GAS (VAPOR) LIQUID SYSTEMS

CONTEMPORARY CHEMICAL ENGINEERING

COMPLEX FORMATION AND STEREOCHEMISTRY OF COORDINATION COMPOUNDS

EDITED BY YU BUSLAEV ISBN 1-56072-186-3

ELECTRON STRUCTURE AND HIGH-TEMPERATURE CHEMISTRY OF COORDINATION COMPOUNDS

EDITED BY YU BUSLAEV ISBN 1-56072-186-1

GAS (VAPOR) LIQUID SYSTEMS

EDITED BY N.N. KULOV ISBN 1-56072-188-X

LIQUID-LIQUID SYSTEMS

EDITED BY N.N. KULOV ISBN 1-56072-189-8

GAS (VAPOR) LIQUID SYSTEMS

EDITED BY

N.N. KULOV

NOVA SCIENCE PUBLISHERS, INC.

Art Director: Maria Ester Hawrys
Assistant Director: Elenor Kallberg
Graphics: Denise Dieterich and Kerri Pfister
Manuscript Coordinator: Roseann Pena
Book Production: Tammy Sauter, Benjamin Fung
Circulation: Irene Kwartiroff and Annette Hellinger

Cover Design: Kerri Pfister

Library of Congress Cataloging-in-Publication Data available upon request

ISBN 1-56072-188-X

6080 Jericho Turnpike, Suite 207
Commack, New York 11725
Tele. 516-499-3103 Fax 516-499-3146
E Mail Novasci1@aol.com

Printed in the United States of America

Contents

Chemistry, Chemical Technology, Progress of Science and Engineering

N.M. Zhavoronkov

> *Science is a light which allows us to see in the darkness. The aim of the science is to forsee and to be useful.*
>
> D.I. Mendeleev

Fundamental Research – The Basis of Engineering Development

The scientific and engineering revolution taking place in today's world is reflected in the rapid development of science and education in an expanding influence of science upon industry, broad penetration in all fields of man's activity and quick changes in the way of life. There have been a number of great discoveries in natural sciences. Revolutionary changes in the development of productive forces started under this influence will become much more significant and promising in the future.

The entity of two opposite tendencies is observed in the development of modern science: on the one side, a deep differentiation takes place, on the other - integration, mutual penetration and obliteration of borders between them. Seemingly different scientific fields and methods of its development are closely interrelated. They influence each other and serve the same purpose: to study and change the surrounding world for the benefit of man. Scientific development is featured by the division accompanied by the cooperation and unification of different trends. Recent scientific history proves that the greatest discoveries take place when different fields of science come together. The appearance of so called "hybrid" sciences, (astrophysics, biophysics, biochemistry, geochemistry, agrobiology and so on), reflects this tendency. The scientific and engineering revolution puts forth many important and complicated problems to be solved by scientists.

The development of science vividly revealed the inadequacy of theoretically unlimited possibilities of human knowledge and the limited possibilities of a human being. Science is now being developed by a collective human mind. Most of the problems of modern science are complex, their solution is impossible for a single scientist, because it requires not only the knowledge of different specialists but the incorporation of various scientific, research, designing and productive organizations.

Such unification and cooperation is of an international character. The character of the problem itself defines the requirements of the specialists involved, and management in this group is carried out by the most interested party. As a rule, large financial and material expenditures are drawn to solve principle tasks of modern science.

The tasks of management, planning and control of science along with its cooperation with the industry are put forward today and require continuous mastering. One of the characteristics and defining features of modern science is the deepening and widening process of its coalescence with industry. Science and industry unite, forming a more closer material and technical basis in society. Thus, the problem of correct correlation between fundamental and purely applied research becomes a reality in modern science. However, interest in the problem has not weakened since science appeared. Today, interest is quite acute when the impetuous growth of industry, transportation, communication, agriculture and public health demand continuous research for the application of fundamental scientific achievements in practice. Scientific development itself proved that knowledge is never exhaustive, any level of knowledge can be surpassed by further development.

The complexity of the problem lies in the fact that it is often difficult to define a boundary line between fundamental and applied research. Presently, the basis of any great technological achievement is profound scientific research. When a big new technical scientific problem arises – whether it's a problem of nuclear energy use, aircraft creation, getting materials for semiconductors and quantum generators started, the elaboration of sea-water distillation methods, or the protection of reservoirs and the atmosphere against harmful industrial waste, there is always the need to begin fundamental research. As an example of fundamental research to solve the problem of atmospheric nitrogen fixation one can quote from the speech F. Gaber gave on June 2, 1920 in Stockholm when he was awarded the Noble prize: "Ammonium synthesis done on a big scale is a reality and may be the most real way to meet the important national economic needs. This practical benefit was not the preconceived aim of my work. I had no doubts that my laboratory work would give nothing more than scientific clarification of the basis and elaboration of the experimental methods and much still has to be done to provide the economic achievements on an industrial scale. However, on the other side, it would be difficult for me to study this problem so profoundly if I were not sure of the productional necessity of chemical success in this field..." Gaber's idea seems to be absolutely impeccable, especially if one bears in mind that by the time he started his work, the general consensus was that ammonium synthesis from elements had no practical significance.

Fundamental research is very important because its results can cause the revolution, in one field or another, of knowledge or production. The fundamental research of F. Gaber, V. Nernst, A. Le-Shatelie and other scientists in the field of ammonium synthesis led to not only the solution of the most pointed problem at the end of last century and the beginning of this one, the problem of air nitrogen fixation and the creation of a powerful nitric industry, but also stimulated the development of new branches of physical chemistry, (chemical thermodynamics and kinetics, catalysis, use of high pressures and recirculation principles in chemical engineering etc.). It is difficult to overestimate their significance for the progress of chemical engineering as a science. The synthetic rubber and polymeric materials industry was also founded on the basis of fundamental research which was carried out by several generations of scientists in the field of chemistry and high-molecular compounds.

All modern technology is the child of science, D.I. Mendeleev who considered the zeal for knowledge to be the motive of a scientist, constantly repeated that knowledge serves the purpose, considering science to be not only a sum of knowledge, but an instrument for social progress. Like Louis Pasteur, he didn't divide science into pure and applied, but instead considered science and its application. The outstanding scientific discoveries of modern time and their application for production processes are examples of mutual influence of practical needs and progress of science. P. Langeven even claimed: "Any academic research, no matter how abstract and "indifferent" it seems, will sooner or later find its application; in other words, no strain of thought is a loss for action." Thus, from the socio-historical point of view, science is utilitarian.

This does not mean that every scientist's activity should only be guided by applicability. The researcher can not always foresee the practical application possibilities of his discoveries. But every true scientist, and especially the leader of a large group of scientists, should be guided by the principle, that the nearest goal of science is in its development, in the perception of the unknown, but the final and most noble goal is in the creative influence on surrounding life and the order of things in our world, and in the benefit it brings directly to the people.

Such understanding of the aims of science was characteristic of D.I. Mendeleev and N.E. Zhukovsky, L.A. Chugaev, N.S. Kurnakov, I.V. Kurchatov, S.P. Korolyov, M.V. Keldysh and many other outstanding scientists. They were not only great scientists who enriched science with their fundamental discoveries, but also great organizers of the basic research programs necessary for the salvation of the most important scientific and technological problems connected with the use of nuclear power, the progress of aviation and space exploration.

One must remember that larger and larger expenditures are required for the realization of all the possibilities opened by a revolution in science and technology. But at every stage of development, the state owned resources can be distributed by the state to achieve these aims. Hence, it is necessary to choose and preferably develop the most important trends of scientific and technological progress, to elaborate and carry out the state industrial policy in succession.

As it is known, fundamental scientific research in the field of natural and social sciences in the Soviet Union is carried out mainly at the scientific institutions of the Academy of Sciences of the USSR, the academies of the Union Republics and at higher educational establishments. Side by side with fundamental research, the importance to the national economy is carried out at these institutions. The main task of the branch scientific-research institutes is the development of applied scientific problems connected with the growth of the national economy, the mastering of industrial production and the creation of new technological processes and techniques. Many research institutes which belong to industry and agriculture also carry out theoretical and experimental research in the field of fundamental science, connected with the development of a particular branch.

One can approximately judge the correlation of fundamental and applied research in our country by the ratio between the number of researches in academic institutes and universities and the number of all researches in the country and by the expenditures.

According to published data, expenditures on fundamental studies in Russia, as well as the USA, England, France and Italy, are about 14-20% of the total expenditures on science and experimental designing projects[*]).

Also of great importance is the problem characterizing the state of sciences, the problem of correlation between the cumulative experimental material and the ability to theoretically explain and foresee it. "Experimentators," - Max Plank wrote[**] , - "are a striking power in science. They carry out decisive experiments and do the most important measurement work. The experiment is the question that the science asks the nature, the measurement is the registration of the nature's answer. But... before asking a question of nature, it is necessary to formulate it. Before the measurement result can be used, it must be explained, nature's answer must be correctly understood. These two tasks are the duty of a theoretician." For a harmonic development of science, it is necessary to balance the experimental and theoretical knowledge.

The division of scientists into experimentators and theoreticians is caused not only by the expediency itself, but also by the existing peculiarities of the abilities. Best results are gained when the two types of abilities are combined in one person. "The experimentator, in order to be worth this title, is to be both a theoretician and a practical person... It would be impossible to divide these two things: the head and the arm, the expert arm without the head guiding it, is a blind tool; the head without the arm, remains forceless," - Clod Bernar said. He also warned against biased conclusions. "Many scientists believe strongly in the theories they have elaborated. Such people cannot only make discoveries, but their observations are completely unsatisfactory. They are guided in their observations by a biased aim, and, making an experiment they try to explain its result in such a way as to prove their

[*] P. Aldac. Scientific-engineering progress and limites of the economic analysis. World economy and international relations. N 5, 1971

[**] M. Plank. Scientific Autobiography and other Papers. London. 1950.

theories. Thus, they misinterpret their observations and sometimes neglect the extremely interesting facts only because these facts do not help them achieve the aim they put forward."

The idea of the significance of the scientific experiment was once strikingly formulated by Francis Bacon: "The axioms, based on logical conclusions cannot be enough to discover new phenomena as the nature is much more clever than any conclusion." I.P. Pavlov expressed the same idea in a more artistic way in his well-known letter to the Soviet youth in February 1936.

THE SIGNIFICANCE OF CHEMISTRY AND CHEMICAL TECHNOLOGY FOR THE MATERIAL AND TECHNICAL PROGRESS

Material and technical progress at all historical periods of mankind was determined by the mutually connected process of creating new energy resources, new technological materials, and new means and methods to influence the material world around us. The main indicator of the material and technical progress is the growth of labor efficiency, which basically depends on the power to weight ratio, mechanization and chemization of the production processes.

The development of power engineering, building, machine-building transportation, communication, all kinds of new techniques, agriculture, health service, food and improvement of people's living conditions requires the production of various materials, substances and preparations with a specific complex of mechanical, physical, chemical and biological properties. Conversion of some matters and materials, (raw materials, semi-finished products), into others with pre-set useful properties, has been and will remain the main task of chemistry and chemical technology. Chemistry can not only process and modify natural matters and produce valuable products and materials from them, but also create new ones which are not available in nature.

The possibilities for meeting the numerous demands of society, to supply people with food, housing, clothes, different materials and power, to preserve one's health and protect the environment from contamination are connected with the progress in chemistry. The development of chemical science is one of the main sources of enriching ideas about the material world. The progress in modern chemistry is vividly manifested in a growing number of natural resources and synthesized by scientists as separate chemical compounds. If about one million of the separate chemical compounds were known in 1930, three million in 1970, and about eight million in 1980, the majority of them were synthesized by chemists.

Chemical methods for processing gain wider application in any production, to say nothing about the chemical industry, metallurgy, by-product coke industry, building materials industry, oil refinery, paper and pulp, pharmaceutical and a greater part of food industries. These branches of industry are basically chemical processes. With the progress of science and technology in industrial production, the role of chemical engineering securing the highest growth of social labor efficiency also increases. One can consider chemization to be the universal way to improve all sides of productive activity. It is difficult, and probably impossible, to find such a sphere of human activity where the application of chemical methods and materials would not revolutionize the processes which had been formed. That is the importance of chemization ideas.

The specific feature of the modern chemical industry is that its progress depends on the development of fundamental research more than other fields of industry. The development of new chemical production methods, (and their ever growing number exceeds ten thousand denominations), creation of new chemical-technological processes in a modern chemical industry principally differs from similar processes in the fields of industry that are connected with mechanics. The progress of the chemical industry in the true sense of the word, that has allowed it to overcome its falling behind in machine-building, was the result of fundamental studies in chemistry, connected with a thorough insight into the microworld aimed at the perceptions of the electronic structure of chemical compounds, the nature of chemical bonds, reaction capability, mechanism of various chemical reactions, etc.

Let us give an example showing the importance of fundamental studies in the creation and development of a very important branch of industry, production of fixed nitrogen. It is common knowledge that the first industrial unit for nitrogen oxidation at 3000°C by air passing over an arc electric furnace by the Birkeland and Eyder method was put into operation in Norway in May of 1905. K.A. Timiryazev's reaction to this event was his enthusiastic article "A New Victory of Science over Nature." The commissioning of the unit completed numerous fundamental investigations carried out by a number of scientists and proved the principle feasibility of industrial production of fixed nitrogen. Nitrogen fertilizers were produced as calcium nitrate, later called Norwegian nitrate in contrast to natural sodium nitrate which is called Chilian after the location of mining. For the next 20 years, plants producing Norwegian nitrate were built in many countries and total production of nitrogen fertilizers using this method by 1925 reached 35 thousand tons per year on conversion to pure nitrogen. The disadvantage of the electric arc method of fixed nitrogen was that high power consumption was the main component in the structure of the product cost price, namely 80000 kW/h per one ton of fixed nitrogen, or on conversion to ideal fuel according to the contemporaneous norms (330 g per 1 kW/h 26.4 tons of ideal fuel (t.i.f.).).

Further progress in the nitrogen industry was conditioned by the desire to lower power consumption and, thus, the cost of the product. The next industrial method of fixed nitrogen introduced to the industrial production at the same time as the arc method, was cyanamide, elaborated by G. Frank and N. Caro. The overall power expenditures, (coke and electric power), per 1 ton of nitrogen as calcium cyanamide were considerably lower than in the arc method, namely about t.i.f. Industrial realization of ammonium synthesis out of hydrogen and nitrogen under a pressure of 20 MPa substituted other methods of nitrogen fixation. The last industrial unit on arc method production was dismounted in 1929, and calcium cyanamide production reached the maximum amounting to 335 thousand tons of conversion to pure nitrogen in 1940, after that, production began to decrease. The production of synthetic ammonium in the world, amounting to about 3 million tons in 1940, reached 60.4 million tons in 1975, and 83.0 million tons in 1987. Such an increase in synthetic ammonium production is surprising if one recalls the pessimistic predictions related to the future of nitrogen production made soon after World War II. The main motive of the nitrogen industry development was the increase of agricultural demand in fertilizers.

In a structure of the synthetic ammonium cost price at plants with an annual capacity of 100-200 thousand tons, the cost of energy amounts to 41-45%, raw material, 33-35% and salary, about 1%. While developing new methods of nitrogen fixation, success is sure to come if power and raw material consumption are decreased. It is not an easy task. As seen in the Table, new units with a capacity of 1200-1500 tons per day consume 1.5 tons of ideal fuel per 1 ton of fixed nitrogen in the form of ammonium, to say nothing of the other advantages of this method. The modern ammonium-producing plant is a power-chemical complex, consuming thermal power supplied by other sources, only about 40 kW/h per 1 ton of fixed nitrogen.

The progress achieved is not the result of some outstanding invention and discovery, but is based on the introduction of the achievements of modern science to the industry causing the modification led to a further decrease in energy expenditures (the first industrial unit, according to Garber and Bosch method, was commissioned in Oppau, Germany, in 1913). More than 80% of the cost price of the synthetic ammonium are the expenditures on hydrogen production. The overall power consumption per 1 ton of ammonium depends on the method of hydrogen production.

The data illustrating power consumption for different methods of hydrogen production is given below (Table 1). For conveniences of correlation, the overall power consumption is converted to tons of the ideal fuel (t.i.f.) per ton of synthetic ammonium.

Table 1. Power consumption per ton of synthetic ammonium*

Source of hydrogen	Overall power consumption, including raw material, t.i.f.
Water electrolysis	4.63
Natural gas by methane conversion with steam	
under air pressure	2.34
under pressure 3 MPa in the unit 600 t/day	1.7
under pressure 4 MPa in the unit 1360 t/day	1.5

* The calculation is made proceeding from ideal fuel consumption at the amount of 0.33 kg per 1 kW/h.

One can see from this data that modern industrial units for ammonium synthesis attained the highest efficiency and improvement of the already known technological processes, elaboration of more efficient equipment, an increase in single unit capacity, change-over for new raw material, introduction of relatively modest inventions and innovations, but in greater number, into practice. In fact, the most modern unit for ammonium synthesis has no new technological process unknown before World War II. It is necessary to note that the improvement of the known technological processes became possible thanks to the progress of science and the creative efforts of engineering thought. The increase of unit capacity became possible thanks to a better understanding of the reaction mechanism of ammonium synthesis, the studying of the kinetic process and the development of the fast heat elimination method from the reaction zone. Achievements in the science of metals, metallurgy and machine-building made it possible to producer larger columns of ammonium synthesis reactors and other equipment working at high pressure than was possible several years ago. As a result, the column structure of ammonium synthesis was developed making it possible to utilize part of the reaction heat to get 0.8 tons of steam with 3 to 4 MPa pressure per each ton of ammonium and providing stable autothermal operation of the unit. In recent years, in all the countries, the ammonium producing plants were gradually converted in order to work on the natural gas and hydrocarbon raw material as a source of hydrogen, based on the results of numerous investigations more perfect methods and highly efficient units to produce hydrogen were elaborated. The most efficient method turned out to be the one of methane conversion with steam under a pressure of 2 to 3 MPa at 800-900°C. This became widespread. The converter is a battery of tubes, filled with a catalyst and heated by a hot gas produced by the burning of the natural gas. To manufacture tubes capable to resist drastic conditions of pressure and temperature, alloys with a high content of nickel and chrome were used. The change for ammonium synthesis units of a larger unit capacity and the use of natural gas conversion under pressure made it possible to substitute a piston compressor for turbocompressors in order to compress the gas up to a final pressure of 30-32 MPa, which is due to some advantages of rotary engines.

The progress of the nitrogen industry in a post-war period vividly illustrates the fact that in the epoch when the role of science as a direct production force reveals itself, the main is not its separate achievements no matter how bright they are, but its high scientific and technical level of all production.

From the socio-historical point of view, science is finally directed to meet the practical tasks put forward by society. However, it does not mean that every scientist in his activity is guided by purely applied tasks. He can not always foresee the possibilities of the practical use of his discoveries. Nevertheless, every real scientist and, moreover, the leader of a scientific staff should not forget that the nearest goal of science is to broaden the knowledge of the surrounding world and its final and the most generous goal is in the benefit it brings to people.

The close interaction of scientific research and industry is a current problem. Moreover, their intrinsic unity is not just a subjective wish or intent, but a serious possibility to realize a practical task which arises from a deep understanding of scientific and technological progress problems. One of these principles is that as far as science and techniques develop the experimental and theoretical, applied and fundamental research are brought closer together and coalesce.

From the end of last century till recently, the period between important scientific discovery and its practical realization has been continuously reduced. The Watt steam engine was created on the basis of principles introduced by the ancient Greek scientist Geron. In this case, the scientific application of a scientific discovery took about 2000 years. Today, society cannot wait for the practical application of scientific results. The process of introducing the scientific achievements into practice is steadily accelerating.

In 1873, a famous work by D. Maxwell "Treatise on Electricity and Magnetizm" was published, which contained a system of equations concluding the existence of electromagnetic waves. Maxwell's theoretical prediction made from 1886-1889 was experimentally proven by German physicist G. Hertz and, in 1985, A.S. Popov, a Russian physicist, was the first to apply the electromagnetic waves to transmit wireless signals at a distance, opening the era of radio communication.

Even less time was required to introduce the fundamental research results to industry, which were carried out by S.V. Lebedev. From 1926-1928, an international competition was announced, aimed at working out an industrial method for synthetic rubber production and Lebedev, along with a group of apprentices worked out the best method of butadiene rubber production at that time. As long ago as 1931, the first batch of synthetic rubber was produced at an experimental plant in Leningrad and in 1932 the Soviet Union started the world's first large-scale industrial production of synthetic rubber by Lebedev's method.

From 1932-1937 four plants were built in Yaroslavl, Kazan, Efremov and Voronezh. During the first post-war years the production of butadiene rubber was mastered at two plants and then under the guidance of B.A. Dolgolosk and A.A. Korotkov work to synthesize steoregular izoprene butadiene rubber and produce it on an industrial scale was completed.

Quantum generators were diversely used soon after the discovery of coherent evolution of light by Soviet physicists N.G. Basov and A.M. Prokhorov and American physicist Ch. Towns. who won the Noble Prize in 1964 for this discovery.

The accelerating tempo of the realization of considerable scientific discoveries witnesses a steady growth in the role of science for the life of society. Scientific progress put forth the problem of increasing scientific research efficiency and organizing scientific activity.

Finances and large material resources are required to solve the fundamental tasks of modern science. In 1940, the expenditures for science in the USSR from the state budget and other sources amounted to 0.3 billion Roubles. In 1970 they reached 11.7, in 1980, 22.3 and in 1988, 37.8 billion Roubles.

The drawing together of science and industry depends not only on the interaction of fundamental and applied research but also on how quickly and efficiently the scientific ideas will be brought to engineering solutions. One of the conditions providing for the fast realization of scientific achievements is to bring scientific investigations and experimental-engineering work into proper correlation. The current practice of financing scientific investigations and experimental-engineering work in industrially developed countries including the Soviet Union, proves that approximately one fifth of the allocations is spent on research and four fifths for experimental-engineering works. A similar practice is in use in other developed countries. These figures are approximate, because it is difficult to differentiate between fundamental and applied research. Besides, they are often different in the various fields of industry.

For individual fields of industry, the share of the capital spent for experimental-engineering work amounts to: 50% for chemical, 55% for ferrous and non-ferrous metallurgical industry, 64% for paper and pulp industry and building materials and 83% for machine-building. These figures can vary, but undoubtedly, one of the most important components in the overall expenditures for science are the expenditures for experimental-engineering work. If fundamental and applied research is not accompanied by experimental, engineering elaborations, the scientific results, especially of applied research would remain unused.

The USSR Academy of Sciences contributed a considerable amount to fundamental research for drawing science and industry together, its 250th anniversary was celebrated in 1975 by Soviet society. The Academy of Sciences played an outstanding role in organizing scientific research, studies and mastering the natural resources of our country, in the development of industry, in the circulation of scientific knowledge, education and culture. On this basis, a large network of scientific-research institutions was developed later. The scientific activity of the Academy became very intense and varied after the Great October Socialist Revolution which opened new, unprecedented possibilities for science.

During a short historical period, a rapid change, incomparable in scale and significance, was made in the development of science. From its very beginning, the Academy of Science took an active part in solving the most important industrial tasks and its scientific researches helped to defeat the fascist troops during World War II and contribute to the creation of highly efficient, scientific and defense potential in Russia and solve the most important problems of material and technological progress. Now, it is one of the largest scientific centers in the world

Scientists from the Academy of Sciences of the USSR carried out fundamental research in mathematics, physics, chemistry, biology, astronomy, sciences about earth and promoting the progress of the national economy in the country.

The growth and development of experimental and theoretical physics, the use of its achievements to solve the most important scientific technical problems in the field of atomic nucleus and nuclear energetics, space and scientific basis of the missile art, the development of the applied mathematics, computers, radio techniques, quantum electronics and the creation of numerous new organic and inorganic materials with preset properties required for technical progress is all connected with the Academy of Sciences. In 1922, the Radium Institute was founded on the basis of the Radiological Laboratory of the Academy of Science, which was organized by academician V.I. Vernadsky which is the study of the phenomenon of radioactivity and radio chemical research.

As early as 1928, the Academy of Sciences began a systematic exploration of the country's territory. The selfless work of the first generation of Soviet scientists gave valuable results. The richest deposits of mineral resources, (iron and polymetal ores, precious and rare metals, coal, oil, gas), were found and explored. An evaluation of water power resources was made, the Northern Sea was explored, the soil and then agrochemical maps were plotted and other numerous discoveries and investigations were made. Since then, the exploration of natural resources were planned. In 1919, scientists began an exploration of Kursk magnetic anomaly. The Academy of Sciences contributed to the exploration and investigation of this richest iron-ore basin. The geological exploration of the Koslk peninsula, made under the leadership of academician A.E. Fersman, was of great importance. Huge apatite-nepheline deposits being the raw materials to produce phosphorous fertilizers were discovered. The practical result of the academic expeditions to explore Kara-Bogaz-Gol which began in the 20s was to begin the industrial extraction of sodium sulfate and other salts.

The works of N.S. Kurnakov and his school on physico-chemical analysis and its application for the study of intermetallic compounds and alloys as starting materials to get new structural materials, to study problems of rare metals, water-salt systems and use of natural salts were also of great importance. They played an important role in the study and exploration of Solikamsk potash salt deposits in particular. H.S. Kurnakov began this work at the laboratories of the Mining and Polytechnical Institutes in Petrograd, it was continued at the Institute of Physical-chemical analysis founded in 1918 at the Academy of Sciences. At the same time, the Academy of Sciences Institute on platinum and other precious metals was founded. It was led by professor Chugaev, who began research in coordination chemistry in Russia.

Research in chemistry of platinum complex compounds, platinoids and gold were carried out on a large scale at the Institute. L.A. Chugaev had many talented students at the Institute, (I.I. Chernyaev, A.A. Grinberg, V.V. Lebedinsky, N.K. Pshenitzin, O.E. Zvyagintzev and others), making up Chugaev's school and continuing his research. The results of platinoids and platinum complex compounds studies formed the basis to set up an industry of precious metals in Russia and works on complex compounds of uranium, plutonium, thorium, circonium and some other elements played a certain role in solving the chemical problems of atomic energy. Recent research on the complexation of some rare earth elements and heavy metal glass-like and liquid matrices led to the creation of new materials for quantum generators. Research in coordination chemistry was wide spread in many scientific-research institutions and higher educational institutes. The main center on synthesis and study of physical-chemical properties of complex compounds is presently the Kurnakov Institute of General and Inorganic Chemistry.

The first scientific research institutions founded under the Academy of Sciences soon after the October Revolution were the Institute of Physical Chemical Analysis and the Institute on platinum and other precious metals studies. They have functioned since May 1918. Academician N.S. Kurnakov and professor L.A. Chugaev headed the Institutes. After the untimely death of L.A. Chugaev in 1922 N.S. Kurnakov became the director of both Institutes. At the same time he was head of the Laboratory on General Chemistry founded by M.V. Lomonosov in 1748.

In 1934, these three scientific institutions were united into the Institute of General and Inorganic Chemistry headed by N.S. Kurnakov. That same year the Institute moved from Leningrad to Moscow and relocated in a building in the Leninsky prospect [31]. After N.S. Kurnakov's death on March 19, 1941, academician I.I. Chernyaev, a pupil of L.A. Chugaev, was appointed director of the Institute (20.01.1893-30.09.1966). From 1962 to 1989, the Institute was headed by the author of this paper. It stood only one story of two rooms in the memorial building of the Lomonosov's Laboratory on Vasilievsky Ostrov in Leningrad and was destroyed by a direct hit of the German bombing during World War II.

The activity of the scientists working at the Academy of Sciences of the USSR in the field of polymers chemistry, electrochemistry, physical chemistry and so on is marked by great success. Their research allowed for not only the important regularities of chemical structure and reactivity to be revealed, but also main principles of chemical compound synthesis of different classes to be developed, which formed the basis of industrial methods to produce different matters and materials. The latter are widely used in all branches of national economy and allowed for the creation of new and improved technological processes.

The works in the field of chemistry of elemental organic compounds which appeared in between organic and inorganic chemistry, were highly successful. They allowed the regularities of organic compound formation of most of the elements of the Periodic system to be found. The success in chemistry greatly influenced other fields of science.

Since the early '20s investigations of chemical engineering began to develop. The pioneers of the theoretical basis, the study of the main processes and apparatus of chemical engineering, were the Lensoviet Leningrad Technological Institute and the Mendeleev Moscow Chemical Technological Institute. Presently, large groups of scientists, engineers and technologists are engaged in solving problems of up-to-date materials study, theory of strength and plasticity, obtaining rare metals, methods of welding, elaboration of new, highly efficient machines for metallurgical and metal processing industries.

Present fundamental studies acquire greater importance as a necessary basis for the technical progress.

Chemistry belongs to those fields of knowledge which began to develop in pre-revolutionary Russia. There are many Russian names among the names of the outstanding chemists in the world. M. Lomonosov began chemical research in Russia. Chemistry in Russia flourished in the second half of the 19th century and was connected with the names of N. Zinin, the founder of the Russian school of chemists organics, A Butlerov, the author of the chemical structure of substances theory, D. Mendeleev, who revealed the fundamental law of nature; the periodic law of chemical elements. Mendeleev's discovery, defining the relationship between all chemical elements of the universe was a milestone in chemistry.

Seventy-six years ago, the president of the London Chemical Society, William Witney said, in his speech "On the importance of the Russian chemists' works for the World science": "If we, appreciating the musical school connected with names like Balakirev, Borodin, (he was a chemist as well), Rimsky-Korsakov, Chaikovsky or writers such as Turgenev, Dostoevsky, Lev Tolstoy and their contemporaries, think that the world would have been poorer without them, then we won't exaggerate that the development of chemistry would have been no less slower if the works of Mendeleev, Butlerov, Vagner and their successors were eliminated for some reasons out of social storage of knowledge.*

The pre-revolutionary years are marked by great achievements of Russian chemists. It is enough to mention the outstanding research of N.S. Kurnakov in the field of physical-chemical analysis, L.A. Chugaev on chemistry of coordination compounds, K.A. Timiryasev and D.N. Pryanishnikov on agrochemistry, V.V. Markovnikov on petro-chemistry, of chromatography discovery by M.S. Tzvet, of G.S. Petrov's works on the synthesis of carbolite, D.P. Konovalov, A.K. Krupsky and I.A. Tischenko on theoretical basis of chemical engineering and many others.

However, the high level of chemical research in pre-revolutionary Russia didn't correspond to its small scale work and practical yield. The chemical industry, which could stimulate scientific research by its needs, was poorly developed. Chemical research was carried out in a few universities.

The Great October Socialist Revolution opened new, unprecedented possibilities for the development of science. Science became of great importance to the state, the object of continuous care of the Soviet Government and people.

The role of the chemical industry in the creation of material and the technical basis of a new society was considered by V.I. Lenin to be major, because only on its basis was it possible to achieve radical and qualitative changes in different spheres of material production of new matters, products and materials, to work out and use new highly efficient methods and production processes and to introduce chemical processes into agriculture.

In April of 1918, V.I. Lenin wrote: "The growth of labor productivity requires the provision of a material basis for the large scale industry first of

* Witney W. on importance of Russian chemists' works for world chemistry. Sci.Chem-tech.Pub.H., 1924, p. 52.

all: to increase the production of fuel, iron, machine-building and to develop the chemical industry.

The selfless labor of the Soviet people transformed economically underdeveloped Russia into a powerful socialist state, with gigantic industry and mechanized agriculture, with high level of education, science and culture.

During a short historical period the Soviet Union changed immensely.

A big scientific and technical potential was created in the country. Soviet science reached a high level. It can solve many tasks connected with the progress of the national economy. Soviet scientists were active participants in working out the GOELRO plan and the first five-year plans. It was especially important because the country had chosen the unknown way and there was no possibility to follow any experience. Soviet scientists took an active part in setting up the main branches of the up-to-date industry, which Tzarist Russia didn't have; in carrying out the cultural revolution, in putting an end to ignorance and in developing higher education.

It was not an easy task. The Soviet Power inherited a country ruined by World War I, Civil War and foreign intervention. A country with an unorganized economy, a destroyed industry and a decaying agriculture. In the period between the end of the Civil War and the fascist invasion efforts have been made to reconstruct and develop industry and restore the economy. Damaged power stations, machine-building, metallurgical and chemical plants were reconstructed and new ones were built and commissioned.

An invaluable contribution was made by the scientists for victory in World War II. During the post-war period, Soviet science actively contributed to the strengthening of Russia's economy and defense. It is enough to give only two examples, a controlling of nuclear energy for military and peaceful purposes and an exploration of space.

December 2, 1942 is considered to be the beginning of the nuclear energy era, when the chain reaction of nuclear Uranium fission was succeeded by Enriko Fermi.

On the 25th of December, 1946 the nuclear reaction was accomplished by I.V. Kurchatov in the USSR. An experimental uranium-graphite reactor which was commissioned by him on the territory of the Institute of Atomic Energy (IAE), now named after him, continues to work today. In June of 1954 the first atomic power plant (APP) with the capacity of 5000 kW was put into operation in Obninsk, the first of its kind in the USSR.

In 1987 there were 427 atomic power plants (units) in the world, producing 15% of the world's power output.

The space era began October 4, 1947 when the world's first Earth satellite was launched into space, and on April 12, 1961, Soviet cosmonaut Yri Gagarin, became the first man in space.

The contribution of Soviet science in the space exploration is well known. E.K. Tziolkovsky's ideas and the selfless labor of the main theorist, academician M.V. Keldysh and the Chief Constructor, academician S.P. Korolov played a major role. Many scientists took part in the elaboration of space materials and fuel for jet-engines.

In 1948 the American magazine "Look" published an article "When will Russia have an atomic bomb?" The same year it was published as a separate booklet in the USSR. * The authors of the article are John F. Hogerton, atomic engineer, one of the leading engineers in atomic plants of Oakridge and Henford, and Ellsworth Raymond, a Russian specialist, the chief consultant on Russian economy in the Military Department in Washington (1943-1946). On the basis of the "detailed" analysis of the Soviet science and industry the authors came to the conclusion that the USSR cannot create the atomic weapon earlier than 1954. However, on August 29, 1949, one year after this article was published, the USSR tested the atomic bomb. The more unexpected fact for the people trying to estimate the scientific and technical potential of Russia was the creation of the thermonuclear weapon. In November of 1952, in the Pacific Ocean, the USA exploded the thermo-nuclear device having the conventional title "Mike" and the dimensions of a small house. On August 12, 1953 the USSR tested the thermo-nuclear bomb.

Chemistry and chemical technology played an outstanding role in solving nuclear energetics problems and the exploration of space.

When retracing the history of Soviet chemistry one should note that it is represented by different trends of modern science. While enumerating them we recall the names of our distinguished scientists who headed the corresponding trends with gratitude:

Inorganic Chemistry (including coordination chemistry)
N.S. Kurnakov, L.A. Chugaev, I.I. Chernyaev, A.A. Grinberg, B.V. Nekrasov, I.V. Tananaev;

Organic Chemistry
N.D. Zelinsky, S.V. Lebedev, N.M. Kizhner, A.E. Arbusov, A.E. Chichibabin, M.Ya. Demiyanov, A.E. Favorsky, I.N. Nazarov, A.N. Nesmeyanov, N.N. Vorozhtzov (junior);

Physical Chemistry
D.P. Konovalov, L.V. Pisarzhevsky, I.A. Kablukov, V.A. Kisyakovsky, A.A. Balandin, S.Z. Roginsky, A.V. Frost, A.V. Rakovsky;

Chemical Physics
N.N. Semenov, Yu.B. Khariton, Ya.B. Zeldovich;

Electrochemistry
A.N. Frumkin, G.V. Akimov, N.A. Izgarishev;

Geochemistry
V.I. Vernadsky, A.E. Fersman, A.P. Vinogradov;

Agrochemistry and Mineral Fertilizers
K.A. Timiryasev, D.N. Pryanishnikov, E.B. Britzke, S.I. Volfkovich, A.V. Sokolov;

---

* Hogerton J. and Raymond E. "When will Russia have an atomic bomb?" M., Foreign Lit.Publ.H., 1948.

Petrochemistry
V.N. Ipatjev, S.S. Nametkin, B.A. Kazansky, A.D. Petrov;

Photochemistry
A.N. Terenin;

Silicate Chemistry
I.V. Grebenschikov, P.P. Budnikov, N.A. Toporov, N.N. Kachalov, V.N. Yung, I.I. Kitaigorodsky, D.N. Poluboyarivov, M.M. Schultz;

Colloid Chemistry
N.P. Peshkov, P.A. Rebinder;

Chemistry of High Molecular Compounds
V.A. Kargin, S.S. Medvedev, G.S. Petrov, S.N. Ushakov, I.P. Losev;

Biochemistry
A.N. Bakh, V.A. Engelgardt, A.I. Oparin;

Bioorganic Chemistry
M.M. Shemyakin, Yu.A. Ovchinnikov;

Chemistry of Carbons and Synthetic Fibers
P.P. Shorygin, Z.A. Rogovin;

The Chemistry of Dyes and Semi-products
V.M. Rodionov, A.E. Porai-Koshitz, N.N. Vorozhtzov (senior);

The Chemistry of Medicinal Substances
A.P. Orekhov, N.A. Preobrazhensky;

Radiochemistry
V.G. Khlopin, B.M. Vdovenko, I.E. Starik;

Theoretical Foundations of Chemical Engineering
N.F. Yushkevich, L.F. Fokin, K.F. Pavlov, A.E. Makovetzky, A.G. Kasatkin, P.M. Lukjyanov, I.A. Tischenko;

Metallurgy
A.A. Baikov, I.P. Bardin, A.M. Samarin.

For the past century the development of chemistry all over the world is featured by a large-scale synthesis of organic compounds, remarkable achievements, especially during recent years are vividly seen. Organic chemistry became the basis to study the fundamental processes and learn the secrets of life.

Inorganic chemistry developed more slowly, mainly as a scientific basis of the traditional branches of industry: mineral acids, alkali and salts, ferrous and non-ferrous metallurgy, cementing materials, ceramics and glass.

However, many new scientific problems, which accelerated the development of inorganic chemistry were raised in the course of scientific and technological progress.

In the middle of the 20th century achievements in physics, mechanics, mathematics, and astronomy cleared the way for nuclear energetics, electronic computers and control tools and allowed man to enter space. It would be impossible without chemistry, which creates new sources of energy and new materials necessary for energetics, electronics, spaceships, and new machinery. At the same time, chemistry raises one cardinal task to be solved by industry, culture, and mode of human life, to substitute old materials and old methods of production and processing for new ones. Chemistry, inorganic chemistry in particular, became the key factor of scientific and technological progress.

Presently, inorganic chemistry goes through an intensive development period. First of all, it is connected with the fact that the development of inorganic chemistry problems extends one's knowledge about the origin of chemical bonds. Wide-spread introduction of physical, namely, quantum and mechanical conceptions for the explanation of chemical bonds characterize and predict elements and their compounds, chemical behavior have a basic significance for the development of theoretical chemistry. Second, inorganic chemistry plays the leading role in the creation of new inorganic materials with valuable properties, first of all wear-, heat- and chemicals- resisting structural materials needed for working under high and low temperature, high pressure and deep vacuum, aggressive media conditions.

Soviet chemists together with physicists and other scientific specialists took an active part in solving the chemical problems of nuclear fission and in the development of nuclear energetics: to work out methods to produce nucleus fuel, to separate Plutonium and products of nuclear fission from unreacted Uranium, to separate isotopes and study their physical and chemical properties.

Great achievements were made in radiochemistry, the synthesis of new transuransium elements and in devising original methods to study their properties. The development of structural research to find the correlation between chemical composition and the structure of the compound on one side and reactivity and physical properties, (optical, magnetic, electrical, mechanical, etc.), on the other side, is characteristic for inorganic chemistry.

At the turn of the century, coordination chemistry appeared as a separate field of inorganic chemistry. The object of its studies became compounds in which the central atom could be isolated, (the metal atom more often), and coupling to it, (coordinated), ligands: atoms, ions, molecules of inorganic and organic origin, as a rule, in these compounds the number of ligands exceeds the classical valency of the central atom defined as a number of uncoupled electrons in its valent shell.

Being in between two fields, inorganic and organic, coordination chemistry became the field of integration, i.e. the process, directly opposite to the differentiation observed during the preceding period of chemical science development. But, contemporary development of any science, (chemistry is only one of the examples), is featured by the unity of opposite tendencies, deep differentiation and specialization on one side and integration and cooperation of different fields of knowledge on the other.

Coordination chemistry, (or chemistry of complex compounds), is a traditional direction of research in Russia. The fundamental works of Soviet scientists in this field are universally recognized. Presently, coordination chemistry is rapidly developing and going through a period of research and discovery. Its success is highly substantial and possibilities are practically unlimited.

The intensive development of research on metal complex catalysis has been witnessed, introduction of coordinational conceptions in biochemistry, and a wide use of complex compounds in medicine and agriculture. Chemistry of rare and transition elements including the coordination chemistry of these elements develops rapidly. Rare elements, being a kind of decoration of the Periodic System, in the past come into our life little by little and soon it will be difficult to do without them as, for instance, without iron or salt.

The greatest achievement of inorganic chemistry of recent years is the obtaining of some inert gases, the inertness of which was explained by the peculiarities of electron structure of their atoms. Outer electron shells of inert gases are extremely stable, that, as per the existed opinion, prevented their interaction with other elements. The myth of chemical inactivity of inert gases prevailed for almost half a century. But in July of 1962 the first publication of an English chemist N. Bartlet on the reaction of platinum hexafluoride with xenon to yield solid chemical compound appeared. Then, xenon di-, tetra- and hexa-fluoride were synthesized directly from xenon and fluorine. Presently, more than 150 xenon, krypton and radon chemical compounds are known. Chemistry of inert gases is developing in Russia.

Scientific and technological progress is defined by an interrelated process of discovery and utilization of new energy resources, creation and use of new materials, elaboration of new instruments and methods and providing higher labor efficiency.

In the process of evolution for more than a million years of intensive purposeful collective labour man acquired the ability to think, created means of communication, (the language), and thanks to this left the rest of the animal world behind, in his development. A great achievement of man was to realize the physical weakness of his body and to create the first primitive labour tools like the stick and the stone hammer to help himself, and later created more complicated instruments called machines. It gave him opportunity to get rid of hard manual labor to increase its efficiency. Our century is marked by a great achievement of man, he realized his limited psychophysiological abilities and created means of getting and processing information, computers. It was preconditioned first of all by the needs of complex modern mass-production. Today, computers provide conditions for the organization and management of such a production, establishment of functional dependence between numerous factors influencing the efficiency of technological processes and their interrelated complexes.

Thus, a new field of applied science appeared, the management science or cybernetics, which importance is far beyond the limits of industrial use. Computers are widely used for scientific research, medicine, weather forecast and in many other fields of human activity.

It is appropriate to mention that the source of developing new natural science tasks and philosophical ideas is finally the means of production of material wealth. K. Marx wrote that "mankind sets such tasks which it can solve because at closer examination it turns out that a task arises only when material conditions for its solution are available or, at least, in the process of being made.

Development of energetics, machine-building, construction, transport, communication, agriculture, public health, food production and improvement of human life require the production of diverse substances, materials, products and preparations with a definite complex of mechanical, physical, chemical and biological properties.

Materials became usual in production and human life. The conception of materials is very comprehensive. It embraces a wide range of natural things such as stone, clay, sand, wood, vegetable and animal fibers, leather and products obtained by processing natural raw materials: metals and its alloys, ceramics, cementing materials, glass, electric conductors, semi-conductors, deelectrics, magnetic materials, paper, cardboard, polymers, elastomers, chemical fibers.

In many cases different materials are required to make products. For example, the modern air-liner consists of tens of thousand of parts, to manufacture them more than 400 metallic and about 600 non-metallic materials are used. We often perceive materials as something understandable in itself. Nylon is known to be stockings and light nylon raincoats rather than polymer with various technical applications for creation of which several generations of scientists and engineers were working.

Some materials, for example, plastic and synthetic fibers led to considerable changes in production and human life. Thus, the result of many years labor of chemists and physicists - luminophores emitting light under electron bombarding, made color TV possible. The possibility to make a machine or instrument often depends on the properties of specific material. For example, to accomplish space flights with a man on board, low cost ablation materials protecting the spaceship against burning when coming through atmosphere, are of great importance.

The transformation of matters and materials into others with the preset properties was and will be the main task of chemistry and chemical technology the possibilities of which are practically unlimited. The role of chemistry and chemical technology in the synthesis of new inorganic and organic matters and the creation of new materials really increased. In recent years, great achievements have been made in this field.

The progress of modern inorganic chemistry is featured by the fact that today almost all chemical elements have this or that industrial or scientific application, but 40-50 years ago many of them except 20-30 were "chemical wonders."

Scientific and technical progress continuously needs new materials. It is not all that necessary to tell what complicated tasks chemistry faces in this connection, how much the variety of starting materials and methods to get new compounds and materials broadens for a researcher. In techniques, for example, many thousands of alloys are presently used, which are composed of more than 50 chemical elements in different combinations.

It would be naive to think that these combinations are found the same way medieval alchemists did mixing and combining different matters, so to say, by methods of testing and mistakes did their guess-work. But is it no less one-sided to assume that today everything in chemistry is opened for preliminary theoretical prediction and mathematical calculation.

Though chemistry is becoming more and more of an exact science, based on the conformities connecting properties of matters and their chemical composition, crystal structure, origin of chemical bond, the skill to synthesize new chemical compounds is and will be one of the most important conditions in its development. As was already mentioned, we now know more than 8 million of the chemical compounds. More of them are synthetic, i.e. the result of creative activity of the human brain and hands and their number is continuously increasing. Let us mention a few of them pertaining to inorganic materials.

First of all, it is necessary to talk about heat-resisting and super wear-resisting materials as well as materials with specific electric and physical properties. In industry and engineering they are widely used and needed in large quantities. Some new infusible matters were made on the basis of rare metal oxides, new methods to synthesize oxide and oxygen-free compounds were worked out, super wear-resisting heat-resisting materials and coatings were developed. The problem of transforming graphite into a diamond was solved. Supersolid material, borazon, was made on the basis of boron and industrially produced. It is not inferior to the quality of a diamond and its thermal resistance is much higher. Borazon does not lose its cutting properties even at temperatures where a diamond burns out. Science today found the way to synthesize rubies, sapphires, amethysts, their properties being better than those of natural.

As it is known, the radio-engineering industry needs piezo electric crystals. But natural resources of piezo-quartz are limited. What to do? Scientists gave an answer. They elaborated hydrothermal methods of growing piezo-quartz monocrystals and organized their industrial production. Nature needed millions of years to grow them. New matters necessary to modulate the laser beam and to create devices of operational memory for computers were synthesized.

Highly solid new glass-crystal and heat resisting materials - sitals and pyrocerams were made. Important success was made in improving properties of ceramics, including fine ceramics, out of which articles can be manufactured by stamping. Usually ceramic is brittle and heavy, but tomorrow it will be different. The main advantage is practically unlimited natural resources.

In the last decade, big quantitative and qualitative changes took place in the production of ferrous and non-ferrous metals, intensification of blast-furnace process, converter melting and continuous casting, elaboration of technology and organization of titanium, circonium, etc. production. Investigations to work out methods of coke-free metallurgy and to create continuous steel smelting aggregate, reduction of iron from ores with coal. Such aggregates should completely change metallurgical plants by removing coke and blast furnaces. It will reduce capital expenditures, increase labor productivity, improve working conditions and reduce environmental pollution.

Metallurgists were successful in getting the so-called amorphous or "glass-like" metals. Liquid metal is cooled so rapidly that it transforms into a solid state without a crystal phase. For example, when melted aluminum hits a cryogen cooled surface at a large speed it becomes solid in the form of flakes within 1 millionth of a second. In another way, which is already being used, melted metal is sprayed in cold inert gas at extremely high speed and becomes solid in the form of fine powder. Properties of such metals are really astonishing. The strength and corrosion resistance of parts made out of amorphous powders increases 10 to 11 times, because amorphous metal has practically no grain limits up to the molecular level. The results gained provide great possibilities for practical use.

Progress in the field of chemistry and technology of super pure materials made notable achievements in the elaboration of methods of getting and industrial production of elementary semi-conductors, silicon and germanium, as well as complex compounds on the basis of gallium, arsenicum, indium, phosphorus and antimony.

Requirements for the purity of substances and materials continuously increases. Atomic technology set a task to create matters and materials containing not more that one thousandth of undesirable impurities. Electronic technology increased the requirements up to one millionth parts. Presently, transmission of information with the help of fiber optics needs glass with a one-billionth content of ferrum and water. Correspondingly, the necessity arises to work out methods and instrumentation to analyze micro-impurities which is an extremely complicated task in itself.

No achievements to create extremely pure materials and even no progress in chemistry is possible without the progress of analytical chemistry.

Gas chromatographic, radio-activation, X-ray fluorescent, X-ray electronic, and mass-spectrum methods of analysis are the combination of achievements in chemistry, electronics, physics of high energies, and crystallography.

Today, different sciences draw together, mutually enriching each other, studying undiscovered secrets of Mother Nature. The content and character of different fields of science changed: biology became more chemical, chemistry, physical and physics, mathematics. Chemistry in alliance with physics and mathematics became the main source to know different biological processes and interpreting biological structures. It is precisely the help of chemical methods that allowed for the interpretation of structures of hundreds of important proteins and nucleic acids, the structure of new antibiotics, vitamins, hormones and other substance-regulators important for every living thing. Thousands of new medicines were prepared by the chemical modification of natural compounds: they are newest antibiotics, biologically active proteins. Chemistry played a special role in interpreting the genetic code and in synthesizing simplest but real gene.

In perspective the synthesis and study of molecules capable to function as individual conductors, resistors, condensors and a combination of which can yield "molecular" boosters, computer memory cells, etc., are of great interest. The development of this trend can cause a revolution in the field of electronic mini-devices, extremely small in size but high-capacity and quick acting. The examples are living organisms: electroconductivity proceeds through electroconductive proteins inside the cells. The understanding of electron transfer mechanism in proteins will allow for the creation of practical organic and inorganic conductors on a molecular level.

Investigations of composition and electron structure of inorganic molecules and substances as well as a search of structural correlation needed for the aimed synthesis of new compounds and materials with the preset properties will play a greater role in the next decade.

Science keeps penetrating deeper and deeper into the microstructure of material and the immense space of the Universe finds their new unexplored properties. No doubt, in the near future science will discover secrets of photosynthesis and ways to control it for a more efficient conversion of solar energy into potential energy of organic substances of plants to increase renewable life resources, will master the controlled thermonuclear reaction and will provide man with practically unlimited energy resources, will make other discoveries increasing man's power over nature to protect and improve it, to eliminate hunger, poverty, diseases and to meet material and spiritual demands of human beings on the Earth.

CHEMICAL TECHNOLOGY AS A SCIENTIFIC DISCIPLINE AND AN EDUCATIONAL SUBJECT

At times the problem of what matters and materials to create, what properties to impart to them, was and will remain inseparable from the problem of how to produce them. Technology, science on industrial production answers this question of paramount importance. But first of all one must be reminded what the term "technology" means. It appeared at the end of the 18th century when the transfer from trade manufacturers based on manual labor to machine industry began. The term "technology" consists of two greek words te´ch ne (art, skill, trade) and *logos* (word, teaching, science) and means "science of trade," or in modern meaning "science on industrial production". This science was first called "technology" in 1777 by I. Beckman, professor of Gettingen University in his work "Introduction to technology." Later he published his first text-books (1779-1788).* In 1795 the two-volume textbook by I.F. Gmelin "The guidance to technical chemistry" was published in Germany and translated into Russian by V.M. Severgin with the title "Chemical bases for trades and plants." Chemical engineering became a compulsory subject in Universities and high technical schools in all the developed countries of Europe at the end of the 18th, beginning of the 19th, century.

In the mid-18th century M.V. Lomonosov in his work "Word about the benefit of chemistry" expressed his idea on the necessity of science and industry correlation by the following: "Science gives way to art, art speeds up the origin of science. Together they bring benefit to the society." The word "art" at that time meant not only art as we now understand it, but the skill to produce materials, products and other objects, i.e. technical trades and industry. M.V. Lomonosov also writes that metallurgy is first among arts which teaches how to find and purify metals and other minerals.

Russia, which, in the second half of the 18th century was not inferior to Western-European countries in industrial development and chemical engineering was given full attention. In Russian commercial and technical schools general technology was introduced as a subject at the end of the 18th century. From 1807 to 1817 the Central Department of schools published the "Technology" textbooks four times. The subject of chemical technology was introduced in Moscow University in 1806 for the first time. The subject was taught by professor I.A. Dvigubsky, who used the textbooks by I. Beckman and I.F. Gmelin, and then in 1807-1808 published his own textbook "Elementary basis of technology or a brief description of the works done at the plants and factories." Since then textbooks on chemical technology were published one by one in Russia during the 18th century. Books by F.A. Denisov, P.A. Iljenkov, M.Ya. Kittar, F. Lesgaft, M.V. Skoblikov, N.N. Lyubavin are especially notable. It is noteworthy that in the first half of the 19th century both in Russia and abroad more textbooks were published on chemical engineering than on general chemistry.

In 1804, in accordance with the Statutes of 1803, the Russian Academy of Sciences began to publish "Technological Journal, or Collection of works and news in technology and the application of scientific discoveries for practical use." In the preface to the first issue, its Editor-in-Chief academician V.M. Severgin wrote about his ideas on the necessity of close connection between science and industry. That preface said that the success of trades and plants is in obligatory connection with the success of science. The same idea stands out in the V.M. Severgin article "A brief review of success and the benefit of technology" also published in his first issue. There, technology is considered to be the applied science on industrial production based on the application of mechanical, physical and chemical knowledge of trades and arts.

So, more than 200 years ago the term "technology" entered European languages as a definition of one of the most important engineering sciences, science on industrial production.

* V. Fratzscher. "Technology, its tasks in connection with material and technical provision of production processes in industry," 1982. Transl. from German.

D.I. Mendeleev defines this term in the Brockgaus and Efron Encyclopedia as: "technology or knowledge on beneficial, (i.e. requiring less manual labor and natural energy), ways of processing natural products into products to be used by people (necessary or useful, or handy). Though technology differs greatly from socio-economic sciences, they have much in common, because the economy, (savings), of labor and materials, (raw materials), as well as time and power is the main task of any production and the essence of science on plant and factory production loses any sense if benefits, (economy), are not taken into account. It is a matter of chemistry, for example, to study methods of getting iron from ores or any other natural materials containing it, but a matter of technology to study the most beneficial ways and choose the most advantageous as per conditions of time and place, to make the product cheap with the desired properties and forms.

"Technology," said D.I. Mendeleev, "is a young science, which appeared only in the 19th century. Systematic summaries, accompanied not only by the description of the existing, (used or proposed), but also by the evaluation of its advantages, i.e. comparison of the achieved and possible, (theoretical on labour and material consumed), appeared only in the 19th century, mainly during the second half, when many specialized technical magazines were published."

Emphasizing the essence of technology as a science, D.I. Mendeleev wrote: "The science of methods used by manufacturers, elucidates scientifically what was practically achieved and not only masters the production but also expands the area of scientific understanding of things and phenomena..."

Though many methods used in the manufacturing route in the mankind experience, a practical combination of particulars will provide general conclusions which will be the basis of technology as a separate, applied science."*

This definition, given by D.I. Mendeleev, fully corresponds to the modern understanding of technology as a science on industrial production which stands on the foundation of natural sciences and sets its head against the economy.

First of all, technology is a science concerned with producing means of production. Its task is the studying, elaborating and non-stop development of general processes and their combinations to produce machines, apparatus, instrumentation and other modern means of production. Technology is a scientific base for machine-building, machine-tool, instrument-making branches of industry. The most important significance for machine building has the technology of metals, the getting and treatment ferrous and non-ferrous metals and their alloys, hot treatment, (casting, moulding, forging, pressing, stamping, rolling, welding, tempering, annealing), cold treatment, (cold rolling, wire-drawing, calibrating), and treatment by cutting (turning, drilling, milling, grinding, etc.).

The biggest and most diverse part of technology is chemical engineering, science on the most economical methods and means of mass chemical process-

* Brockgaus-Efron Encyclopedia. 1901. V. 33, p. 132.

ing of raw natural resource including agricultural production into industrial and consumer goods and intermediate products, used in other branches of material production.

Chemical engineering is a scientific basis of chemical, petrochemical, coke by-product, pulp and paper, food industries, building materials, electronics and many other branches of industry (Table 2.).

In recent years chemical engineering processes are widely used in all branches of industrial production. Development of machine-building, energetics, building, transport, communication, agriculture, health protection, and food is connected with the utilization of different materials, substances, products and preparations with the definite complex of mechanical, physical, chemical and biological properties. The development and production of these substances and materials were and will always remain the main task of chemistry and chemical engineering.

The aim of any production, including chemical is the obtaining of a final product by minimal capital investments and operating costs. This is achieved by choosing the proper technological processes, necessary equipment and rational production chart, as well as by automation control, technological process and production regulation.

All multiformity of chemical engineering processes are divided into five main classes: 1) hydromechanical, 2) thermal, 3) diffusional (or mass transfer), 4) mechanical, 5) chemical.

The class of chemical processes is the most numerous due to wide nomenclature of substances and materials produced under modern conditions on the basis of chemical transformations, for example: oxidation and burning, the burning of solid, liquid and gaseous fuel, sulfur, sulfur pyrite for sulfuric acid production, steel production from cast iron in Martin furnaces, electric furnaces and converters; reduction, the obtaining of iron and other metals from ores; electro-chemical, the decomposition of water and aqueous solutions of salts by electrolysis, electrolytic precipitation of metals, electrolysis in melt media, aluminum, magnesium and other metal production; electrothermal, obtaining calcium carbide, electric sublimation of phosphorous, electric welding of metals; heat treatment of metals, the reduction of metals from their compound, (oxides, halogenes, etc.), by more active metals (aluminum, magnesium), as well as silicon, accompanied by heat liberation, production of titanic sponge, carbon free alloys; plasmochemical, chemical reactions in a low-temperature plasma, getting powder and non-metallic materials from high-melting metals; synthesis of simple inorganic compounds, acids, alkalis, salts, metallic alloys and other inorganic materials and substances; complexing, separation and purification of platinum and other rare metals, gas purification from carbon oxide, obtaining hyperpure metals by synthesis and decomposition of complex compounds; roasting and clinkering, high temperature synthesis of silicon, obtaining cement clinker, aluminum oxide from nepheline, ceramic materials and products, thermal dissociation, obtaining lime, calcination of sodium bicarbonate, aluminum hydroxide; chemical enrichment of ores of nonferrous metals being in isomorphic state in the crystal lattice of the associated minerals, for example, by chlorinating or sulfating roasting with the aim to convert metal into volatile or easy to leach out by water compounds; oxidation-reduction-gasification of solid and liquid fuels, hydrocarbon conversion with water vapor, partial hydrocarbon oxidation; hydrogenation - synthesis of ammonium, methanol, and higher alcohols, hydrocracking and hydropurification of oil and refining products from sulfurous compounds, hydrogenation of fats; pirogenous processes, coking and low temperature carbonization, pyrolysis and cracking of liquid and gaseous petroleum products, dry distillation of wood; synthesis of organic compounds on the basis of carbon oxide, oleins, paraffins, aromatic hydrocarbons, acetylene and other organic compounds; chemical decomposition of complex organic compounds, boiling of wood chips in chemical solutions, (caustic soda, calcium bisulfite, etc.), aimed at lignification during cellulose production; polymerization and condensation, obtaining high-molecular compounds and their bases synthetic resins, plastics, synthetic rubber and other elastimeters, polymer films, paint and varnish coatings glues, synthetic fibers; synthesis of complex organic compounds dyes, intermediate products, medicines, pesticides, detergents and others; ammonolys, oxidation of hydrocarbons together with ammonia with the aim to get prussic acid, acrylonitrile and other compounds; aminating - obtaining carbamide and organic amine compounds of aliphatic-aromatic rows; nitrating, obtaining nitrate derivatives of organic compounds, powders, explosives; chlorination, obtaining chlor-organic compounds, chemicals to protect crops against pests, titanium tetrachloride; sulfonation, obtaining sulfoorganic substances, synthetic detergents; cyclization, conversion of organic compounds with the opened cycle into cyclic or aromatic compounds aiming at the improvement of motor-fuel quality; biochemical processes, fermentation in wine, beer, bread production, microbiological synthesis of protein, biological purification of industrial and domestic waste waters, etc., hydrolysis, exchange decomposition of different substances and water, hydrolysis of salts, obtaining metal hydroxides, decomposition of cellulose and other polysaccharides, which waste products of wood processing and agricultural production continue, to get glucose and other monosaccharides by treatment with mild solution of mineral acids; nuclear-chemical processes, obtaining fuel out of natural ores, extraction of pure transuranic elements and unreacted urane from fission products, use of radiation in chemical fusion, etc.

Presently, a new and promising branch of technology, biotechnology, is in the making. Technology of products obtained in the combined complex process of fine synthesis by biochemical methods, (amino acids, prostaglandins, interferon, etc.), is of great importance.

Industry uses the catalysis to accelerate chemical reactions being the basis of many of the above-mentioned processes.

Table 2

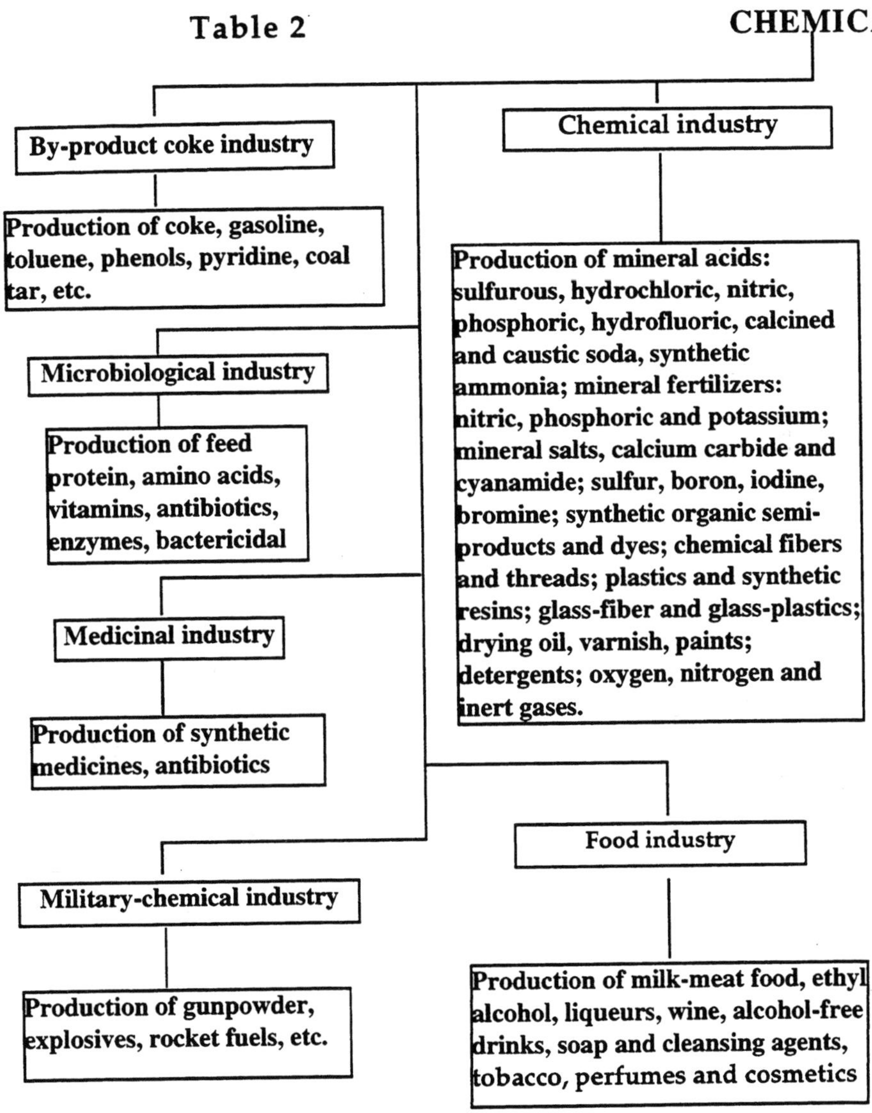
CHEMICAL
By-product coke industry
Production of coke, gasoline, toluene, phenols, pyridine, coal tar, etc.
Microbiological industry
Production of feed protein, amino acids, vitamins, antibiotics, enzymes, bactericidal
Medicinal industry
Production of synthetic medicines, antibiotics
Military-chemical industry
Production of gunpowder, explosives, rocket fuels, etc.
Chemical industry
Production of mineral acids: sulfurous, hydrochloric, nitric, phosphoric, hydrofluoric, calcined and caustic soda, synthetic ammonia; mineral fertilizers: nitric, phosphoric and potassium; mineral salts, calcium carbide and cyanamide; sulfur, boron, iodine, bromine; synthetic organic semi-products and dyes; chemical fibers and threads; plastics and synthetic resins; glass-fiber and glass-plastics; drying oil, varnish, paints; detergents; oxygen, nitrogen and inert gases.
Food industry
Production of milk-meat food, ethyl alcohol, liqueurs, wine, alcohol-free drinks, soap and cleansing agents, tobacco, perfumes and cosmetics

Table 2 (continued)

TECHNOLOGY

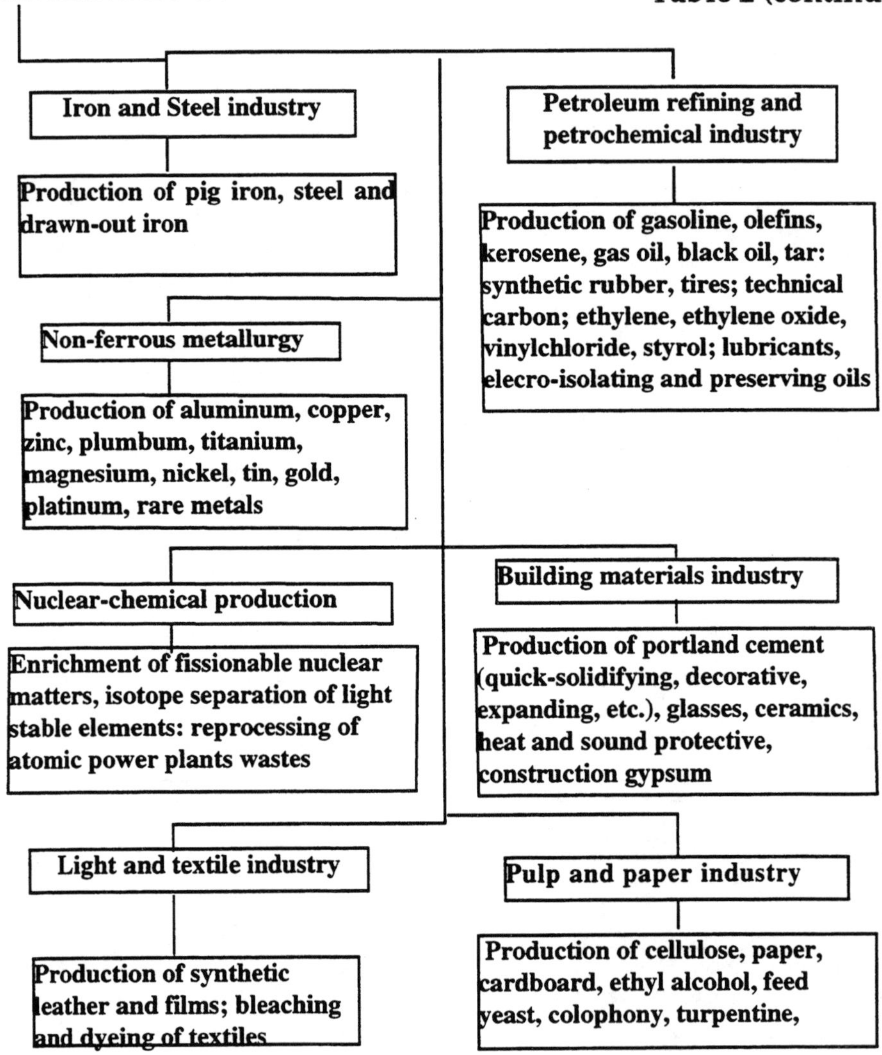

A quantitative description of the chemical engineering processes is based on the laws of chemical thermodynamics and chemical kinetics combined with the momentum- heat- and mass-transfer laws. Theoretical foundations of chemical engineering is the variety of tasks dealing with the description of phenomena on a microlevel, when physical and chemical effects are investigated for the estimation of the rate of process on a molecular level and in a local volume of the apparatus, and on a macrolevel, when hydrodynamic, heat, diffusion phenomena of a large scale character are added to those of a microlevel phenomena, the structure of which is explained by the design of the industrial apparatus, character of external energy supply, etc.

The diversity of chemical and physico-chemical processes used in practice explains various designs of apparatus for chemical processes, making highly efficient apparatus consuming minimum energy is the main task of chemical machine-building, to be solved on the basis of up-to-date achievements of theoretical foundations of chemical engineering. The design of such apparatus must provide the possibility to achieve and maintain favorable (from the point of view of chemical engineering) operation conditions and simultaneously correspond to modern machine building requirements. Evidently, to carry out an optimum number of chemical processes, an extremely large choice of chemical equipment is needed. However, to expedite elaboration and manufacture this equipment industrial engineers must have a limited choice of standard solutions. This contradiction can be solved by making chemical apparatus as a combination of the unified units and assemblies. To provide maximum possible number of combination of the machine units for making this or that version of apparatus or machine, these units, manufactured according to the common technology must be unified. Large capacity equipment must be both extremely reliable and efficient. Low-capacity chemical producing equipment must be easy to re-adjust and to reserve the provision of a wide-range application.

The chemical reactor is that element of the technological scheme, the perfection of which makes industrial production possible. The general theory of chemical reactors is comparatively young. Application of mathematical modelling for solving the task of scaled transfer from the results of a laboratory experiment to industrial conditions proved to be highly efficient. Successful studies of chemical kinetics heat- and mass- transfer accompanying chemical reaction, application of mathematical modelling allow industrial reactors to be designed more precisely. Some new efficient designs of reactors having high unit capacity were elaborated and the best conditions to carry out some important processes were found.

Modern chemical engineering processes must be uninterrupted and at high rates, self-adjusting to the most favorable operating conditions, meeting complex raw materials and energy utilization demands, combination and cooperation of different production, excluding possible air and water pollution. Chemization is one of the most important factors promoting intensification of economy as a whole. That is why chemical industry is and will be developing rapidly. Serious measures are required to make proper use of natural resources, fuel and energy. Qualitative changes of technological processes are required, including elaboration of new generation of catalysts, introduction to industrial production of such promising processes and apparatus as membrane processes, extraction by compressed gases at supercritical pressure and other new methods influencing the technological process with the aim of intensifying it. As an example, the influence of surface tension gradients promoting the appearance of surface convection and sharp intensification of mass-transfer at absorption with chemical transformations, rectification, extraction, creation of transient conditions by cyclic processes in chemical reactors, extractors, absorbers, utilization of electric and magnetic fields, pulsations, vibrations, etc.

Prospects for chemical engineering development are ascertained by the necessity to elaborate new economical technological processes, a highly efficient apparatus and a system of automation control, operation and optimizing not only separate processes but their interdependent complexes, i.e. complete computerization of production.

These are the tasks of a chemical engineer technologist. Many engineering problems are of a fundamental scientific character. The chemist investigator continuously receives new knowledge, synthesizes new substances, studies their properties and reaction kinetics. The chemical engineer creatively "materializes" this knowledge, combines it with a knowledge of other fields of science, such as physics, mathematics, mechanics, biology, economics or cybernetics. This unification of different knowledge allows the chemical engineer to elaborate new methods of production, new technological processes and apparatus, that is suitable for practical use. Figuratively speaking, the chemist-investigator perceives the new methods, the chemical engineer-technologist creates the new methods. The highest success is achieved only when the efforts of the investigators and engineers are combined, and they can equally divide this success.

The development of chemical industry training for the engineering and scientific staff is of paramount importance. The first technological institute in Russia was founded in St. Petersburg in 1828. The first clause of Status on the foundation of the institute reads, "...the aim of the Practical Technological institute is to prepare people with enough Theoretical and Practical knowledge for factories and shop management." By order of the curriculum drawn up by academician G.I. Gess, chemistry, mechanics and general technology were to be taught in detail. Study of educational subjects was divided into the mechanical and the chemical. The first was subdivided into solid and fiber materials treatment aimed at getting threads and textiles. The second included chemical processes required for threads and textiles treatment, first of all bleaching, dyeing and pattern printing. In the last century the institute became a higher educational institution to teach engineers-technologists two specialists, mechanical and chemical proceeding from the requirements of the developing industry.

In 1830, the Moscow Higher Technical School was founded to teach technologists. Later, in the 20th century, Russian technological and polytechnical institutes for the training of chemical engineers were founded in other big cities.

In the 19th century chemical engineering was mainly a descriptive science. The stage of chemical engineering development is related to the first quarter of our century, when knowledge of general processes and unit operations began to form. Nevertheless, it is interesting that early in the last century, the author of one of the first textbooks on chemical technology (1828), F.A. Denisov, professor of the Moscow University, noted the unity of many processes and apparatus for different chemical production or, as he wrote: "... means, instruments and machines used in various chemical arts." The basic role in the development of modern chemical engineering as an independent branch of science belongs to D.I. Mendeleev. He anticipated the idea of general processes and apparatus as the theoretical foundation of chemical engineering.

D.I. Mendeleev's ideas were developed in pre-revolutionary Russia by famous scientists-engineer A.K. Krupskii at the Petersburg Technological Institute and I.A. Tischenko in Moscow. Professor A.K. Krupskii, the specialist in the field of glass technology and industrial furnaces and fire-chambers was one of the first to publish, (1909), a Process and Apparatus of Chemical Technology course, describing industrial chemical processes based on the principles of physical chemistry.

I.A. Tischenko, a specialist in sugar production as well as in thermodynamics and heat- transfer almost simultaneously and independently, (1911), read a course "General Processes and Apparatus of Chemical Technology) to the students of chemical faculty of Moscow Higher Technical School. This course was published from 1913-1914. I.A. Tischenko thought that the main difference between higher chemical engineering and university education must be in the basic and specific character of engineering and chemical subjects and, most of all, the course of general processes and apparatus of chemical engineering and the course of general chemical technology. Moreover, the course "General Processes and apparatus of chemical technology covered the first four groups of processes - hydromechanical, thermal, diffusion and mechanical, i.e. physical and physico-chemical, but the course, "General Chemical Technology," covered chemical processes. Based on general chemical knowledge, mainly physical chemistry, on the one side, and general-engineering-technical thermodynamics, theoretical mechanics, strength of materials, machine parts, heat engineering and electrical engineering, on the other, chemical engineering subjects, reflecting the requirements of the specific branches of the industry. I.A. Tischenko realized the importance of a new subject to be taught to future chemist engineer-technologists. In the course of Processes and Apparatus of Chemical Technology he saw the missing section which accomplished the preparation of a new type of engineers and technologists and united general scientific and general engineering with specialized technical teaching. I.A. Tischenko put his ideas into practice, having been appointed the Rector of the Mendeleev Moscow Chemical Technological Institute in 1923. At that time, his disciple and colleague, academician D.P. Konovalov, contributed greatly to the development of Mendeleev's ideas at the Petrograd Technological Institute. He published the fundamental monograph "Materials and processes of chemical technology" in 1924.

The important stage of chemical technology development as a science was a book by F. Fokin, the professor of the Petrograd Technological Institute, "Methods and instruments of chemical technics," published in 1923, the same time a famous monograph of American scientists Walker W. , Lewis W. and McAdams, W. "Principles of Chemical Engineering" was published. For several years both were the main textbooks for teaching engineer chemists. A textbook by A.G. Kasatkin, the professor of the Mendeleev Moscow Chemical Technological Institute, "General processes and apparatus of chemical technology," was published from 1935 to 1937 and had 9 editions. The last two editions published after A.G. Kasatkin's death, were essentially added by his colleagues (1970 and 1973).

The textbooks of P.M. Lukjanov, a course in two volumes which was written by authors under the guidance of S.I. Volfkovich, a textbook under I.P. Mukhlenov's editorship and others, which played an important role in teaching engineers, chemists anechnologists, should be mentioned among numerous textbooks on general chemical technology. During the post-war years, many textbooks and monographs by Soviet and foreign authors were published both on the general problems of chemical technology its important parts.

Many textbooks, manuals and monographs on the processes and apparatus of chemical technology and general chemical technology evidencing the achievements of the international cooperation of scientists in this field of knowledge have been published from the 30s to now. Confirmation is large but it's not a complete list of References.

The national chemical societies of engineer-chemist societies organized in the 19th-29th centuries played a major role in the formation and propaganda of chemical technology and, during the post-war years, became the basis of the main engineering-technical international Societies; the European Federation of engineering chemistry, the Inter-American Confederation of engineering chemistry and the Asian-Pacific Confederation of engineering chemistry.

Scientific and technical progress assume an interrelated and interdependent development of science industry and education. At the same time, education, as the most long-range factor laying the foundation for success in future, requires scientifically stipulated long-term program of development. Training of personnel, indisputably, is an inherent part of scientific and technical progress and one of the most essential factors of its acceleration. Higher education and personnel training are the production branch of national economy. Academician V.I. Vernadskii said: "Higher education is an instrument in the world struggle for existence and is much stronger than dreadnought". That is why engineering chemical education requires new different measures to strengthen it. First of all, these measures are the supply of higher educational institutions with modern scientific equipment and computers. It is also expedient to introduce systematic columns dealing with the questions of improvement of engineering chemical education to some scientific and technical journals on chemistry.

We are deservedly proud of higher and especially of engineering education progress, which made the achievements of our science and industry possible. Under the present situation, further improvement of higher chemical engineering education is necessary with due regard for the newest achievements in science and technology, a realistic evaluation of the vast possibilities opened for the studying and application of chemical engineering processes in industry using cybernetics and computer methods.

Positive success has already been achieved in teaching general engineering and chemical engineering subjects, a new course of main processes and apparatus of chemical engineering has been adopted and introduced. Some new courses meeting modern requirements were introduced into the curriculum. "Chemical processes and reactors," "Heating engineering of chemical production," "Computerisation in chemical technology" are among such courses. However, the level of chemical engineering training is lagging behind the requirements of modern technology. Tomorrow's engineers need new knowledge. But new knowledge must make use of the methods developing scientific engineering intellect, engineering thinking based on a system analysis. System approach to the education expects, in particular, indissolubility of both courses, "Processes and apparatus of chemical engineering" and "Chemical processes and reactors". At the same time the course of "General chemical engineering" must present broader knowledge of complex raw materials processing, efficient arrangement of chemical technological processes, specific production in complex with power and ecological problems.

A chemical engineer needs interdisciplinary knowledge. It is apt to quote the aphorism of ancient Greeks, paraphrased by Bernard Shaw: "Narrow specialist learns much more of much less and so, until he knows everything about nothing and nothing about everything."

Chemical technology as an engineering science is responsible for the elimination of harmful discharges, because methods of processing and utilization of waste are mainly chemical and physico-chemical. One of the most important ways to avoid an "ecological spasm" caused by human activity is to qualitatively improve higher chemical engineering education. Its mission is to reform the attitude of engineers and chemists and to point out the general directions to elaborate ecologically pure production.

The main role among general engineering and chemical engineering subjects belongs in the course, "Processes and apparatus of chemical engineering." The aim of the course is to study the profoundly basic appropriateness of calculation, research and modeling of main processes and apparatus. The program must provide for teaching separate parts of the course in close system interconnection with one another. Special attention is given to the possibility to use the knowledge acquired for practical, design and research work.

A very important problem is the training of highly qualified engineers. Presently, large chemical concerns were and are being created to produce various products applying different methods and technological processes. That is why specialists that are engaged there should have a profound knowledge of the main processes and apparatus of chemical engineering. Chemical engineering subjects are not only the basis to teach a chemical engineer technologist, but they also play an important role in the building up of chemical thinking and the general outlook of the graduates.

No matter how important specific knowledge is the students obtain in the course, "Processes and apparatus of chemical engineering," the main task of education is the development of creative character of engineering chemical thinking. It is possible to achieve if education is the only problem: science on processes and apparatus must be taught in its development, its achievements and unsettled problems and analysis of its possible solution.

An explanation of some specific problems of processes and apparatus improvement should be subordinated to the solution of the most important problem of the economical development necessary to improve the quality and efficiency of production. Problems of efficiency and quality improvement in chemical engineering can only be solved in deep dialectical interdependence. Thus, the transition from the relatively small-scale production to the large-scale production units not only leads to a quantitative increase of production units but to the improvement of qualitative index. Reduction of capital and operation expenditures, improvement of quality control for technological processes results in the improvement of quality of products as it is, for instance, in ammonium production. Evolution of nitric industry, beginning with the first industrial unit of nitrogen fixation by the arc-method (1905) until now is a striking example of a struggle to reduce power consumption and the cost of production.

Chemical engineering education can contribute a lot. In a book by D. Ross, "Energy of waves,"* it is said of G. Goodwin, the scientific leader of the Ministry of Power of Great Britain: "He is a chemical engineer and asserts that such education is an ideal basis for being a specialist because it covers any engineering field and allows to talk the same language with mechanics, designers and electricians."

In conclusion, let us touch upon the classification and order of scientific terminology in the field of chemical engineering.

The classification of the main processes and apparatus played a positive role in the formation of chemical engineering as a science and subject. Strict classification of any branch of science requires precise terminology, which is typical for this field of knowledge.

Among all the languages of science, (logical, physical, mathematical), the chemical language is the most suitable auxiliary means of international scientific communication, (especially in written form), because of its distinct features as a ramified chemical nomenclature and its conventional symbolism. Nevertheless, to put scientific terminology in order in the field of chemical engineering, it is impossible to simply add physical, mathematical and other terms without touching upon the essence of the described processes and apparatus. The most important problems of modern scientific terminology is, first of all, putting the terminology itself in order, and, second, its standardization. These problems are closely connected with automation, storage and the search for scientific technical information and the necessity for international communication and the fruitful cooperation of specialists for interrelated branches of industry.

* D. Ross. Energy of waves. D.: Gidrometizdat, 1981, p. 52.

Much work was done by the chemical department of the Institute on the standardization for the "Classification and terminology of Main processes of chemical engineering." This work was the first attempt to lay the groundwork for cooperation between countries in chemical engineering and machine building when making out the unified lists for purchased equipment and designing. Classification of the main processes of chemical engineering was made on the basis of the general process appropriateness, their purposefulness and a way of accomplishment. It can also be the basis for chemical equipment classification.

The results were published by P.G. Romankov and M.I. Kurochkina* Five standard classes of chemical technology processes were the basis for this classification. The classes are divided into sub-classes, with the essence of their physico-chemical processes being different. Sub-classes then are divided into groups of processes, which differ in the way or purposefulness of their accomplishment. Classification provides the possibility for further details.

A reasonable classification of chemical processes is extremely complicated and the work has not yet been completed. More attention must be paid to further ordering the terminology to provide for a mutual understanding. It should be noted that thoroughly classified and accepted terminology is the maturity sign of any branch of science.

Presently, work is under way to accept and check the effectiveness of different standards, norms, work out classifiers, descriptive dictionaries and terminological recommendations. But much has yet to be done in the field of chemical technology. The introduction of international standards on terms will unify them, and exclude the professional slang from terminology. Descartes said: "Define the meaning of the words, and you will rid humanity of many fallacies." This idea is true for fields of science during the tempestuous development period, the same chemical technology enters at present.

ROLE OF CHEMISTRY AND CHEMICAL TECHNOLOGY IN SOLVING THE GLOBAL PROBLEMS OF MANKIND

Mankind is faced with four uneasy questions due to the growth of the population, rapid industrialization, concentration of population in cities and the ever-growing use of automobiles:

How to feed a rapidly-expanding population?

How to prevent natural resources from exhaustion and to provide industry with raw materials?

How to meet people's demands for energy?

How to prevent the contamination of the environment (man's habitat)?

* P.G. Romankov, M.I. Kurochkina, Zhurn. Prikl. Khimii. 1972, V. 14, No. 11, p 2971. Theor. Found. of Chem. Techn., 1974, V. 8, No. 3, p. 331.

These global problems provoke many pessimistic prognoses. Let us try to give short answers from the point of view of "historical optimism."

Yes, the world population grew quickly over the last decades. Nearly eight millenium back, when primitive man began to cultivate land and change his way of life to a settled one, about 5 to 10 million people lived on the Earth. By the beginning of this era, the population increased to 200 to 250 million, in the year 1000 it was more than 300 million, and in 1500, about 450 million people. At the beginning of the 19th century, the world population increased to one billion and mankind faced the 20th century as a family of 1.65 billion people! The century of high speeds became very speedy indeed in the area of rapid population growth. In the middle of 1987, the population of the planet numbered 5 billion people, and in 50 years it is predicted to reach 8 billion. In 1972, P.L. Kapitza, the academician, judging by the world population of 3.7 billion people and growth rates of 2% a year, calculated that in 700 years our planet would be populated so densely, that every man would have only 1 sq. meter of surface.* Lately, a decrease in growth rates is observed; presently it is about 1.73% a year. UN experts forecast a further decrease of growth rates and estimate that by the year 2100 growth will stop. By that time about 11 billion people will live on our planet.

According to the great physiologist I.P. Pavlov, care for daily bread prevails over the other phenomena of human life. Bread is the main product on the Earth. The health, physical condition and development of a human being depend first and foremost on nourishment. To provide people with the necessary amount of food is most important. The primary source of nutrition for men and animals are photosynthesis products, produced by green plants with the help of sunlight from carbon dioxide, water and mineral salts. Photosynthesis is based on chemical reaction:

$$H_2O+CO_2=O_2+1/6\ C_6H_{12}O_6$$

The increase of free energy $+\Delta F$ for this reaction makes up 120 Kcal. It can be accomplished only by supplying external energy. The reverse thermodynamic probability process, oxidation of sugar, is accomplished in living organisms when breathing uses various ferments called biocatalysts.

The majority of organic matter, produced by plants is polysaccharide-cellulose. About 10^{11} tons of cellulose is formed every year on Earth, the majority of which is drawn in a biological cycle of oxidation and decomposition.

The green leaf of a plant and its most typical organ, chlorophyll grain, plays the cosmic role of mediator between the main source of energy in our planet system, the Sun, and life on Earth. The organic resources of the Earth consumed by man in the form of food, clothes, fuel and other materials are filled with the help of green plants and the sun's energy is stored in the form of chemical energy of organic matter. The cultivation of land is the factor linking man and his activity with the sun that sends radiant energy to the Earth. Cattle-breading transforms the energy of primary plant matter to

* Kapitza P.L. Experiment. Theory. Practice. M., "Nauka", 1987, p. 418.

more valuable farm products, meat, milk, eggs, etc. However, this transformation is not efficient, because a major part of the forage energy is spent on the animal's activity. For example, the transformation of forage energy into animal produce for beef amounts to 10%, and for pork 20%. Recent experience shows that the assimilation of feed can be increased, the sickness rate can be lowered, the productivity of cattle-breading can be increased by proper composition of fodder rations, balanced protein, vitamins, chemical biostimulators and rich mineral matters. In fact, modern farms for raising broilers spend about 2 kg of feed per 1 kg of broiler weight.

Can the Earth feed all people living on it?

Yes! According to the UN Food and Agricultural Organization (FAO) data, the current overall production of food in the world in calorie and protein content exceeds the required level (on the average, per capita). Nevertheless, a great number of people, mainly in developing countries, suffer from malnutrition. According to FAO statistics, this number doesn't look like it is going to decrease, instead it keeps growing: from 1969 to 1971 it was 360 million and from 1974 to 1976 it was already 460 million people. Now the number of people suffering from hunger has almost doubled. 1/5 to 1/4 of the world's population eats much less than the admitted norms and 30 to 40 million people starve to death.

What are the maximum food resources of our planet? Possible data on the primary productivity of greens in the Earth's land and ocean differ greatly. In his article "Evolution of the biosphere and ecological prediction"* Academician S. Schwarts estimates the annual primary biomass production at 380 billion tons. Here, 300 billion tons of CO_2 is extracted from the air, 5 billion tons of nitrogen and 10 to 15 billion tons of other mineral elements from the soil.

Larcher W.** and many other scientists consider that H. Lieth data *** are more reliable. He estimates the pure productivity of the Earth's green cover to be 155.2 billion tons of dry biomass per year, including 100.2 billion tons on land and 55 billion tons in the ocean. It is possible that these data are underestimated, especially concerning ocean productivity. According to the later estimations**** 200 billion tons of carbon is bound by photosynthesis annually. It is estimated that approximately 1.5 tons of dry primary biomass is required every year to feed and to meet other vital needs of one man. Thus, if Lieth's data are accepted, the productivity of the Earth's green cover is enough for 100 billion people.

This figure has more principle than practical meaning. A number of population isn't defined just by food resources. According to demographer predictions, as was already said, the world population should stabilize at approximately 10-12 billion people. We are obliged to a timely and effective means

* See: World of Science. 1976, No. 2, p. 2.

** Larcher W. Physiological Plant Ecology. Springer Verlag Berlin-New York, 1975.

*** Lieth H. Angewandte Botanik, 1972, B.46, No. 1, p. 37.

**** Les biotechnologies. Vue d'ensemble Le Potentiel des biotechnologies. "Problemes Politiques et sociaux," 1980.

for solving the most important food problems which arise from the ever growing population, a lack of food rich in calories for a considerable part of population and its uncertainty and its influence upon the harvest. A steep rise in food resources is presently the biggest problem.

What are the ways to increase food production?

The main way is to increase the output of fertilizers and use them properly. It is necessary to remember, that highest efficiency can be reached if an expression from an old peasant's proverb is followed: 'Live, as if you will die tomorrow, and cultivate the land as if you will live hundred years.' Of course, the use of mineral fertilizers can be highly effective if it is combined with other agrotechnical measures and improvements in agriculture. But, on the other hand, no agrotechnics can compensate for the shortage of elements in the soil to feed the plants, most of all nitrogen, phosphorous and potassium and in many cases sulphur and microelements (B, Mn, Fe, Co, Cu, Zn, Mo).

The application of chemical fertilizers in the volume that fully meets the needs of agriculture, in combination with proper agrotechnics, as well as irrigation, land-improvement and selection of high-yielding varieties of plants and prevention of soil erosion, all this could result in a productivity increase equal to the West European level, that would mean, at least, doubling or even tripling the agricultural production of the world's cultivated land. It must provide 9 to 10 billion people with food rich in calories.

The number of chemical problems in agriculture are of great concern. The most important of them is the development of the most effective methods of fertilizer use, i.e. the solution to the problem of where, when and how to feed plants. A lack of nutrients in various soils, and the removal of them from the soil increases with the growth of productivity, and must be compensated by applying fertilizers. The dose of one or another element to be applied must be determined by its content in the soil and the amount removed with the harvest. The application of fertilizer prior to soil analysis is considered to be a flagrant violation of national economy interests.

We should not fail to mention the biological production of fixed nitrogen and the importance of such plants as clover, lucerne and other leguminous plants which make the soil fertile and increase albuminous feed production. In K.A. Timiryazev's opinion, all the tasks of agriculture come to the studies as well as creation of necessary conditions for plants nutrition. The correctness of this statement is illustrated by the fact that the achievements of the so-called "green revolution," in particular, the raising of high-yielding short-stalk cereals by selection, cannot be accomplished due to the lack of nutrients in the soils, particularly in old agricultural areas, and lag behind the needs of agriculture in mineral fertilizers production.

The full provision of plants with mineral nutrients is the main task of agro-chemistry. While working out ways to solve this problem the richest scientific inheritance of K.A. Timiryazev, D.N. Pryanishnikov and other scientists, founders of agro-chemical science, should be used.

Possibilities open up through the application of mineral fertilizers, illustrated by the growth of crop yield in some countries. In his book, D.N.

Pryanishnikov* provides a diagram of average wheat yield changes in Western Europe for 150 years and analyzes the reasons for these changes. At the end of the 18th century the average wheat yield in Western countries was about 7 centners/ha, called medieval by D.N. Pryanishnikov. The transition from a three-field to a multi-field system and the introduction into the crop rotation of clover, lucerne and other grasses enriching soil with nitrogen gradually led to a stable average yield of 15 to 16 centner/ha by the mid-19th century and was kept at that level for 50 years. Application of mineral fertilizers since the end of the 19th century resulted in an average yield increase of up to 30 centners/ha from 1936 to 38 and 40 to 50 at the present time. Comparing the conditions of land cultivation in Belgium, Holland, Denmark and other Western countries with conditions of Russia's Non-Chernozem Zone, D.N. Pryanishnikov emphasized that the soil in Western countries was originally poorer than Russia's, its fertility is secondary, it is connected with the application of labor and knowledge, and if there are no summer droughts in the Volga region or severe winter frosts in Holland and England, if there is no sun in Kiev or even a summer sun in Moscow.

Approximately 60 million people or 44% of the population of the Russian Republic live in a Non-Chernozem Zone, comprising 29 regions and autonomous republics. Agricultural areas occupy 52 million hectares, 32 million hectares of which are arable land.

The Non-Chernozem Zone has favorable possibilities for the intensive development of agriculture. Mineral fertilizers are of paramount importance for improving the productivity of that soil. Simple liming of the soil reduces its acidity and increases its productivity. During the pre-revolution period the average crop yield was 6 to 7 centners/ha in this area. The average productivity of soil in this area considerably increased due to the application of mineral fertilizers, liming of soils and land-reclamation measures (Table 3).

The power Russian fertilizer industry makes it easier to solve the problem of productivity increase.

World production of mineral fertilizers continues to grow rapidly. In 1980, it was 123 million tons on conversion to 100% of nutrients (N, P_2O_5, K_2O) against 30.2 million tons in 1960. During that period in the USSR it increased from 3.3 to 24.8 million tons, i.e. almost 7.5 times. In 1988 the USSR produced 37.1 million tons of fertilizers on conversion to 100% of nutrients.

Through photosynthesis all plants assimilate roughly 0.1 to 0.2% of the solar energy reaching the Earth. Agricultural plants assimilate solar energy better than plants in general: cereals, 0.5 to 1.5% and highly productive crops like rice, sugar-cane, sugar-beet, corn and some others, 2 to 4%. There is good reason to believe that the discovery of a photosynthesis mechanism in its full scale and mastering its control will allow for the increase of its efficiency of 2 to 3 and more times.

---

* Pryanishnikov D.N. Nitrogen in the life of plants and land cultivation. USSR. M.-L. AS USSR. 1945, p. 136.

Table 3.

	Average consumption of mineral fertilizers per 1 ha of cultivated land (on conversion to 100% of nutrients, kg)			Crop capacity of wheat in the USSR and some foreign countries (100 kg per 1 ha)		
	1970	1986	1987	1980	1987	1988
USSR	47	118	122	16.0	17.8	17.6
Bulgaria	159	199	180	39.7	38.2	39.9
Hungary	150	261	146	47.6	43.7	54.4
Eastern Germany	319	331	339	43.8	54.0	48.3
Poland	160	233	245	26.0	37.2	34.8
Romania	67	150	–	28.6	28.7	–
Czechoslovakia	230	328	303	45.0	50.8	52.6
China	–	–	–	18.9	30.4	–
Yugoslavia	77	134	–	33.6	36.2	–
USA	84	103	106	22.5	25.3	22.9
Great Britain	226	359	380	58.8	59.2	67.4
France	219	292	298	51.7	55.5	61.3
West Germany	379	427	427	48.9	54.9	60.5
Italy	82	170	–	26.9	30.3	29.4
Belgium	648	569	566	46.8	55.8	–
Denmark	220	243	–	46.9	57.5	–
Holland	759	772	–	62.1	69.3	–
Austria	–	–	–	9.6	13.8	–
Japan	390	382	–	–	–	–
Canada	19	54	–	17.4	19.5	16.0
Norway	242	263	271	–	–	–
Finland	174	211	234	28.8	20.2	–
Sweden	155	143	139	41.4	47.8	–
Argentina	–	–	–	15.5	20.4	–
Brazil	–	–	–	8.7	16.7	–
India	–	–	–	14.4	20.0	–
Mexico	–	–	–	37.7	41.8	–
Turkey	–	–	–	18.6	20.2	–

In the article "Biological transformation of solar energy" Academician A.A. Krasnovsky* shows, on the basis of experimental data on measuring quantum photosynthesis output for unicellular algae, that the maximum possible photosynthesis efficiency, i.e. transformation of the absorbed solar energy into potential chemical energy reaches 30%. The other 70% of sunlight energy absorbed by chlorophyll, is finally transformed into heat. This surplus energy is used to overcome potential barriers in intermediate reactions, for back reactions to active photosynthesis products and for other internal needs of a green cell. However, nearly a half of the solar radiation energy

* Vestnik of the USSR Academyof Sciences. 1979, No. 1, p. 83-85.

reaching the Earth, is in the visible range of spectrum and photosynthetically presents active radiation (PAR). Therefore, one can assume that the maximum possible use of solar energy in photosynthesis is no more than 15%. However, even this figure evidences the large reserves for a more intensive use of solar energy both on land and in the oceans of Earth.

Replacement of food and other agricultural products used for technical tasks by synthetic chemical products obtained on the basis of mineral raw materials has great importance for the growth of food resources. Thus, areas occupied by growing technical crops, will be used for growing food crops.

Traditional methods of food production are not the only way to increase it. The extraction and concentration of food matters, mainly proteins, from green plants, wood or sea-plants, as well as the chemical or biochemical transformation of unedible organic mass into foodstuffs, in other words, a more complex use of plants, this problem must be the first to be solved. In principle, all plant matters are good enough to feed, if not animals, at least microorganisms. The latter can be used as concentrated feed for agricultural animals.

The modern level of chemical technology, biotechnology and microbiology allows for the production of monosaccharides, ethyl alcohol, glycerin, feed yeast, amino-acids, and protein-vitamin preparations in a large industrial scale. Raw materials for them are polysaccharide containing plants in the form of lumbering, wood-working wastes, industrial processing of cobs, sun-flower peel, cotton seeds and wild grass and bushes.

The main method of plant waste processing is the hydrolysis with hydrochloric or sulphuric acids to get monosugars and using aqueous solutions of the latter as a nutrient medium to grow food and feed yeast. The microbiological industry was set up in Russia on this basis and produced about 1.8 million feed yeast in 1988. The synthesis of food matter is expedient, most likely, in cases when the necessity of certain food components arises (amino acids, vitamins, some protein preparations, etc.).

Care should be taken with the scientific research development which will provide a more effective progress of agriculture, medicine, and the food stuff industry. One of the most important tasks is the study of chemical structure and the properties of those compounds which are the primary basis of living matter, as well as a functioning of different physiologically active compounds in biological systems. These studies will become the material bases to know and understand the processes of vital functions. The discovery of structure and the mechanism of biological catalyst action can cause not only the progress of biology itself but chemical industry as well, because it allows one to create an active artificial industrial and specific catalyst.

Putting the possibilities of the living resources into practice will require great efforts. However, man possesses such knowledge and technology which opens the prospects for the improvement of mankind's well-being and most of all for providing him with full-value food.

Modern industry consumes a huge and ever increasing amount of natural mineral raw materials to produce metals, cement, ceramics, concrete, asbestos-cement materials, mineral fertilizers, acids, alkalis, salts, etc. Mined and processed rock mass amounts to many billion tons. In 1987, cement output reached 1051 million tons, steel, 737, mineral fertilizers, (on conversion to 100% nutrients), 146, but actually about 500, sulphuric acid -135.9* , synthetic ammonium, about 83, soda ash, about 29, caustic soda, 33 million tons.

Metals and their alloys are the basis of civilization. Consumption of metals and alloys will increase in the future. Therefore, a high rate of metal production growth is able to deplete the usual natural sources of raw materials, because rates of consumption growth may take the lead over the possibility of finding new deposits. (Table 4).

Table 4. Index meaning of metal resources (IMR) (years)*

Metals	Annual growth of consumpt. %	World capacity mln. ton	Index of metal resources (IMR)	IMR 10 times increas. capacity	IMR 50% recycling of metals	IMR 100% recycling of metals from 1990
Iron	1.3	1x105	109	267	319	598
Aluminum	5.1	1170	35	77	91	135
Copper	3.4	308	24	76	95	170
Zinc	2.5	123	18	16	101	212
Molybde-mun	4.0	5.4	36	87	104	165
Silver	1.5	0.2	14	82	117	328
Chrom	2.0	775	112	222	256	416
Titanium	2.7	147	51	127	152	255
Uranium	10.6	4.9	44	66	–	–

*Tien J.K., Arons R.M., Clark R.W. J. Metals. 1979, V. 28, No. 12, p. 26-28.

As it appears in the Table, should raw material deposits increase 10 times, (which is unlikely because the planet is almost fully explored), then the provision with raw materials required for metal production would increase only 2.5 to 3 times. Raw materials for iron production at average statistical rates of growth is left for 109 and aluminum, for 35 years. For all the conditional character of these figures, one naturally makes a conclusion that the raw materials crisis stands out against the critical situation in the economy, ecology etc.

Russia is rich in natural resources, but unlike the US meets their needs in raw materials by exploiting their mineral wealth. There are more than 50% of world reserves of iron, rich in ores and other metals. Nevertheless, despite huge natural resources, the mineral wealth of the Earth is not all that infinite. A lack of high-quality ores for some metal production becomes a real threat. The vital problem to be solved in the fight against this threat is the rational utilization of ores, in particular, complex raw materials processing aiming at extraction of all the useful elements, (considering the economic ex-

* In 1988.

pediency of course). The important reserve of natural resources is the secondary utilization of metals (Table 5). In the case of 50% metals recycling between the consuming and producing spheres, the provision of the world requirements in iron would increase from one to three centuries and in the case of full recycling up to six centuries. Known, but not fully utilized possibilities of the metal economy are of great important due to the decrease of metal consumption in machine-building, building and other spheres, corrosion, wear and tear losses, and an increase of metal utilization factors in the manufacturing industry.

Table 5. Content of the most important metals in the earth-crust*

Metals	Metals content, tons		Their ratio
	in the earth-crust**	in ores***	
Iron	$8.8x10^{15}$	$1.0x10^{11}$	$8.8x10^{4}$
Aluminum	$1.5x10^{16}$	$5.0x10^{8}$	$3.0x10^{7}$
Zinc	$2.4x10^{13}$	$8.2x10^{7}$	$2.9x10^{5}$
Copper	$1.8x10^{13}$	$1.8x10^{8}$	$1.0x10^{5}$
Tin	$7.0x10^{12}$	$5.1x10^{6}$	$1.4x10^{6}$
Plumbum	$3.0x10^{12}$	$5.1x10^{7}$	$1.6x10^{6}$
Molybdenum	$2.7x10^{12}$	$1.6x10^{6}$	$1.7x10^{6}$

*Zhavoronkov N.M., Evdokimov V.I. Complex use of mineral raw materials. 1978, No. 1, p. 29.
**Callot "Ann mines," 1971, V. 11-20, No. 1,3, p. 101.
***Falkie Thomas V. World minerals: the US perspective. "Skill Mining Rev.," 1975, V. 64, No. 42, p. 18-22, 24-25.

Bringing down the losses of metals while mining and enriching ores, in the production of base metal is a great reserve. The losses for different metals are estimated at 25 to 30%, in some cases they reach 50%. During the stage of enrichment losses are at their highest. The application of traditional methods of enrichment to obtain rich selective concentrates becomes less effective due to the increase in the utilization of ores with complex mineral compositions or poor ores. In these cases, the metal to be enriched is often a part of a crystal lattice of mineral rock in an isomorphic form and to extract it chemical processes are needed. Chemical technology here has a large field of practical application.

The development of economical processes of chemical enrichment of the poor for the metal production ores can make raw materials resources practically unlimited. The content of many important metals in the two kilometer stratum of the Earth exceeds its content in the ores many times over (Table 5).

As seen from the Table, for most of the metals their content in the Earth exceeds its content in the ores 100 thousand to 1 million times. It goes without saying that the extraction of metals from poor ores requires higher power expenditures. Of particular importance is the substitution of highly scarce metals and their alloys for less scarce or non-metallic materials, if possible. Aluminum takes the first place among metals in the consumption growth rates. By now, aluminum alloys and mainly magnesium aluminate are the basic structural material for the aircraft industry and rocket sciences. In perspective, aluminum alloys as structural materials will supplane iron. They have the advantage of lightness and are highly corrosion-resistant. Aluminum takes the first place in raw materials resources: its content in the earth-crust is 8.05%, but iron only 4.65%. Aluminosilicates of calcium, magnesium, sodium and potassium constitute more than 90% of the earth-crust weight. The main raw material to produce aluminum is bauxite (aluminum ore), the world reserves of which will be depleted by the first quarter of the 21st century. The mining, transportation and processing of aluminum ores into aluminum oxide become more expensive as the quality gets worse. To provide for aluminum production growth, it is necessary to expand its raw materials sources by involving poor-quality bauxites, nephelines, alunites, kaolin clays and the mineral part of coal having infinite deposits. But to process by them applying the known alkali method is inefficient.

The original method to obtain aluminum oxide from natural nephelines and its waste was developed in Russia. It makes it possible to obtain 0.3 additional tons of potassium, 0.7 tons of soda ash, 7 to 8 tons of portland-cement and 40 grams of gallium per each ton of aluminum oxide. There are three plants currently applying this method in Russia.

The electrolysis method applied by the aluminum industry today has some disadvantages, namely: electrolysis requires pure enough aluminum oxides, its production is rather complicated; fluoride sales and carbon electrodes are needed in large amounts; high power consumption for electrolysis, 16 to 18 thousand kW/h per one ton of aluminum. That is why the search for a more economical, less power consuming processes is the most important task of chemical technology.

Chemistry and chemical technology face important tasks in the elaboration of new materials and aluminosilicate based methods. New materials would be able to be substituted for highly scarce ones. The above examples show that difficulties arising from the provision of industry with natural mineral raw materials can be overcome, though supreme efforts should be exerted by scientists and engineers.

Power engineering can become the only real limit for the further development of material culture in the world. Coal became the basis for industrial revolution. Earlier this century, its share among energy sources was 95%. Presently, it is cut down to 30% and the share of oil and gas increased to 65%. The remaining 5% is the energy produced by nuclear power and hydropower plants. Energy consumption increased 20 times during last century. It exceeds the growth of the population: growth rates of population=1.73%, production of primary energy resources=3.9% a year.

Production of world energy resources in 1987* was: coal and brown coal - 4566 mln. tons, oil (including gas condensate - 2795 mln. tons, electric power - 10400 billion kW/h overall = 10.36 billion tons of ideal fuel.

The USA share is 2225 million tons, Russia's share is 1961.2 million tons of ideal fuel.

How are natural energy resources used? Academician A.P. Alexandrov in his article "Energy prospects"** quotes the following world average data: 20% of energy resources are used for energy generation; 20% to provide aviation, automobile and railway transport, sea and river fleet, and small-scale power diesel units with fuel; 30% for heating and industrial technological processes, consuming low-potential heat; 30% is consumed by industry as high-potential heat at 500 to 1500°C and higher temperatures in chemical processes, metallurgy, building materials industry, etc.

The second half of this century was marked by ever-growing oil and gas consumption for power engineering. They are easy-to-use fuels. Oil and gas consumption for chemical and petro-chemical industries as starting raw materials, for important products, has also increased.

What are the energy resources of the planet?

World reserves are estimated at 12800 billion tons of ideal fuel (7000 Cal/kg), comprising 11200 billion tons of coal, 740 billion tons of oil and 630 billion tons of natural gas.

World prognosticated energy reserves to be extracted are estimated at 3800 billion tons of ideal fuel, coal, 2900, oil, 370 and gas 500 billion tons***. According to some prognoses, mineral fuel reserves are higher and the reserves of liquid fuel contained in bituminous sands and shale exceed oil reserves. According to scientific estimates the reserves of coal, peat, oil and gas will suffice for the next 200 to 300 years if the rates of consumption would remain at present level. Discovered reserves of U_{238} used in nuclear energy generating reactors, exceed the overall resources of organic fuel.

Water power resources on our planet should not be underestimated. Should water power be fully used a greater part of the world demand in electric power would be satisfied. The report of "World Watch" Institute in Washington* says, that water power promises to become the basis of economy in the post-oil era and hydro-power plants will generate electric power years after the last oil-well is run dry and the last coal mine is depleted.

Overall world reserves of water power amount to 73 trillion kW/h annually. Presently, only 2.75% of its potential is used. In 1979 electric power generated by all the world's hydro-power plants is equivalent to 542 million tons of ideal fuel. West European countries, Japan, USA and South Canada did all that was possible to use the most water energy. Europe uses only 60%

* National economy in the USSR in 1988. Annual Statistics. M., p. 673. Finance and statistics. 1989.

** Izvestia, 10 April, 1979.

*** Kirillin V.A. Energy engineering. Today and tomorrow. Publ. House "Pedagogica," 1983.

* "Za rubezhom," 1990, No. 29, p. 21.

of its water power potential. Though it is only one quarter of the Asian continent potential, hydro-power plants of Europe generate 2 times more electric power than Asia. Africa uses only 5% of water power.

The first hydro-power plant (HPP) in Russia was built in 1892 on the Berezovka river in Rudnyi Altai to pump out water from the Zyryanovsky mine. The four-turbine HPP had 150 kW total capacity. In 1917 there were 24 hydro-power plants in Russia with a total capacity of 15 thousand kW**.

The first HPP built after the October Revolution was Volkhov, which was commissioned in 1926 with the capacity of 58 thousand kW. In 1932, Dneprovskaya HPP with a capacity of 560 thousand kW was built and commissioned in accordance with the GOELRO Plan***. In 1975, its capacity increased to 820 thousand kW. In 1967, Bratskaya HPP on the Angara River with the capacity of 4.5 mln kW, and in 1972, Krasnoyarskaya the largest in the world on the Enisei river with the capacity of 6 million kW, then the Sayano-Shushenskaya with an installed capacity of 6.4 million kW. Construction of cascade type hydro-power plants on the Volga river with an overall capacity of 10 million kW began before World War II and was completed after the war. Nureckskaya in Tadzhkistan on the river Vakhsh, Toktogulskaya on the river Naryn in Kirghizia, Ingurskaya in Georgia and less powerful hydro-power plants should also be mentioned.

It is impossible to use the entire potential of water power. In flat country areas construction of hydro-power plants requires flooding large territories and in the highlands construction problems arise. Hydro electric power generation will grow in the near future. Direct solar, as well as wind, tidal, geothermal and other renewable energy sources should be widely used, (wherever expedient), but we cannot expect them to play an important role in power engineering. First of all, solar energy should be directed for the intensification of land cultivation.

Thus, estimations of world energy resources made by different scientists confirm the groundlessness of the impending energy crisis. But we cannot deny that energy resources are not infinite. Actual solution of this problem is controlled nuclear fusion. Nuclear fusion, (gellium being a chemical element), was first discovered in the Sun and other Stars of the Universe. Even in 1929, physicists P. Atkinson and F. Hautermans concluded that energy emitted by the Stars is the result of nuclear fusion that takes place in the entrails of Stars. The theoretical works of G. Gamov, G. Bete, K. Vartszeker and others accompanied this discovery. Studies showed that nuclear fusion of gellium from hydrogen nuclei-protons, is a basic source of energy. The Sun, granting life on our planet, is a natural thermonuclear reactor, in which hydrogen is already "burning" for 5 billion years, is converted to gellium, the "burning" being under stable controlled conditions.

Thermo-nuclear reaction was first carried out on the Earth during fusion bomb tests. But these tests, as well as subsequent ones, were uncontrolled. The

** "Nauka i Zhizn", 1990, No. 4, p. 37.

*** The first long term plan of the national economy based on the electriciation, elaborated in 1920, on the instruction and under V.I. Lenin's guidance.

energy generated is of a great destructive force. Control over thermonuclear reaction for peaceful power generation would provide humanity with practically an inexhaustible source of energy.

In 1955, at the 1st International Conference on the peaceful use of nuclear power in Geneve, a famous Indian physicist Homa Baba expressed his opinion that the problem of a controlled thermonuclear reaction would be solved in the next 20 years. Though this prediction did not come true, works on nuclear fusion still advance. It is highly probable that a controlled thermonuclear reaction will be accomplished in a test-reactor at the end of this century, which will create the necessary prerequisites for the emergence and construction, early next century, of the first thermonuclear power plant.

Thermo-nuclear fusion is based on the interaction of a preliminary heated mass of light element nuclei. To provide for the corresponding kinetic energy required for over-coming the electrostatic repulsion of the nuclei, positively charged, the temperature should be high enough. The so-called "conditional ignition" needs power generated by reacting masses exceeding its unavoidable losses. It is usually expressed by minimal temperature, density, (number of nuclei per volume unit), and time for the retaining of plasma. To carry out a thermonuclear reaction plasma should be heated up to 100° million C. While heated molecules and atoms are ionized and primary mass is transformed into plasma, which consists more or less of free moving electrons and electrically charged ions. Owing to that the fact that plasma can be "locked" in magnetic field. The required conditions of success in a controlled nuclear fusion is the extending of knowledge in the field of plasma physics.

More nuclear fusion reactions are possible between the first five elements of the Mendeleev Periodic Table, namely: hydrogen, gellium, lithium, berillium and boron. In the near future, the so-called deuterium-tritium reaction is important. In the course of this reaction deuterium and tritium, (heavy and super-heavy hydrogen isotopes), are transformed into gellium nuclei:

$$D + T \rightarrow {}^{4}He + n + 17.6\ MeB$$

The neutron bombardment of natural isotopes of lithium results in trinium. Thus "starting fuel" for this cycle consists of deuterium and lithium:

$$n + {}^{6}Li \rightarrow {}^{4}He + T + 4.8\ MeB$$

or total:

$$D + {}^{6}Li \rightarrow 2\ {}^{4}He + 22.4\ MeB$$

The advantage of thermonuclear fusion as compared with uranium fission reactions is that fusion products are not radioactive. Elements of reactor structure undergoing neutron irradiation, will be radioactive in connection with that special choice of material which is required.

Impressive progress is achieved in controlled nuclear fusion under the "Tokamk" program, the research in thermonuclear physics began in the '50s by I.V. Kurchatov and L.A. Artsimovich.

A famous nuclear physicist engineer, H. Knoepfel, in his lecture at Ispra Courses on Engineering Aspects of Thermonuclear Fusion Reactors*, showed the ratio of mineral fuel energy resources and thermonuclear energy.

The deuterium cycles present only inexhaustible sources of energy. Actually, deuterium energy in 1 liter of water is equivalent to 300 liters of petrol, there are $14 \cdot 10^{20}$ liters of water on the Earth. Reserves of deuterium in sea water are equivalent to 10^{21} tons of oil according to present-day estimates. For comparison, let us recall that world consumption of energy resources in 1987 amounted to 10.36 million tons of ideal fuel. As per Knoepfel's opinion, the deuterium-tritium cycle will be, most likely, the basis for the first commercial reactors. It has the lowest working temperature and reaction rate 100 times higher than that in competing nuclear fuels. Nevertheless, tritium fuel can be treated as an intermediate stage. The main task - the reactor working on pure deuterium or proton fuel, the sources of which are infinite, radioactivity minimal and no complex process to obtain tritium.

As predicted by academician Styrikovich, ** energy consumption will grow slower in the near future than in the past decades and at the end of the 21st century, when the world population stabilizes and the slow growth of actual energy consumption will be compensated by higher efficiency, practically permanent levels of energy consumption would be easily maintained.

The prerequisites being what they are, by the growth of energy resources, consumption by the year 2000 will reach 20 billion tons of ideal fuel per year and by 2100, 60 to 80 billion tons. It is impossible to allow the non-renewable energy sources to be exhausted before power engineering will be converted to practically inexhaustible energy sources. It requires the elaboration of optimum strategy for power engineering development for a longer period.

During the transition period, the use of nuclear power, coal and hydropower will grow. Oil and gas should be spared as raw materials for the chemical industry, their use in power engineering should be minimized. In this connection, but on a new basis, an old problem arises for the production of synthetic liquid fuel from coal to meet the demands of auto- and air-transport. Chemical technology has great opportunities in this area. It also faces great tasks to save energy resources by substituting high power consuming technological processes for low power consumption. Here is an additional source of energy. Much can be, and must be, done or this.

Another global problem rivets everyone's attention. This is the evolution of biosphere and environmental protection. Creative forces of Nature considerably exceed the productive focuses of a human being. Plants, microorganisms and animals of the planet carry out geochemical work, exceeding the possibilities of the world industry many times over. But the productive forces of man grow rapidly. Today, many already upset the balance of Nature on large territories. Large industrial cities and towns face problems of fresh water bodies exhaustion and fresh air deficiency. The removal and utilization of wastes and sewage from large cities become complicated and expensive.

* Knoepfel, H. Lecture at Ispra Courses, June 23-27, 1980.
** World of Science. 1980. No. 1, p. 2.

Different forms of environmental contamination, (air, soil, water), of one country often affects other countries' interests. In this respect, the atmosphere is most affected.

Pollution of the world's oceans is of great concern. Oil tanker accidents become more frequent, causing the contamination of the sea-coast, the death of fishes, oysters and other sea life, and does much damage to health resorts. Some countries have made international waters the permanent burial place of radioactive waste.

At the same time, the contamination of rivers with everyday waste finally results in the contamination of the ocean. All of this has hazardous side-effects and naturally worries millions of people.

Thus, a new aspect, ecological, became another concern of future problems. New predictions related to rapid environmental pollution and ecological balance disturbances between man and nature were added to the rapid growth of the population forecast and from that arises the threat of hunger and the depletion of natural resources. Many scientists picture a gloomy future for mankind and call to save the planet from the harmful consequences of progress. There was even an expression: "industry attack on nature." In his article "In defense of DDT and other pesticides" Nobel Prize winning scientist Norman Berlough remarks that the seeds of the present-day storm boosted by newspapers, radio and TV, aimed at environment protection were planted in 1962 by the book "Silent Spring" written by the talented scientist Rachel Karson, which sold many copies[*].

One recalls the involuntary warning expressed by F. Engels in "Dialectics of nature"... "we do not rule over nature as a conqueror rules over foreign people, we do not rule as someone staying away from nature... on the contrary, we with out flesh and blood and brain belong to it and are inside of it... all our supremacy over it is that we, unlike other creatures, can perceive its laws and apply them correctly."[*]

The last lines of this warning shows that the way to solve the problem is the thorough research of all processes taking place in the biosphere and the study of industrial activity of a man over the evolution of biosphere. Rational principles of modern technology should be based on that knowledge proving for the reasonable use of natural resources, not just its passive "protection." It would be incorrect to think that man's interference in nature should be restricted, that industrialization must be stopped. On the contrary, modern science and technology must be used as the most important means of rational consumption of raw materials, increase of renewable raw materials and improvement of the biosphere.

The development of industry is required for the progress of humanity and there must be an appeal to protect nature, (as if nature needs protection, instead of the people), often becomes indifferent to people's destiny. To be fair, let us note that the emotional approach to the evaluation of possible con-

---

[*] UNESCO Express. February 1972.

[*] Engels F. Dialectics of nature. Marx K. Engels F. Coll. Works. V. 20, p 496.

tamination of the environment and its consequences began to change for the sober reckoning in solving this problem.

For thousands of years the problem "Man and Nature" was perceived by social consciousness as the problem of the most intensive exploitation of natural resources, industrial development of vast territories and the substitution of natural "chaos" for man's "order." Today, the problem is raised so, as to cause no spontaneous distortion or break in the natural cycle during public production but to be included in it as a necessary link and become a condition of its further development.

Academician S. Shwarts in the above mentioned article "Evolution of biosphere and ecological prediction" notes, the relationship between man and nature which created him, the relationship between the biosphere and the sphere of the brain and labor, noosphere, which becomes one of the most important and difficult problems to solve. Conflicts that arise out of it, the complexity of which is sometimes treated as an ecological crisis, are numerous and multiform. But at closer consideration it appears that many of them have a common origin, the crux of which is in contradictory relations of two self-regulation systems the biosphere and human society.

The ecological crisis is not in the destruction of biological natural resources as a result of an unreasoned activity of a man, but in the fact that the self-regulating ability of natural complexes is injured or begins to "work" against man and humanity. There are objective prerequisites to establish a new, harmonious unity of man and nature. The realization of these prerequisites should be based on the knowledge of laws of biosphere development. The basic ideas in this field were formulated by academicians V.I. Vernadsky and V.N. Sukachev.

Vernadsky's main idea is that the superior form of material development on Earth determines life, and subordinates other planetary processes. This is the difference between the Earth and other planets in the solar system. A man transforms nature by the cultivation of millions of hectares of land to grow farming plants, by processing billions of tons of rock to extract minerals and by regulating the flow of rivers.

Man's influence on nature became one of the most important problems of natural science. The task is to foresee for certain the possible consequences of this interference, to be able to neutralize undesirable results and enrich the environment. It requires a broad approach to the natural processes and a comprehension of them. The scientific basis for such an approach is the study of the biosphere and of course, to study matter and energy circulation on the planet. This circulation includes food chains of biogenocenosis, metabolism between animate and inanimate nature, global circulation of carbon dioxide, oxygen, nitrogen and other biogenic elements. The cyclic recurrence of biospheric processes proves their recurrence, stability, and reproductivity. To perceive these cycles is the task of natural science.

Environmental protection and the rational utilization of natural resources were of great concern to V.I. Lenin, the founder of the Soviet State. In its policy on environmental protection Russia acknowledges the vital importance of this problem not only for the Soviet people but for humanity as a whole.

One should not underestimate the danger of the damage of environmental contamination, which can deprive humanity of normal living conditions on our planet. The decisive role in the fight with this danger belongs to science. Even now, science provides the means for air and water pollution prevention, wind and water erosion of soils, irrational use of forests, etc. It is necessary to continuously improve and develop new effective ways and means to purify industrial exhaust. This decisive role belongs to chemistry and chemical technology. The efforts of scientists all over the world should be united on the basis of international division of labor.

Modern chemistry makes it possible to recirculate wastes. D.I. Mendeleev pointed out that the main goal of advanced technology is finding the way to produce something useful out of something useless. One of the most important tasks of modern technology as a science is to elaborate processes while excluding harmful exhaust to air and water bodies.

It should be remembered that in the past there was no lack of pessimistic predictions. So, in 1887, English scientist Thomas Hexly predicted the end of civilization in 50 years due to nitrogen hunger, nitrogen exhaust in the soil and complete depletion of natural resources good for mineral fertilizer production (Chile saltpeter). The same idea was repeated by William Cruks, a famous physicist from the last century. In 1898, K.A. Timiryasev read a public lecture to the British Royal Society: "Is mankind threatened with destruction?*, which became answers to W. Cruks' words about the danger of the world nitrogen hunger and Lord Kelwin's prophecy of mankind's death caused by suffocation due to the mineral fuel consumption growth resulting in the gradual decrease of oxygen in the air and increase of carbon dioxide. In his lecture, K.A. Timiryasev not only revealed scientifically proven erroneousness on the grounds of prominent scientists, but expressed confidence in a solution to the problem of production of fixed nitrogen out of fairness, let us note, that W. Cruks being a real scientist gave his gloomy prospects a ray of hope for science.

The problem of the production of fixed nitrogen was solved at the beginning of this century, and the creation of the nitrogen industry was a great achievement of science and technology.

K.A. Timiryasev ended his lecture: "Until the sun shines and scientists' ideas persist mankind should not be afraid of the future." These words remain actual even today. In 1944, during the war, when the problem of "man and nature" seemed to be not so urgent, academician V.I. Vernadsky in a "Few words about noosphere" wrote: "For the first time in the history of mankind the interests of each and every person and people's ideas determine the life of humanity and are the criterion of his ideas about justice..."

Ideas of democracy act in unison with spontaneous geological processes and with the laws of nature. Russians can face their future with confidence. It is in our hands, and we will not let it go.*

---

* K.A. Timiryasev. Sel. Works. V. 2, M., Selkhozgiz. 1948. p. 283.

* Vernadsky V.I. Progress of modern biology. Part 8, 1944.

As science and technology progress, the role of chemistry and chemical technology in the growth of labor productivity increases. We can speak about the chemization a universal way of mastering production activity. It is difficult and may be impossible to find such a field of human activity in which the application of chemical methods and materials would not revolutionize the established processes. Here is the importance of chemization, as one sees it, not only the development of chemical industry and the broadening of the fields of its application, but also the introduction of chemistry and chemical methods and processes into other branches of industry, agriculture, medicine and everyday life.

The progress of the modern chemical industry depends more on the development of fundamental research than in any other branches. It is in a wide sense the result of fundamental research, connected with deep penetration into microworld, knowledge of electronic structure, nature of chemical bond, reactivity and mechanism of chemical reactions. Chemists are forced to draw various physical methods of research into the sphere of their interests.

Modern chemistry has many sides and is omnipresent. The innermost secret is that of animate and inanimate nature, the more necessary is the interference of chemistry to discover it. Science reaches out its power to learn the most surprising phenomena of nature from the processes of the living cell to the transformation of matters taking place deep in the Earth and ocean, on the planet and in space.

The progress of science and technology in Russia as well as industrial progress, is not isolated. Soviet scientists and engineers study and appreciate the achievements of other scientists and engineers in the world. The scientific and technological revolution put forward many new and complex problems. Many of them are of a global character, the main task is, indisputably, to keep peace on Earth. That is why cooperation between scientists in the world is of great importance. Their efforts to solve the specific problems of science and increase the effectiveness of scientific research should be united.

A list of some books in the field of Chemical Engineering, (section 3. Chemical technology as a scientific discipline and an educational subject).

1. L.F. Fokin *Methods and Instruments of Chemical Technology*, L.-M.: Gosizdat, 1923.
2. D.P. Konovalov, *Chemical Technology. Materials and Processes of Chemical Technology*, L.-M.: Gosizdat, 1925.
3. A.A. Kirov, *Apparatus and Fundamental Processes of Chemical Engineering*. M.-L. Gosizdat, 1927.
4. W.H. Walker, W.K. Lewis, W.H. McAdams, *Principles of Chemical Engineering*. New York-London: McGraw-Hill Book Co., J.N.C., 1927.
5. M.L. McGabe, J.C. Smith, *Unit Operations of Chemical Engineering*. New York-Toronto-London: McGraw-Hill Book Co., J.N.C., 1956.
6. W.L. Badger, J.T. Banchero, *Introduction to Chemical Engineering*. Ibid., 1955.
7. Kasatkin, A.G. *Main Processes and Apparatus of Chemical Engineering*, M.: Khimiya, 1973.

8. Ya. Tsiborovsky, *Processes of Chemical Engineering*. Ed. P.G.L. Romankov: Gos. nauch.-tekhn. Izd. Khim. Lit., 1958 (in Polish).
9. D.A. Frank-Kamenetsky, *Diffusion and Heat-transfer in Chemical Kinetis*, M.: Nauka, 1977.
10. V.B. Kogan, *Theoretical Foundations of the Main Processes of Chemical Engineering*, L.: Khimiya, 1977.
11. A.N. Planovsky, P.I. Nikolaev, *Processes and Apparatus of Chemical and Petro-chemical Technology*. M.:Gos. nauch. -techn. Izd. petrochem. min.-fuel industry, 1960.
12. J. Perry, *Handbook of Chemical Engineer*. Ed. N.M. Zhavoronkov, P.G.L. Romankov: Khimiya, 1969 (in English).
13. S. Bretshnaider, *Properties of Gases and Liquids*. Methods of Technical Calculation. Transl. Ed. P.G.L. Romankov: Khimiya, 1966 (in Polish).
14. P. Reed, J. Prausnitz, G. Sherwood, *Properties of Gases and Liquids*, L.: Khimiya, 1982.
15. P. Benedek, A. Laslo, *Scientific Principles of Chemical Technology*, Trans. from German. Ed. P.G.L. Romankov: Khimiya, 1970.
16. N.M. Zhavoronkov, *Hydraulic Principles of Scrubbering Process and Heat Transfer in Scrubbers*. M.: Sovetskaya Nauka, 1944.
17. M. Lewa, *Tower Packing and Packed Tower Design*. U.S. Stoneware C., 1953.
18. G.A. Morris, J. Jackson, *Absorption Towers*, Butterworth Sci. Publ., 1953.
19. P.V. Danckwerts, *Insights into Chemical Engineering*, Oxford: Pergamon Press, 1981.
20. *Process Heat Exchanger*. Y. Cavaseno, Ed. New York: McGraw Hill Publication, 1979.
21. *Handbook of Separation Techniques for Chemical Engineering*. P.A. Schweitzer, Ed. New York: McGraw Hill Book Co., 1979.
22. *Plant Design and Economics for Chemical Engineering*. M.S. Peters Ed., K.D. Timmerhaus, Tokyo: McGraw Hill Kogakusha Ltd., 1980.
23. K.F. Pavlov, P.G. Romankov, A.A. Noskov, *Course on Processes and Apparatus in Chemical Engineering. Examples and Problems*. L.: Khimiya, 1976.
24. P.G. Romankov, M.I. Kurochkina *Hydromechanical Processes in Chemical Engineering*. L.: Khimiya, 1982.
25. Ya.L. Averbuch, F.I. Zaostrovsky, L.I. Matusevich, *Processes and Apparatus in Chemical Engineering. Hydraulic and Mechanical Processes* K.N. Shabalin Ed. Sverdlovsk: Ural. Politechn. Inst., 1969.
26. T.A. Milinovskaya, I.A. Kobrinsky, O.S. Kirsanov, V.V. Reinfart, *Separation of Suspended Matters in the Chemical Industry*, M.: Khimiya, 1983, p. 263.
27. D.G. Pazhi, V.S. Galustov, *Principles of Liquid Spraying*, M.: Khimiya, 1984, p. 254.
28. L.I. Braginsky, V.I. Begachev, V.M. Barabash, *Mixing in Liquid Media*, L.: Khimiya, 1984.
29. V.A. Zhuzhikov, Filtration. *Theory and practice of Suspended Matter Separation*. M.: Khimiya, 1980, p. 398.

30. M.A. Mikheev, *Principles of Heat Transfer*. M.-L.: Gosenergoizdat, 1947.
31. S.S. Kutatelidze, *Principles of Heat Transfer Theory*, M.: Atomizdat, 1979, p. 415.
32. T. Hobler *Heat transfer and Heat exchangers*. Transl. from Polish. Ed. P.G.L. Romankov, Goskhimizdat, 1961.
33. P.G. Romankov, V.F. Frolov, *Heat exchange Processes of Chemical Engineering*, L.: Khimiya, 1982.
34. *Heat transfer in Two-phase Flow*, D. Batterworth, G.M. Hewitt, M.: Energia, Eds. 1980, p. 326.
35. E. Gaudsbrandt, M. Girsh. Evaporation, *Condensation and Cooling*. A.B. Pakshver Ed., M.: ONTI, 1936 (in German).
36. E.I. Taubman, *Evaporation*, M.: Khimiya, 1982, p. 327.
37. J. Botterhill, *Heat-exchange in Pseudo-liquid Layer*, M.: Energia, 1980, p. 345.
38. V.I. Tolubinsky, *Heat exchange at Boiling*. Kiev: Nauk. dumka, 1980, p. 315.
39. A.M. Kutepov, L.S. Sterman, N.G. Styushin, *Hydrodynamics and Heat exchange at vapor-formation*, M.: Vyssh. shkola, 1983.
40. A.V. Lykov, *Heat mass exchange*, M.: Energia, 1972.
41. *Hydro-mechanics and Heat Mass exchange in Wieghtlessness*. M.: Nauka, 1982, p. 263.
42. J. Delaie, M. Gio, M. Ritmuller, *Heat exchange and Hydrodynamics in Nuclear and Thermal power engineering*. M.: Energoatomizdat, 1984, p. 422.
43. P. Berd, V. Stewart, E. Lightfood, *Transfer Phenomenon* Ed. N.M. Zhavoronkov, V.A. Malyusov, M.: Khimiya, 1974.
44. N.V. Dankwerts, *Gas-Liquid Reactions*, M.: Khimiya, 1973.
45. T. Shervurd, P. Pigford, *Wiklee Ch. Mass transfer*, Ed. V.A. Malyusov, M.: Khimiya, 1982 (in English).
46. V.V. Kafarov, *Principles of Mass Transfer*, M.: Vyssh. shkola., 1972.
47. L.G. Voroshnin, B.M. Husid, *Diffusion Mass Transfer in multi-component Systems*. Minsk: Nauka i Tekhnika, 1972, p. 225.
48. *Heat and Mass exchange. Experiment*. Ref. book V.A. Grigorjev, V.M. Zorin, M.: Eds Energoixdat, 1982, p. 510.
49. A.E. Berlges, T.G. Collier, J.M. Delhaye, G.F. Hewitt, F. Majinger, *Two-phse flow and heat transfer in the power and process industries*. Washingtin-New York-London: Hemisphere Publ. Corp., 1981. p. 707.
50. L.N. Matusevich, *Crystallization from solutions in Chemical Industry*, M.: Khimiya, 1968.
51. A.V. Bamfort, *Industrial Crystallization*, Ed. L.N. Matusevich M.: Khimiya, 1968 (in English).
52. E.V. Khamsky *Crystallization in Chemical Industry*, M.: Khimiya, 1979.
53. O.M. Todes, V.A. Seballo, A.D. Goltsiker, *Mass Crystallization from Solutions*, L.: Khimiya, 1984, p. 232.
54. B.S. Sazhin, *Principles of Drying*, M.: Khimiya, 1984, p. 319.
55. A.M. Tregubov *Theory of Distillation and Rectification*. Baku: Gos. nauchn. -tekhn. izd. neft. igorno-topl.lit., 1946.

56. N.I. Gelperin, *Distillation and Rectification*. M.-L.: Goskhimizdat., 1947.
57. T. Hobler, *Mass Transfer and Absorption*, Ed. P.G. Romankov, L.: Goskhimizdat, 1964 (in Polish).
58. V.M. Ramm, *Gas absorption*, M.: Khimiya, 1977, p. 655.
59. G. Astarita, D.W. Savage, A. Bisio, *Gas Treating with Chemical Solvents*, N.Y.: J. Wiley and Sons, 1983.
60. P.O. Protodiakonov, S.V. Siparov, *Mechanics of adsorption process in Gas Solid Systems*. L.: Nauka, 1985.
61. N.V. Keltsev, *Principles of Adsorption Technic*, M.: Khimiya, 1976.
62. G.A. Yagodin, et al. *Principles of Liquid Extraction*. M.: Khimiya, 1981.
63. K. Khanson, *Latest Achievements in Liquid Extraction*. M.: Khimiya, 1974.
64. G.M. Ritch, A.V. Eshbruk, *Extraction. Principles and Application in Metallurgy*. M.: Metallurgy, 1983, p. 407.
65. P.G. Romankov, M.I. Kurochkina, *Extraction from Solid Materials*. L.: Khimiya, 1983, p. 256.
66. Yu.I. Dytnersky, *Back Osmos and Ultrafiltration*. M.: Khimiya, 1978.
67. Yu.I. Dytnersky, *Baromembrane Processes*. M.: Khimiya, 1986, p. 271.
68. S.-T. Khvang, K. Kammermeier, *Membrane Separation Processes*. M.: Khimiya, 1981, p. 464.
69. A.M. Rozen, et al. *Scale Transfer in Chemical Engineering. Elaboration of Industrial Apparatus by Hydro-dynamic Modelling*. M.: Khimiya, 1980.
70. V.V. Kavarov, I.I. Dorokhov, *System Analysis of Chemical Engineering Processes*, M.: Nauka, 1976.
71. T.J. Williams, *Designing of Chemical Engineering Processes by System Technics*, M.-L.: Khimiya, 1967.
72. S. Roters, *Dynamic Programming in Chemical Engineering Processes and Control Methods*, Ed. V.V. Kafarov. M.: MIR , 1965 (in English).
73. I.O. Protodjakonov, O.V. Muratov, I.I. Evlampiev, *Dynamics of Chemical Engineering Processes* L.: Khimiya, 1984.
74. M.S. Peters, *Plans, Design and Economics for Chemical Engineers*, New York-Toronto-London: McGraw-Hill Book Co., 1958.
75. I.P. Usyukin *Units, Machines and Apparatus of Creogeneous Technics*. M.: Food Industry, 1976.
76. O.M. Todes, O.B. Tsitovich, *Apparatus with boiling Layer*. L.: Khimiya, 1981.
77. S.M. Karpacheva, B.E. Pyabchikov, *Pulsating Apparatus in Chemical Engineering*. M.: Khimiya, 1983, p. 223.
78. S.A. Krapivina, *Plasmochemical technological Processes*. L.: Khimiya, 1981, p. 246.
79. S.I. Volfkovich et al. *General Chemical Engineering*, In 2 volumes. M.-L. *Gos. nauchn. tekhn. izd. khim. lit.* 1959.
80. I.P. Mukhlenov et al. *General Chemical Engineering*, M.: Vyssh. shkola, 1977.
81. A.G. Amelin, A.I. Manakhov, I.E. Zubova, V.N. Zaitsev, *General Chemical Engineering*, M.: Khimiya, 1977.

82. S. Bretsznajder, W. Kawecki, J. Leyko, R. Marcinkowski, Podstawy *Technologii Chemicznej*. Warszawa, 1973.
83. A.M. Kutepov, T.N. Bondareva, M.T. Berengarten, *General Chemical Engineering*, M.: Vyssh. shkola, 1985.
84. S.D. Beskov, *Technological Calculations*, M.-L.: Goskhimizdat, 1950.
85. K. Denbig, *Theory of Chemical Reactors* Ed. N.M. Zhavoronkov M.: Nauka, 1968 (in English).
86. P. Aris, *Analysis of Processes in Chemical Reactors*, Ed. I.I. Ioffe L.: Khimiya, 1967 (in English).
87. S.M. Korsakov-Bogatkov, *Chemical Reactors as object for Mathematical Modelling*, M.: Khimiya, 1967.
88. H. Kramers, K. Westerterp. *Chemical Reactors. Calculation and Operation*. M.: Khimiya, 1967.
89. A.L. Leonov, F.Ya. Revdin, *Chemical Processes Automation and Ways to Develop it*. M.: Khimiya, 1967.
90. O. Levenshpil, *Engineering description of chemical processes*. Ed. Slinko M.G. M.: Khimiya, 1969 (in English).
91. S. Veilas, *Chemical Kinetics and Industrial Reactors Calculations*. M.: Khimiya, 1967.
92. G.M. Ostrovsky, Yu.M. Volin, *Methods of Chemical Reactors Optimization*. M.: Khimiya, 1967.
93. A.L. Leonov, F.Ya. Revzin, *Automation of Chemical Processes and Ways of its Development*. M.: Khimiya, 1967.
94. M.G. Slinko, *Chemical Reactors Modelling*. Novosibirsk: Nauka, 1968.
95. A.Yu. Zakgeim, *Introduction into Chemical Engineering Processes Modelling*, M.: Khimiya, 1982.
96. B.I. Kuznetsov, Z.A. Zaitseva, *Chemistry and Chemical Engineering. Evolution of Interelation*. M.: Nauka, 1984.
97. *Mathematical Modelling of Chemical Reactors* G.I. Marchuk, Ed., Novosibirsk: Nauka, 1984.
98. *Industrial Gas separation*. (ACS Symp. Ser. No. 223), E. Jhaddeus, Ed., Washington, Amer. Chem. Soc., 1983, p. 292.
99. *Chemical Reaction Engineering. The 8th Intern Symp. on Chemical Reaction Engineering*. Pergamon Press. 1984, p. 826.
100. *Condensers: Theory and Practice*. Pergamon Press. 1983, p. 510.
101. *Chemical and Catalytic Reactor Modelling* (ACS Symp. Ser. No. 237) Ed. P. Milord. Washington: Amer. Chem. Soc., 1984, p. 426.

Onset of Instability Due to Marangoni Effect

V.V. Dil'man, N.N. Kulov and V.I. Naidenov

Abstract—Within the framework of a linear theory, a mechanism of the onset of Marangoni instability in quiescent and flowing liquid films has been studied. The unstable state arises for various reasons, including vertical and horizontal temperature gradients, heat evolution upon the physical dissolution of gases, formation of surface-active substances and heat evolution during the chemical reactions occurring in an interfacial diffusion layer. The surface tension of a liquid was assumed to depend linearly on temperature and concentration, and temperature dependence of viscosity was assumed to be exponential (thermocapillary convection in a quiescent liquid layer).

We have obtained analytical expressions representing the critical Marangoni number, (at which the steady-state regime loses its stability) as a function of wavenumber, and hydrodynamical parameters of the problem.

In the range of subcritical Marangoni numbers, a model of interfacial diffusion layer, taking the volume chemical reactions into account, was used to study the effects of surface tension gradient on the mass transfer rate. Analytical relations between the Sherwood and Marangoni numbers have been found. The surface tension has been shown to affect the accumulation of reaction products in the surface layer.

NOTATION

C_A	absorbed gas concentration in liquid;
C_B	reaction product concentration;
c_P	specific heat of a liquid;
D_A, D_B	molecular diffusion coefficients for the absorbed gas and reaction product, respectively;
g	acceleration of gravity;
ΔH	dissolution heat (thermal effect of reaction);
k	rate constant of chemical reaction;
$N=-\beta \dfrac{dT}{dy}$	dimensionless temperature gradient (β is the coefficient in Eq. (1.6));
p	pressure
Q	wetting density
T	temperature
t	time

$W=(N_2Pe)/(NPe_B)$; $N_1=(\varepsilon_1/\mu U_i)$; $N_2=(\varepsilon_2 C_{Ai}\Delta H)/(T_G \rho c_p \mu U_i)$	dimensionless parameters
V_j	velocity components of liquid (j=1,2,2); $V_1=U$ the longitudinal velocity component, $V_2=V$ the transverse velocity component;
v_j	dimensionless velocity components (j=1,2,3), $v_1=u$ the longitudinal velocity component, $v_2=v$ the transverse component (the v_j values are defined differently in different sections);
X_j	rectangular Cartesian coordinates (j=1,2,3), $X_1=X$ the longitudinal coordinate; $X_2=Y$ the transverse coordinate, $X_3=Z$;
$x_j=X_j/h$	dimensionless coordinates (j=1,2,3), $x_1=x$ the longitudinal coordinate; $x_2=y$, ε the transverse coordinate, $x_3=z$ $x=X/\delta$; η, $\varepsilon=Y/\delta$; $z=Z/\delta$; $x=kX(2\nu/g)(3\nu Q/g)^{-2/3}$; $y=Y(k/D)^{1/2}$;
α	wavenumber;
α_G	coefficient of mass transfer from liquid film into air;
δ	thickness of interfacial diffusion layer;
$\varepsilon;\varepsilon_1;\varepsilon\_$	activity coefficients determining the dependence of surface tension on concentration and temperature;
θ	dimensionless temperature (defined differently in different sections);
$\eta_h=h/\delta$	dimensionless thickness of liquid layer; $=\lambda/(\rho c_p))$ thermal diffusivity
λ	heat conductivity;
μ,ν	dynamic and kinematic viscosity coefficients, respectively
ρ	density of liquid;
σ	surface tension
$\Lambda=k\delta/U_i$	dimensionless rate constant of chemical reaction;
$\tau=t\mu_w/\rho h^2$; $\tau=(tU^0)/\delta$	dimensionless time;
Ω	perturbation build up coefficient
$Bi=\alpha_G h/\lambda$	Biot number;
$Le=D/æ$	Lewis number
Sh	Sherwood number
Re	Reynolds number (defined differently in different sections);
Ma	Marangoni number (defined differently in different sections);
$Pe_A=(U_i\delta)/D_A$ $Pe_B=(U_i\delta)/D_B$	Péclet diffusion number;
$Pe=(U_i\delta)/æ$	Péclet thermal number;
$\Pi=\dfrac{\varepsilon\Delta H\, C_{Ai} Le}{\delta C_p \mu U_i}$	dimensionless complex
$Pr=\nu/æ$	Prandtl number;

indices:

"0"	steady-state values,
'	perturbed (oscillatory) values;
–	dimensionless values
~	modulus of perturbed value (function of a coordinate normal to the free liquid surface);
w	values pertinent to solid walls;
i	values pertinent to free liquid surface;
x	critical values;

INTRODUCTION

An investigation into the problem of surface convection is aimed mainly at the search for novel ways to increase the mass transfer rate in the gas (vapor)-liquid and liquid-liquid systems, whereby the surface tension forces driven by compositional and temperature nonuniformities produce local tangential stresses at the system interface. Systematic experimental studies of mass transfer processes complicated by chemical reactions occurring in streams and flowing liquid films [1-10] allowed for the establishment of several important facts which have yet to receive adequate theoretical explanation. Among these facts, the most interesting are the anomally high values of coefficients of mass transfer across the gas-liquid interfact observed in several experiments. Thus, it has been found [1] that the desorption rate of inert gases, (nitrous oxide, helium, xenon), from a flowing film of an aqueous monoethanolamine solution, (with a simultaneous chemisorption of carbon dioxide from a gas phase), is considerably affected by the carbon dioxide concentration (see Figure 1).

It should be noted that the lowest curve, (No. 1), in Figure 1 corresponding to the zero carbon dioxide concentration in a gas phase agrees with the known experimental data on the rates of simple physical absorption for both laminar-wave and turbulent regimes of the film flow [11-13]. However, in all the other cases, the chemical reaction between the carbon dioxide and monoethanolamine occurring in the liquid phase resulted in a pronounced increase in the desorption rate of nitrous oxide. This can only be attributed to the onset of an intense surface convection[1] whose effect is retained within the entire range of the Reynolds numbers studied.

An increase in the desorption rate in the conditions studied was found to exceed the values that were expected to provide a growth in the Reynolds number alone. Therefore, it could be anticipated that the case under consideration involved a loss of hydrodynamical stability in the near-surface layers of the liquid and the onset of a surface convection.

It may be suggested that the effect of the surface convection will decrease upon the flow turbulization because the mixing produces a leveling of the surface tension gradients. Thus, the bubbling which results in strong turbulization of the medium shows a much weaker manifestation of the surface convection effects on the mass exchange rate as compared to the case of the film flowing down.

The flow turbulization favors the exchange processes by stimulating oscillatory mechanisms of mass and energy transfer which are more efficient than the molecular diffusion mechanism. However, on approaching the inter

[1] Sometimes, this phenomenon is referred to in literature as the "interfacial turbulence." However, the processes considered here are characterized by a spontaneous onset of stationary unequilibrium states with a high degree of ordering, so that the "surface convection" seems to be a more correct term.

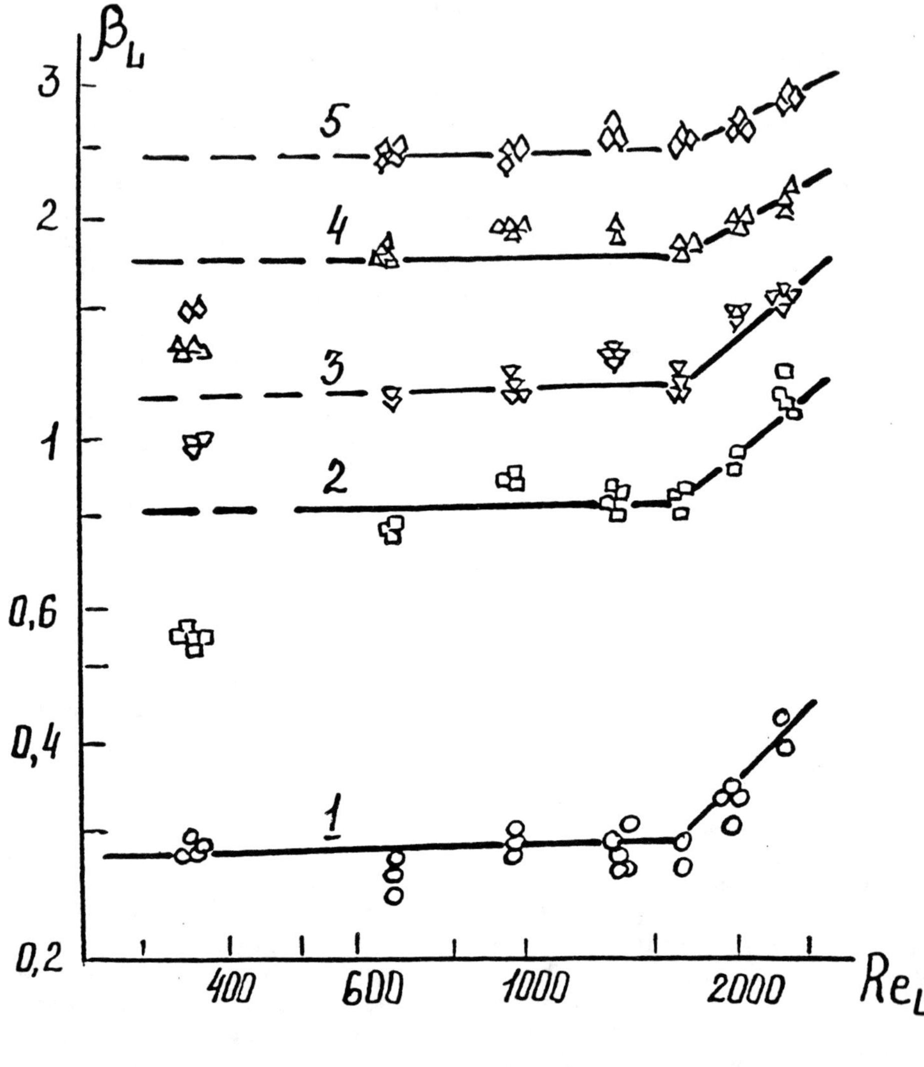

Figure 1. Coefficient of mass transfer in a liquid film for the desorption of nitrous oxide from 1M aqueous monoethanolamine solution with simultaneous absorption of carbon dioxide vs. the Reynolds number for different carbon dioxide concentrations: (1) 0%; (2) 10%; (3) 20%; (4) 50%; (5) 100% (1m wetted-wall tube).

face, the turbulent oscillations exhibit a rapid decay and a purely diffusional retardation of the mass transfer occurs even for very high Reynolds numbers.

An increase in the desorption rate in the conditions studied was found to exceed the values that were expected to provide a growth in the Reynolds number alone. Therefore, it could be anticipated that the case under consideration involved a loss of hydrodynamical stability in the near-surface layers of the liquid and the onset of a surface convection.

It may be suggested that the effect of the surface convection will decrease upon the flow turbulization because the mixing produces a leveling of the surface tension gradients. Thus, the bubbing which results in strong turbulization of the medium shows a much weaker manifestation of the surface convection effects on the mass exchange rate as compared to the case of the film flowing down.

The flow turbulization favors the exchange processes by stimulating oscillatory mechanisms of mass and energy transfer which are more efficient than the molecular diffusion mechanism. However, on approaching the interface, the turbulent oscillations exhibit a rapid decay and a purely diffusional retardation of the mass transfer occurs even for very high Reynolds numbers. For this very reason, the interfacial flux of distributed substance A will be always proportional to the coefficient of molecular diffusion, $j_A \sim D_A{}^n$ $(n>0)$. However, this statement is only valid for a passive interface. The interfacial instability which arises under certain conditions favors a growth in the rate of transfer processes. In such cases, the laws of mass transfer kinetics are no longer described by any of the existing mass transfer theories [14].

These considerations are well illustrated in Figure 2 which presents the data on the desorption of various purely soluble gases from an aqueous monoethanolamine solution [1]. In the absence of carbon dioxide in the gas phase, the value of exponent n in the relation of $j_A \sim D_A{}^n$ amounts to approximately 0.5, which is in good agreement with other researchers' experimental data and with the theoretical concepts [15,16]. As the carbon dioxide concentration grows and, surface convection increases, the mass transfer rate enhances, while the magnitude of n diminishes. As can be seen from Figure 2, the mass transfer coefficient for the curves 2 to 7 is virtually independent of the molecular diffusion coefficient while this last varies nearly ten-fold. Thus, the experiments pertinent to a co-desorption of nitrous oxide and helium showed practically equal mass transfer coefficients for both gases under the surface convection conditions, in contrast to a nearly two-fold difference between these coefficients observed on the absence of chemical reactions.

These findings are qualitatively explained by the disturbing influence of the surface tension gradients on the flowing film. Here, a determining role in the loss of hydrodynamical stability belongs to the sign of the $d\sigma/dC_B$ gradient value. Indeed, the reaction products formed upon the carbon dioxide chemisorption on aqueous monoethanolamine and alkali solutions and accumulated near the surface of a flowing down film produce different effects on the surface tension, increasing its magnitude $(d\delta/dC_B>0)$ in the former, and decreasing $(d\delta/dC_B<0)$ in the latter. In all other respects, the two systems behave in a similar manner. In particular, they exhibit the same mechanism of chemical reactions with close rate constants and activation energies [16].

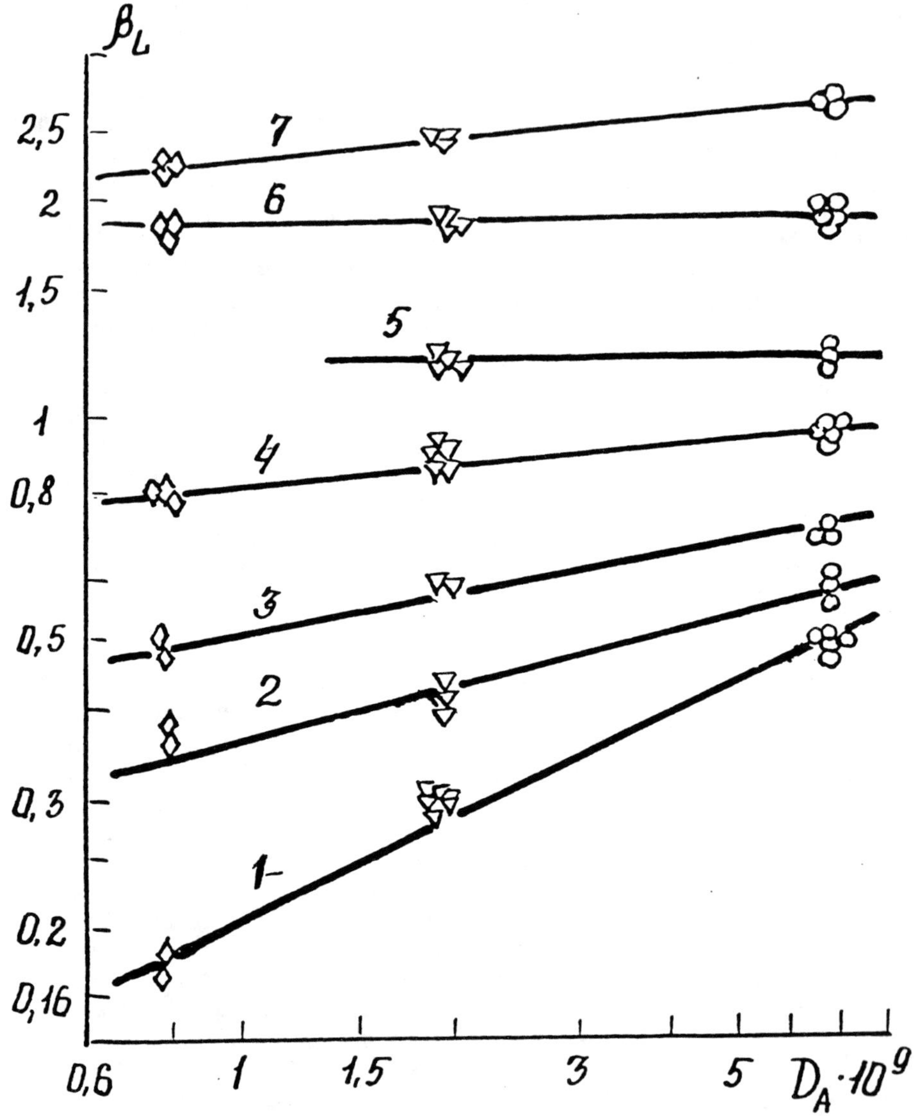

Figure 2. Coefficient of mass transfer for the desorption of nitrous oxide, helium, and xenon from 1M aqueous monoethanolamine solution with simultaneous carbon dioxide absorption vs. thermolecular diffusion coefficient for different carbon dioxide concentrations: (1) 0%; (2) 2.5%; (3) 5%; (4) 10%; (5) 20%; (6) 50%; (7) 100% (1m wetted-wall tube).

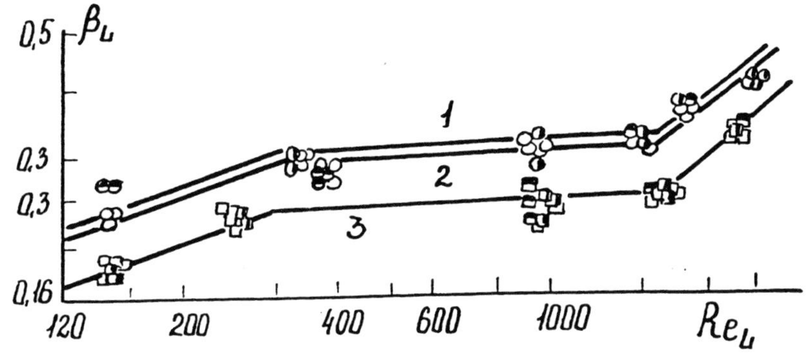

Figure 3. Coefficient of mass transfer in a liquid film for the desorption of nitrous oxide vs. the Reynolds number: (1) desorption from water; (2) desorption from 0.5 M aqueous NaOH solution, with different carbon dioxide concentrations in the gas phase (0 to 100% CO_2); (3) desorption from 2 M aqueous solution, with different carbon dioxide concentrations in the gas phase (0 to 100% CO_2).

Moreover, even with intentionally created equal concentrations of chemisorbents in the flowing films, the experimentally measured carbon dioxide adsorption rates showed qualitatively different trends. As can be seen in Figure 3, the absorption rate has increased noticeably in experiments on monoethanolamine, (with positive surface tension gradient), and decreased in an alkali solution (negative surface tension gradient).

It should be noted that, depending on whether the mass-transfer induced variation in the surface tension during the absorption, rectification, or extraction is positive or negative, the systems are often classified into three groups. We may distinguish between the positive, negative, and neutral systems, where the interfacial tension grows, diminishes, or remains unchanged during the mass transfer, respectively. Thus, experiments have shown the mass transfer rate during rectification in positive systems to be noticeably higher as compared to that in negative systems.

An interfacial hydrodynamical instability results in a system that becomes capable of transforming a part of its free surface energy into the kinetic energy of convective liquid flow, hence, increasing the mass transfer rate. If the hydrodynamical instability arises due to the action of capillary forces, it is referred to as the Marangoni instability. Determination of critical conditions favoring the onset of the Marangoni instability is one of the most important problems of the hydrodynamics of interfaces.

In a general case, the Marangoni effect may arise due to the dependence of interfacial tension not only on the concentration, but also on the temperature, surface electric charges, and other physico-chemical factors. Convective motions of a liquid driven by surface forces may essentially change the hydrodynamical conditions in the vicinity of an interface and considerably enhance the transfer rates (as illustrated in Figures 1-3).

A full theory of Marangoni instability can be constructed only when taking into account the essential nonlinear effects inherent in the particular class of problems considered. Nevertheless, a linear theory still retains its value in allowing for the elucidation of some important details of the mechanism of interfacial instability and serving as a basis for a more profound investigation in this mechanism within a nonlinear theory.

Presently, among the completed studies are the problems of the linear theory of stability in quiescent liquid layers with the temperature or concentration gradients affecting the magnitude of surface tension. These problems were developed in the classical works of Pearson [17] and Scriven and Sternling [18] which had originally formulated the spectral problem concerning the thermal (diffusional) and hydrodynamical perturbations and determined the critical Marangoni numbers separating the stable and unstable regimes.

As for moving liquid films, the corresponding problems are still far from completely studied, the linear theory of stability for the interrelated diffusion and hydrodynamical problems is still in the making, and many experimentally observed phenomena [4, 14, 19-24] are waiting for theoretical explanation.

Obviously, a liquid flow and an intense convective and turbulent mass transfer in the flow core lead to new mechanisms which are responsible for the onset and development of the surface convection. Investigation in this convection process is necessary for the construction of a correct mass-transfer theory.

1. Thermocapillary Convection in a Quiescent Liquid Layer with Variable Viscosity

The Marangoni instability in a liquid with constant viscosity has been studied in numerous works for moderate Prandtl numbers [17, 18, 25-33]. However, it can be expected that chemic-technological processes, e.g., those involving (a non-isothermal chemisorption of gases by liquid films, will employ viscous liquids with high Prandtl numbers [32, 33]. The viscosity of such liquids usually decreases exponentially with the growing temperature, which could noticeably affect the conditions for the onset of thermocapillary convection. It should be noted that taking the variability of the physical properties of liquids into account is also of importance in studying the hydrodynamics and heat exchange during the crystal growth [34-38].

Consider a quiescent liquid layer of thickness h and let a heat exchange with an ambient medium occur on the free surface (Y=h) according to the Newton law:

$$-\lambda \frac{dT}{dY} = \alpha_G (T - T_G), Y = h \tag{1.1}$$

where T and T_G are the temperatures of liquid and are over the free surface, respectively; α_G is the coefficient of heat transfer from liquid into air; and Y is the coordinate measured along the normal to free liquid surface.

At the bottom of the layer (Y=0), the boundary condition is specified either as a constant temperature,

$$T = T_w, Y = 0 \tag{1.2}$$

or as a constant heat flux

$$-\lambda \frac{dT}{dY} = q_w, Y = 0 \tag{1.3}$$

The steady-state temperature distribution is described by the equation

$$\frac{d^2T}{dY^2} = 0 \tag{1.4}$$

The solution of Equation (1.4) for the boundary conditions (1.1), (1.2) or (1.3) has the form

$$T = T_w - (T_w - T_i)\frac{Y}{h} \tag{1.5}$$

The values of characteristic temperatures, T_W and T_i, for the various boundary conditions are presented in Table 1 where $B_i = \alpha_G h/\lambda$ is the Biot number. In what follows we shall assume that the wall temperature exceeds the temperature of gas in contact with the free liquid surface, so that the temperature gradient across the liquid layer is negative ($T_W > T_G$).

Table 1. Temperatures of solid wall and free liquid surface.

Problem conditions	Wall temperature (Y=0)	Temperature of free surface (Y=h)
Wall temperature, T_w, specified by Equation (1.2)	T_w	$T_i \frac{T_w + BiT_G}{1 + Bi}$
Heat flux, q_w, specified by Equation (1.3)	$T_w = T_G + \frac{q_w h}{\lambda}\frac{1 + Bi}{Bi}$	$T_i = T_G + \frac{q_w h}{\lambda Bi}$

Let the viscosity of the liquid depend exponentially on the temperature, and let the coefficient of surface tension be a linear function of the temperature [39]. Then we have

$$\mu = \mu_w e^{-\beta(T - T_w)}, \sigma = \sigma_w - \varepsilon(T - T_w) \tag{1.6}$$

where $\beta > 0$ and $\varepsilon > 0$ are the approximating parameters; σ_W and μ_W are the surface tension and viscosity at the wall temperature, respectively.

The steady-state solution (1.5) remains valid for relatively small temperature gradients across the layer thickness. An increase in the gradient results in the loss of stability and the development of convective motions.

Near the equilibrium, the distribution of velocity and temperature fields can found from a set of equations of motion

$$\delta \frac{\partial V_i}{\partial t} = -\frac{\partial P}{\partial X_i} + \frac{\partial}{\partial X_j} \mu \left(\frac{\partial V_i}{\partial X_j} + \frac{\partial V_i}{\partial X_i} \right) \tag{1.7}$$

(with the inertial terms omitted as the second-order quantities), a discontinuity equation

$$\frac{\partial V_j}{\partial X_j} = 0 \tag{1.8}$$

and an equation of convective heat conduction

$$\frac{\partial T}{\partial t} + V_j \frac{\partial T}{\partial X_j} = æ \frac{\partial^2 T}{\partial X_j^2} \tag{1.9}$$

where X_j are the rectangular coordinates (i,j=1,2,3), and V_j are the corresponding velocity components.

Let us study the stability of a stationary solution (1.5) with respect to small perturbations in Liapunov's sense. It is convenient to introduce the quantities

$$T = T^\circ + T'; \; p = p^\circ + p'$$

where the superscript "°" denotes the stationary, and the prime indicates the perturbed values of the corresponding variables.

Let us also introduce the dimensionless temperatures

$$\theta = \frac{T - T_G}{T_w - T_G}; \theta' = \frac{T'}{T_w - T_G} \frac{1 + Bi}{Bi}$$

for the problem with a fixed wall temperature, and

$$\bar{\theta} = \frac{T - T_G}{q_w h} \lambda \frac{Bi}{1 + Bi}; \theta' = \frac{T' \lambda}{q_w h} \frac{1 + Bi}{Bi}$$

for the problem with a fixed heat flux at the wall.

Taking into account the fact that liquid in a stationary state is quiescent, the velocity components in Equations (1.7) and (1.8) should be considered as perturbed quantities.

By doubly applying the vortex operation to Equation (1.7), we may exclude the pressure and arrive at an equation for the velocity component, $V_2=V$, normal to the liquid surface.

Linearization of the convective heat exchange equation (1.9) in the vicinity of a stationary solution yields a set of equations for the perturbed velocity $v=Vh/æ$ and temperature θ'

$$\frac{\partial}{\partial \tau}\Delta v = f\Delta\Delta v - \frac{d^2 f}{dy^2}\Delta v + 2\frac{df}{dy}\frac{\partial}{\partial y}\Delta v + 2\frac{d^2 f}{dy^2}\frac{\partial^2 v}{\partial y^2} \Pr\frac{\partial \theta'}{\partial \tau} - v = \Delta\theta' \qquad (1.10)$$

where $f=\exp Ny$; $N=-\beta(dT/dy)$; $y=Y/h$ is the dimensionless coordinate along the normal to the liquid surface; $\tau=(t\mu_w)/(\rho h^2)$ is the dimensionless time; $\Pr_w$ is the Prandtl number for the temperature of a solid wall (at a bottom of the liquid layer); Δ is Laplacian

$$\Delta = \partial^2/\partial x^2 + \partial^2/\partial y^2 + \partial^2/\partial z^2 -$$

and $x_j=X_j/h(x_1=x,\ x_2=y,\ x_3=z)$ are the dimensionless coordinates.

Let us now assume that the free liquid surface is non-deformable, that is, the parameter of the curvature is small and the effect of capillary forces is reduced to the production of surface tension gradients equilibriated by tangential stresses. An analysis of the interfacial curvature effects has been given in References 40-45. The temperature perturbation will be described by usual boundary conditions specifying the heat exchange at the free surface and on the solid wall:

$$\begin{aligned} &v=0,\ \partial v/\partial y = 0,\ \theta' = 0\ (\text{or}\ \frac{d\theta'}{dy} = 0)\ \text{for}\ y=0 \\ &v=0,\ \partial^2 v/\partial y^2 = Ma\Delta{\cdot}\theta',\ \frac{\partial\theta'}{\partial y} = -Bi\theta'\ \text{for}\ y=1. \end{aligned} \qquad (1.11)$$

The Marangoni number Ma entering in (1.11) characterizes the stability of the steady-state temperature distribution (1.5) and is defined as

$$Ma = \frac{\varepsilon(T_w - T_G)h\,Bi}{\mu_i\,æ(1+Bi)}$$

for the specified temperature conditions (1.2), and

$$Ma = \frac{\varepsilon q_w h^2\,Bi}{\lambda\mu_i\,æ(1+Bi)}$$

for the specified heat flux (1.3), where $\mu_i=\mu_w \exp N$ is the liquid viscosity at a temperature of its free surface.

The first and second boundary conditions given by (1.11) follow from the sticking hypothesis and discontinuity equation (1.8).

It should be noted that in the case of constant viscosity N=0, f=1 we arrive at a known Pearson's problem [17] on the onset of thermocapillary convection in a quiescent liquid layer.

Following a well known hypothesis of Pellew and Southwell [46-50], the solution of Equation (1.10) with the boundary conditions (1.11) will be found in the form

$$\begin{aligned} v &= -F(x,z)\tilde{v}(y)e^{\pi\tau} \\ \theta' &= F(x,z)\tilde{\theta}(y)e^{\pi\tau} \end{aligned} \tag{1.12}$$

where F(x,z) is the function of horizontal structure obeying the equation

$$\Delta_1 F = -\alpha^2 F$$

$\tilde{v}(y)$ and $\tilde{\theta}(y)$ are the amplitudes of variation of the perturbed velocity and temperature values along the coordinate y, respectively; α is the wavenumber; $\Delta_1=(\partial^2/\partial x^2)+(\partial^2/\partial z^2)$ is the Laplacian for the plane; and Ω is the perturbation build up coefficient.

Using Equations (1.10-(1.12) we can formulate a spectral problem which is reduced for the neutral perturbations (Ω=0) to differential equations with constant coefficients

$$\begin{aligned} &\frac{d^4 v}{dy^4} - 2\alpha^2\frac{d^2 v}{dy^2} + \alpha^4 v - N^2\left(\frac{d^2 v}{dy^2} - \alpha^2 v\right) + 2N\left(\frac{d^3 v}{dy^3} - \alpha^2\frac{dv}{dy}\right) + 2N^2\frac{d^2 v}{dy^2} = 0 \\ &d^2\theta / dy^2 - \alpha^2\theta = v \end{aligned} \tag{1.13}$$

$$v = 0; \frac{dv}{dy} = 0; \theta = 0 \text{ (or } \frac{d\theta}{dy} = 0, y = 1) \text{ for } y = 0,$$

$$\frac{dv}{dy} = 0; \frac{d^2 v}{dy^2} = \alpha^2 Ma\theta; \frac{d\theta}{dy} = -Bi\theta \text{ for } y = 1$$

Let us construct Green function G for the operator

$$\frac{d^2\tilde{\theta}}{dy^2} - \alpha^2\tilde{\theta} = 0$$

with boundary conditions

$$\tilde{\theta}=0,\, y=0;\, \frac{d\tilde{\theta}}{dy}=-Bi\tilde{\theta},\, y=1$$

and calculate its values at the interface:

$$G(\xi,\alpha,Bi)=\frac{e^{\alpha\xi}-e^{-\alpha\xi}}{e^{-\alpha}(\alpha-Bi)+e^{\alpha}(\alpha+Bi)}$$

Using the Green function, we can obtain an equation for the neutral stability curve, Ma(α), in the form

$$Ma=\frac{d^2\tilde{v}/dy^2}{\alpha^2\int_0^1 G(\xi,\alpha,Bi)\tilde{v}(\xi)d\xi} \tag{1.14}$$

where $\tilde{v}$ is the solution of hydrodynamical equation (1.13). From Equation (1.14) it follows from that Ma→∞ at α→0 or α→∞. With α→∞ the influence of parameter N is insignificant and Equation (1.14) has the same asymptotes as do the curve of neutral stability in the Pearson problem [17].

In the limit α→0, the hydrodynamical equation (1.13) simplifies to

$$\frac{d^4\tilde{v}}{dy^4}+2N\frac{d^3\tilde{v}}{dy^3}+N^2\frac{d^2\tilde{v}}{dy^2}=0 \tag{1.15}$$

Equation (1.15) has a fundamental system which can be used to derive the expressions for the Marangoni numbers:

$$Ma=\frac{1+Bi}{\alpha^2}\varphi_1(N)$$

for the specified temperature at the bottom of the layer, and

$$Ma=\frac{Bi}{\alpha^2}\varphi_2(N)$$

for the specified heat flux at the wall. The functions φ_1 and φ_2 are obtained by solving the hydrodynamical equations:

$$\varphi_1 = \frac{6N^4 e^{-N}(N^2 - 2N + 2 = 2e^{-N})}{(N^3 - 18N + 12) + (2N^4 + 5N^3 + 12N^2 + 12N - 24)e^{-N} + (6N^2 + 12)e^{-2N}}$$

$$\varphi_2 = \frac{2N^3 e^{-N}(N^2 - 2N + 2 + 2e^{-N})}{N^2 - 4N + 2 + (N^3 + N^2 - 4N - 4)e^{-N} + 2e^{-2N}}$$

It can be shown that for $N \to 0$ we have $\varphi_1=80$ and $\varphi_2=48$, that is, the above expressions are reduced to the Pearson curve for a liquid layer with constant viscosity.

Introducing the functions $\chi_1=\varphi_1/80$ and $\chi_2=\varphi_2/48$, we can write

$$\frac{Ma}{Ma_i} = \chi_1(N), \frac{Ma}{Ma_i} = \chi_2(N) \tag{1.16}$$

where Ma_i is the critical Marangoni number for the case of constant viscosity calculated at a temperature of liquid on the free surface.

Equation (1.16) reflects the influence of temperature-dependent viscosity on the critical conditions for the onset of thermocapillary convection in a quiescent liquid layer. Numerical values of the functions χ_1 and χ_2 are presented in Table 2.

Table 2. Numerical values of functions χ_1 and χ_2 for different values of parameter N

N	0	0,5	1	2	3
χ_1	1	0,86	0,65	0,41	0,26
χ_2	1	0,83	0,62	0,38	0,23

It is interesting to note the strong reduction in the critical Marangoni numbers with decreasing viscosity. For example, in glycerol we have $\beta=0.07$ deg^{-1}, so that the Marangoni number decreases by a factor of 1.5 for a temperature drop of 15°C.

Let us construct the curve of neutral stability for the wave number range $0<\alpha<\infty$. For this purpose, it is necessary to solve the hydrodynamical equation (1.13). The solution will be found in the form $\tilde{v}=\exp(\beta_n y)$. Substituting this expression into Equation (1.12), we obtain an algebraic equation with respect to β_n:

$$\beta_n^4 - 2\alpha^2\beta_n^2 + \alpha^4 - N^2(\beta_n^2 - \alpha^2) + 2N(\beta_n^3 - \alpha^2\beta_n) + 2N^2\beta_n^2 = 0 \quad (1.17)$$
$$n = 1,2,3,4.$$

The roots of Equation (1.17) are

$$\beta_n = Z_n - \frac{N}{2},\ Z_n^2 = \alpha^2 + \frac{N^2}{4} \pm \alpha Ni$$

where z_n is determined by the root of square equation, i is the imaginary unit.

Let $\tan \varphi_0 = (\alpha N)/(N^2/4+\alpha^2)$, $k=[(N^2/4+\alpha^2)+\alpha^2N^2]^{1/4}$. Then, the roots of Equation (1.17) are

$$\beta_1 = K\left(Cos\frac{\varphi_o}{2} + iSin\frac{\varphi_o}{2}\right) - \frac{N}{2},\ \beta_2 = K\left(Cos\frac{\varphi_o}{2} - iSin\frac{\varphi_o}{2}\right) - \frac{N}{2}$$
$$\beta_3 = -K\left(Cos\frac{\varphi_o}{2} + iSin\frac{\varphi_o}{2}\right) - \frac{N}{2},\ \beta_4 = -K\left(Cos\frac{\varphi_o}{2} - iSin\frac{\varphi_o}{2}\right) - \frac{N}{2}$$

The fundamental system of solutions for Equation (1.13) can be written as

$$\tilde{v}(y) = \left\{e^{K_1y}CosK_2y;\ e^{K_1y}SinK_2y;\ e^{K_3y}CosK_2y;\ e^{K_3y}SinK_2y\right\}$$

where

$$K_1 = KCos\frac{\varphi_o}{2} - \frac{N_2}{2},\ K_2 = KSin\frac{\varphi_o}{2},\ K_3 = -KCos\frac{\varphi_o}{2} - \frac{N}{2}$$

Using this system and calculating the corresponding integrals, we can obtain the curve of neutral stability, minimum critical Marangoni numbers, critical Marangoni numbers, and wavenumbers. Figure 4 presents an example of constructing the neutral stability curves for Bi=0.01 and various N values.

It should be noted that the obtained solution does not take the temperature dependence of the density of a liquid into account and, hence, is valid only for small Rayleigh numbers, i.e., for sufficiently thin liquid layers.

It is known [5,46] that a critical thickness exists in the liquid layer, h_x, such that a natural convection is predominant at $h \gg h_x$, while at $h < h_x$ an essential role is played by the thermocapillary mechanism of the loss of stability of a liquid layer heated from beneath.

Let us evaluate a critical value of the film thickness, h_{xx}, at which the temperature dependence of viscosity should be taken into account. It is convenient to introduce a dimensionless parameter $K=Ne^N/Ma=\mu_w bæ/\varepsilon h$ which characterizes the relative contribution of two instability factors, the temperature dependence of viscosity and that of surface tension. Consider the dependence $Ma_x(N)$ constructed using the calculated critical Marangoni num

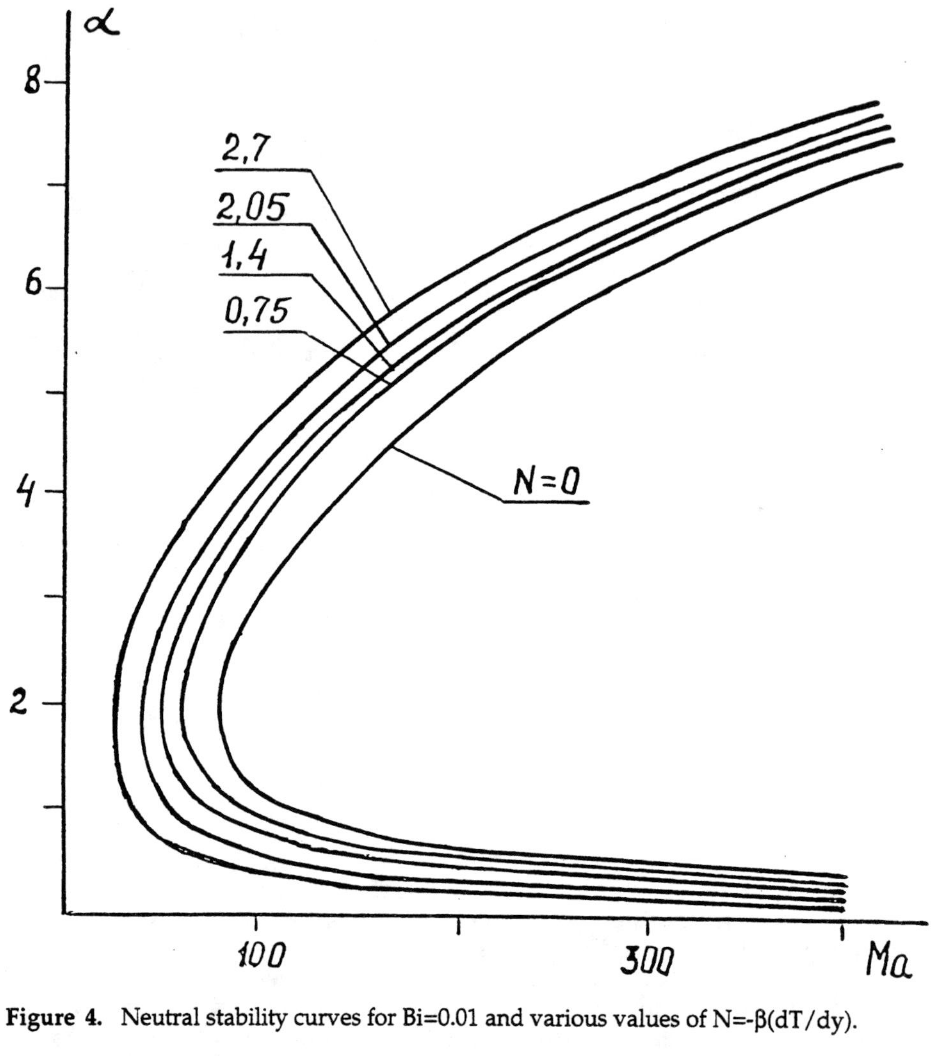

Figure 4. Neutral stability curves for Bi=0.01 and various values of N=-β(dT/dy).

bers (see Figure 5). An analysis of this curve shows that a considerable decrease in the critical Marangoni number occurs already at K=0.05, this range of K values corresponding to film thicknesses of the order of $h_{xx}=\mu_w b æ/\varepsilon K$. This relation suggests that an increase in the viscosity of liquid and the parameter β, (proportional to the activation energy of viscous flow), leads to a growth of the critical film thickness. Therefore, a considerable decrease in the minimum Marangoni number accompanying a decrease in the liquid viscosity with growing temperature is characteristic of liquids with high Prandtl numbers. The critical film thicknesses for various liquids are presented in Table 3.

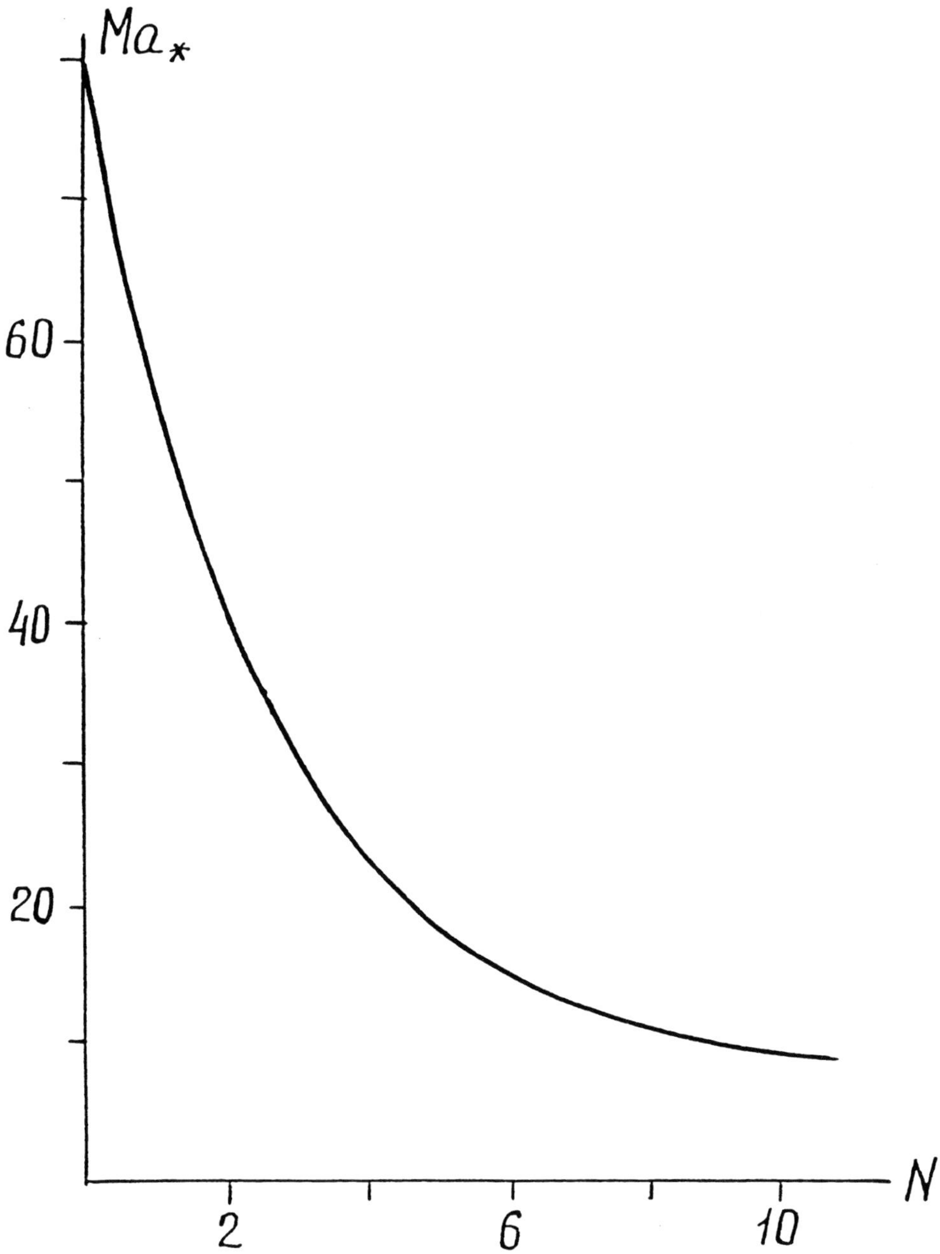

Figure 5. Critical Marangoni number vs. the parameter N.

Table 3. Critical film thicknesses for various liquids

Liquid	T_w °C	μ_w Pa s	æx 10^7 m^2/s	εx 10^3 $N/(m^2 deg)$	β 1/deg	h_{xx} mm	Prω
Glycerol	20	1,48	0,94	0,04	0,07	4,8	12490
MS-20 type oil	30	0,466	0,0725	0,07	0,05	1,7	7300
Methanol	25	0,553◊10	1,03	0,08	0,02	0,28◊10	6
toluene	25	0,552◊10	0,9	0,124	0,02	0,16◊10	9
water	20	0,894◊10	1,44	0,17	0,02	0,3◊10	6

As can be seen from Table 3, the h_{xx} values do not exceed 5 mm. In terrestrial conditions, the mechanism of decreasing stability due to viscosity reduction is essential only in liquids with large Prandtl numbers (Pr_w>100). In liquids with moderate Prandtl numbers, (organic liquids, water), the destabilizing effect of variable viscosity is insignificant.

2. Heat-Diffusion Instability in a Quiescent Liquid Layer

It is usually believed that the regimes of physical absorption in cases of gases possessing no surface activity are stable and involve no surface convection. However, there are cases that encounter mutually coupled processes of convective diffusion and heat exchange which may lead to Marangoni instabilities.

Consider a quiescent horizontal liquid layer, with a free surface featuring gas absorption and the evolution of the corresponding dissolution heat [12]. Assume the viscosity of liquid to be constant. The system of equations (1.7)-(1.9) for μ=const will be complemented by the equation of convective diffusion [15]:

$$\frac{\partial C_A}{\partial t} + V_j \frac{\partial C_A}{\partial X_j} = D_A \frac{\partial^2 C_A}{\partial X_j^2} \tag{2.1}$$

where c_A is the gas concentration in liquid, and D_A is the coefficient of molecular diffusion.

A stationary solution of the above problem is found from the set of equations

$$\begin{gathered} \frac{d^2T}{dY^2} = 0; T = T_w, Y = 0; -\lambda \frac{dT}{dY} = \alpha_G (T_i - T_G) - \Delta H D_A \frac{dC_A}{dY}, Y = h \\ \frac{d^2C_A}{dY^2} = 0; C_A = C_{Aw}, Y = 0; C_A = C_{Ai}, Y = h \end{gathered} \tag{2.2}$$

where c_{Ai} and c_{Aw} are the concentrations of an absorbed component at the interface and at the wall, respectively, (in the general case, the walls are considered to be similar to porous membranes in that they are permeable for the transferred substance A); ΔH is the heat of dissolution.

An exact solution of Equation (2.2) has a simple form:

$$T = T_w - mY, m = -\frac{dT}{dY} = -\left[\frac{1}{1+Bi}\frac{\Delta H D_A(C_{Ai}-C_{Aw})}{\lambda h} + \frac{Bi}{1+Bi}\frac{T_G - T_w}{h}\right]$$
$$C_A = (C_{Ai} - C_{Aw})\frac{Y}{h} + C_{Aw} \tag{2.3}$$

Consider the case of $c_{Ai}>c_{Aw}$, $dc_A/dY>0$, where the concentration of gas at the surface is greater than that in the adjacent layers, just as it must be for physical absorption.

The thermal condition of Equation (2.2) takes the balance of heat fluxes at the interface due to convection into account and the heat of dissolution reflects a relation between the convective diffusion and heat exchange processes. As in Section 1, let us separate the dimensionless perturbed velocity component v normal to the interface and linearize Equations (1.7), (1.9), (2.1) in the vicinity of stationary solution (2.3). Then the system of dimensionless equations for the perturbed values of velocity, concentration, and temperature can be reduced to the form

$$\frac{\partial(\Delta v)}{\partial \tau} - \Delta(\Delta v) = 0$$
$$pr\frac{\partial c_A'}{\partial \tau} - Le\Delta c_A' = -v \tag{2.4}$$
$$pr\frac{\partial \theta'}{\partial \tau} - \Delta\theta' = v$$

where $\theta'=T'/(|m|h)$ and c_A' are the oscillatory values of temperature and concentration, respectively, (the concentration rendered dimensionless by scaling to c_{Ai}-c_{Aw}); $Le=D_A/æ$ and $Pr=\nu/æ$ are the Lewis and Prandtl numbers, respectively.

Let us specify the boundary conditions for Equations (2.4) without taking the deformation of a free liquid surface into account

$$v = 0, \frac{\partial v}{\partial y} = 0, \theta = 0, C_A' = 0, for\ y = 0$$
$$v = 0, \frac{\partial^2 v}{\partial y^2} = Ma\Delta_1\theta, -\frac{\partial\theta}{\partial y} = Bi\theta - K\frac{\partial C_A'}{\partial y}, C_A' = 0, for\ y = 1 \tag{2.5}$$

where

$$Ma = \frac{|m|h^2\varepsilon}{\mu æ}\quad K = \frac{\Delta H D_A (C_{Ai} - C_{Aw})}{|m|\lambda h}$$

The Marangoni number describes a condition for the loss of stability of a stationary solution (2.3), and parameter K takes the effect of the temperature gradient created by the heat of dissolution into account.

Let us search for the solution of Equations (2.4) and (2.5) in the form of Equation (1.12) for the velocity and temperature perturbations and in the form

$$cA' = F(x,z)\tilde{C}_A(y)e^{\Omega\tau}$$

for the disturbed concentration.

For neutral perturbations (Ω=0) we arrive at the following spectral problem:

$$\frac{d^4\tilde{v}}{dy^4} - 2\alpha^2\frac{d^2\tilde{v}}{dy^2} + \alpha^4\tilde{v} = 0$$
$$\frac{d^2 g}{dy^2} - \alpha^2 g = \tilde{v} \tag{2.6}$$
$$g = 0,\ y = 0;\ \frac{dg}{dy} = -Big,\ y = 1$$

where

$$g = \frac{\tilde{\theta} - K\tilde{C}_A}{1 + K/Le}$$

The boundary conditions for the normal velocity component of the liquid will remain the same (see Section 1).

Using Equations (2.5) and (2.6), we obtain an expression for the critical Marangoni numbers

$$Ma = \frac{\dfrac{d^2\tilde{v}}{dy^2}}{\alpha^2(1 + \dfrac{K}{Le})g},\ y = 1$$

The spectral problem, written as Equation (2.6) has been originally solved by Pearson [17], and the solution has been analyzed in detail in Reference 47. Using this solution, we obtain the neutral stability curve

$$Ma = \frac{Le}{Le+R} \bullet \frac{8\alpha(\alpha Ch\alpha + Bish\alpha)(\alpha - sh\alpha \cdot Ch\alpha)}{\alpha^3 Ch\alpha - Sh^3\alpha} \tag{2.7}$$

where

$$R = \frac{1+Bi}{1+\gamma Bi}, \gamma = \frac{(T_G - T_w)\lambda}{\Delta H D_A (C_{Ai} - C_w)}$$

and α is the wavenumber of three-dimensional perturbations.

Let us construct the plot of function γ. Consider two regions: $-\infty<\gamma<0$ where the temperature gradient is negative ($dT/dy>0$), and $0<\gamma<(1+Bi-Le)/(BiLe)$ where the temperature gradient is positive ($dT/dy<0$). In the first region, the heating of the liquid from beneath results in the onset of thermocapillary convection, as described by Pearson. In the second region, which has not been considered in the literature, the Marangoni instability arises due to the heating (at the expense of dissolution heat) from above. This last instability will be referred to as the heat-diffusion Marangoni instability [48,49]. The physical meaning of this phenomenon will be discussed in more detail in Section three where the mechanism of flowing down will be treated.

An interfacial convection driven by the dissolution heat will arise provided that $Le<1+Bi$, that is, if the coefficients of diffusion and thermal diffusivity obey certain restrictions. For most liquids and gases involved in the commercial chemical processes $Le\sim10^{-2}$, so that the above instability is highly probable.

For $Bi=0$, the minimum critical Marangoni number occurring at $\alpha=\alpha_x=2$ equals

$$Ma_x = \frac{80Le}{1-Le} \tag{2.8}$$

Examples for various systems. For the absorption of carbon dioxide by organic liquids and water, the critical conditions of diffusion-thermal Marangoni instability are determined by Equation (2.8) and presented in Table 4. In the methanol-carbon dioxide system, the Lewis number is $Le=3.66\times10^{-2}$, and the temperature gradient $m_0=\Delta H(c_{Ai}-c_{Aw})D_A/(\lambda h)=2.64$ deg/m. The corresponding quantities in tholuene-carbon dioxide and water-carbon dioxide are $Le=5.13\times10^{-2}$, $m_0=1.84$, and $Le=1.27\times10^{-2}$, $m_0=0.468$, respectively. The physical properties of liquids were taken at 20°C.

The data in Table 4 were obtained for cAi-cAw=8 mole/m3. As can be seen from this table, the 1mm deep layers of methanol and toluene will be unstable with respect to small perturbations, (because the Marangoni number exceeds the critical level), while a water layer with the same thickness is stable. The stability of thin water layer is explained by its relatively high viscosity and heat conductivity. The extreme sensitivity of organic liquids with respect to the heat-diffusion instability should be noted: the temperature variations of the order of 10^{-3}°C are capable of producing the Marangoni convection.

Table 4. Critical Marangoni numbers for the absorption of carbon dioxide by various liquids.

Liquid	ΔH kJ/mole	D_A x10^9 m^2/s	æx 10^7 m^2/s	ε$\times10^3$ N/(m deg)	μ$\times10^5$ Pa s	λ W/(m deg)	Ma	Ma_x
Methanol	17,6	3,75	1,03	0,08	0,553	0,2	3,7	3,04
Toluene	6,7	4,62	0,9	0,124	0,552	0,134	4,5	4,32
Water	19,2	1,83	1,44	0,17	0,884	0,602	0,61	1,03

A simple analysis reveals that heat removal from the evolution region to the wall increases the liquid viscosity in the surface layer and stabilizes the heat-diffusion Marangoni instability.

3. HEAT-DIFFUSION INSTABILITY IN A FLOWING DOWN LIQUID FILM

In the preceding sections we have discussed the heat-diffusion instability in a quiescent liquid layer. Now, the linear theory is used to establish the critical conditions for the onset of Marangoni instability driven by temperature gradients due to the absorption heat evolved in a flowing down liquid film.

Consider a planar non-isometric motion of viscous liquid, taking the convective diffusion processes into account:

$$\begin{aligned}
&\frac{\partial}{\partial t}\Delta\psi + \frac{\partial\psi}{\partial Y}\frac{\partial}{\partial X}\Delta\psi - \frac{\partial\psi}{\partial X}\frac{\partial}{\partial Y}\Delta\psi = \upsilon\Delta\Delta\psi \\
&\frac{\partial C_A}{\partial t} + \frac{\partial\psi}{\partial Y}\frac{\partial C_A}{\partial X} - \frac{\partial\psi}{\partial X}\frac{\partial C_A}{\partial Y} = D_A\Delta C_A \\
&\frac{\partial T}{\partial t} + \frac{\partial\psi}{\partial Y}\frac{\partial T}{\partial X} - \frac{\partial\psi}{\partial X}\frac{\partial T}{\partial Y} = \text{æ}\,\Delta T
\end{aligned} \tag{3.1}$$

where Δ is Laplacian and ψ is the flow function.

In Equation (3.1), the velocity of the liquid is expressed via the flow function:

$$V = \frac{\partial\psi}{\partial Y}, V = -\frac{\partial\psi}{\partial X}$$

where the coordinate Y is directed inward from the free surface.

As a stationary approximation of Equation (3.1), we shall treat the process of ordinary physical absorption accompanied by the heating of the liquid phase due to the heat of dissolution. The velocity field in a flowing down liquid film is determined by a parabolic profile

$$V = \frac{\rho g}{2\mu}(h^2 - Y^2), V_2 = \frac{\rho g}{2\mu} h^2 \tag{3.2}$$

where g is the acceleration of gravity.

For simplicity, we shall restrict the consideration to a waveless flow regime known to occur at low Reynolds numbers.

Taking the proposed constancy of velocity in diffusional and thermal interfacial layers into account [50], the system of equations for the stationary approximation acquires the form

$$\begin{aligned} &V_i \frac{\partial C_A}{\partial X} = D_A \frac{\partial^2 C_A}{\partial Y^2}; C_A = C_{Ai}, Y = 0; C_A = 0, Y = \infty. \\ &V_i \frac{\partial T}{\partial X} = æ \frac{\partial^2 T}{\partial Y^2}; \lambda \frac{\partial T}{\partial Y} = \Delta H D_A \frac{\partial C_A}{\partial Y}, Y = 0; T = T_H, Y = \infty \end{aligned} \tag{3.3}$$

Equation (3.3) represents an automodel system with respect to the transverse coordinate

$$\eta = \frac{Y}{\delta}, \delta = 2\sqrt{\frac{D_A X}{V_i}}$$

and can be solved by standard techniques [51,52]:

$$\frac{dC_A^0}{d\eta} = -\frac{2}{\sqrt{\pi}} e^{-\eta^2}, \frac{d\theta^\circ}{d\eta} = -\frac{2}{\sqrt{\pi}} e^{-Le\eta^2} \tag{3.4}$$

where $c_A{}^0 = c_A/c_{Ai}$, $\theta^0 = (T - T_1)\rho c_p / æ\Delta H c_{Ai}$ are the dimensionless concentration and temperature values for the stationary state, respectively.

It should be noted that the $\Delta H c_{Ai}$ value determines the heat evolution at the interface and is proportional to the dissolution heat and Henry's constant [12].

In the problem considered we have $d\theta^0/d\eta < 0$, that is, the temperature at the interface is greater than that in the adjacent liquid layers. According to Pearson [17], such states are usually stable and do not lead to an interfacial convection. In what follows, we shall assume a linear dependence of the surface tension on the temperature.

Let us study the stability of stationary physical absorption regime determined by the solution of Equation (3.3), with respect to infinitesimal thermal and hydrodynamical perturbations. Consider the perturbed solutions of Equation (3.3): $\psi^0 + \psi'$, $\theta^0 + \theta'$, and $cA^0 + c_A{}'$, and let the boundary conditions for ψ', T', and $c_A{}'$ be written in a usual form, without taking the effects related to the surface deformation into account, (i.e., assuming the smallness of the curvature parameter, $\mu D_A/\sigma\delta \ll 1$) and surface viscosity:

$$\frac{\partial\psi'}{\partial X}=0,\ \frac{\partial^3\psi'}{\partial X\partial^2 Y}=-\varepsilon\Delta T',\ \lambda\frac{\partial T'}{\partial Y}=\Delta HD_A\frac{\partial C_A'}{\partial Y},\ C_A'=0 \text{ for } Y=0,$$
$$\frac{\partial\psi'}{\partial X}=0,\ \frac{\partial^2\psi'}{\partial X\partial Y}=0 \text{ for } Y=h;\ T'=0,\ C_A'=0 \text{ for } Y=\infty \tag{3.5}$$

Substituting the perturbed solution into the set of Equations (3.1), (3.3) and linearizing it in the vicinity of a stationary state (3.4), we arrive at a linear system of equations. The solution is conveniently found in the form

$$\psi'=\varphi(\eta)e^{i\alpha(\bar{x}-C_\omega\bar{\tau})},\theta'=\theta(\eta)e^{i\alpha(\bar{x}-C_\omega\bar{\tau})},C_A'=\tilde{a}(\eta)e^{i\alpha(\bar{x}-C_\omega\bar{\tau})}$$

where c_A', θ', ψ' are the dimensionless values of the perturbed concentration, temperature, and flow functions (the definition of c_A' and θ' are given above (see Equation (3.4)), while the dimensionless flow function is obtained by dividing with $U_i\delta$); a is the wavenumber $c_w=c_r+ic_i$, c_r is the phase velocity of waves; c_i is the perturbation build up coefficient; $x=X/\delta$, $\tau=U_it/\delta$ are the dimensionless coordinate and time, respectively.

One should also assume that within a thin interfacial diffusion layer ($0\le Y\le\delta$) the hydrodynamical perturbations can expanded in a series in powers of h, and the first terms of these expansions can be used in an analysis of the spectral problem studied. The stationary field of velocities is represented in a similar manner.

On accomplishing the above outlined procedure, we obtain the spectral problem:

$$\frac{d^4\varphi}{d\eta^4}-2\alpha^2\frac{d^2\varphi}{d\eta^2}+\alpha^4\varphi=i\alpha\,\mathrm{Re}\left[(1-\mathrm{Cw})\left(\frac{d^2\varphi}{d\eta^2}-\alpha^2\varphi\right)+\frac{2\varphi}{\eta_\delta^2}\right]$$
$$\frac{d^2\theta}{d\eta^2}-\alpha^2\theta=i\alpha\mathrm{Pe}\left[(1-\mathrm{Cw})\theta-\varphi\frac{d\theta^\circ}{d\eta}\right]$$
$$\frac{d^2\tilde{a}}{d\eta^2}-\alpha^2\tilde{a}=\mathrm{idPe_A}\left[(1-\mathrm{C_w})\tilde{a}-\varphi\frac{d\tilde{a}}{d\eta}\right] \tag{3.6}$$
$$\varphi=0,\ i\alpha\frac{d^2\varphi}{d\eta^2}=-\mathrm{N}\alpha^2\theta,\ \frac{d\theta}{d\eta}=\frac{d\tilde{a}}{d\eta},\ \tilde{a}=0 \quad \text{for } \eta=0;$$
$$\varphi=0,\ \frac{d\varphi}{d\eta}=0 \text{ for } \eta=\eta_h;\ \tilde{a}=0,\ \theta=0 \quad \text{for } \eta=\infty$$

where

$$\mathrm{Re}=\frac{\mathrm{Vi}\delta}{\upsilon},\ \mathrm{Pe}=\frac{\mathrm{Vi}\delta}{æ},\ \mathrm{Pe_A}=\frac{\mathrm{Vi}\delta}{\mathrm{D_A}},\ \eta_h=\frac{h}{\delta}$$

The spectral problem presented by Equation (3.6) has a trivial solution $\bar{\varphi} \equiv 0$, $\theta \equiv 0$, $a \equiv 0$, hence, the existence of nonzero eigen-solutions of Equation (3.6) will imply the instability of physical absorption regime determined by Equation (3.4).

It is suggested that $\alpha Re \ll 1$ and the solution of Equation (3.6) for the neutral perturbations $c_i=0$ will be found in the form $c_r=1$, $\varphi=i\varphi_i/\alpha$. Assuming the quantities φ_i, α, and θ to be real we obtain:

$$\frac{d^4\varphi_i}{d\eta^4} - 2\alpha^2 \frac{d^2\varphi_i}{d\eta^2} + \alpha^4\varphi_i = 0$$
$$\frac{1}{Pe}\left(\frac{d^2\tilde{\theta}}{d\eta^2} - \alpha^2\tilde{\theta}\right) = \varphi_i \frac{d\theta^{\circ}}{d\eta}$$
$$\frac{1}{Pe_A}\left(\frac{d^2\tilde{a}}{d\eta^2} - \alpha^2\tilde{a}\right) = \varphi_i \frac{dC_A^{\circ}}{d\eta} \tag{3.7}$$
$$d^2\varphi_i / d\eta^2 = \Pi\alpha^2\tilde{\theta},\ \varphi_i = 0 \text{ for } \eta = 0;$$
$$d\varphi_i / d\eta = 0 \qquad\qquad \text{for } \eta = \eta_h,$$

where

$$\Pi = \frac{\varepsilon \Delta H C_{Ai} Le}{\delta C_p \mu V_i}$$

The boundary conditions for $\tilde{\theta}, \tilde{a}$ will, naturally, be the same as before. Within the assumptions adopted, it is not difficult to obtain the asymptotic (for small η) solution for hydrodynamical perturbations

$$\varphi_i = Bz,\eta,\ \frac{d^2\varphi_i}{d\eta^2} = Bz_2$$
$$z_1 = \frac{sh^2k - k^2}{2sh^2k},\ z_2 = \frac{k(2k - sh2k)}{2\eta_h sh^2k},\ k = \alpha\eta_h,\ B = Const \tag{3.8}$$

Taking the above expressions into account for the velocity field and Equation (3.4) for the temperature and concentration gradients, we obtain a boundary problem for the difference $L = \tilde{\theta} - \tilde{a}$

$$\frac{1}{Pe_A}\left(\frac{d^2L}{d\eta^2} - \alpha^2 L\right) = Bz_1\left(\eta e^{-\eta^2} - Le\eta e^{-Le\eta^2}\right) \tag{3.9}$$

Let us construct the Green function for the operator $d^2L/d\eta^2 - d^2L = 0$ with the boundary conditions $dL/d\eta=0$, $L=0$; $L=0$, $\eta=\infty$ and calculate its value at the interface:

$$G=-\frac{1}{2}e^{-\alpha\xi}$$

The solution of Equation (3.9) and L(0) can be written in the form

$$L(0)=-BPe_A\frac{z_1}{\alpha}\int_0^{\infty}e^{-\alpha\xi}(e^{-\xi^2}-Le\,e^{-Le\xi^2})\xi d\xi$$

On accomplishing the necessary calculations, we obtain the neutral stability curve:

$$Ma=2\frac{sh2k-2k}{sh^2k-k^2}\left[\frac{\alpha}{\sqrt{Le}}e^{\frac{\alpha^2}{4Le}}erfc\frac{\alpha}{2\sqrt{Le}}-\alpha e^{\frac{\alpha^2}{4}}erfc\frac{\alpha}{2}\right]^{-1} \tag{3.10}$$

The asymptotes of Equation (3.9) can be written as

$$\underset{\alpha<<1}{Ma}=2\frac{\sqrt{Le}}{\alpha^2\eta_h(1-\sqrt{Le})},\ \underset{\alpha>>1}{Ma}=\sqrt{\pi}\frac{\alpha^2}{1-Le} \tag{3.11}$$

The expressions (3.10), (3.11) suggest that a minimum critical Marangoni number Ma_x exists, whereby at $Ma>Ma_x$ the physical absorption process will lose its stability. The relation (3.10) allows us to derive simple analytical formulas which are valid for $Le<<1$ and $\eta h>>1$. For $Le=0$ (i.e., when the molecular diffusion coefficient is much lower as compared to the heat-diffusion one), Equation (3.10) yields

$$\begin{gathered}Ma=\frac{sh2k-2k}{sh^2k-k^2}\frac{1}{F}\\ F=\frac{1}{\sqrt{\pi}}-\frac{1}{2}\alpha e^{\alpha^2/4}erfc\frac{\alpha}{2}\end{gathered} \tag{3.12}$$

where the function F exhibits the following properties: $F=1/\sqrt{\pi}$, $dF/d\alpha=-1/2$ for $\alpha=0$. By differentiating (3.12) with respect to α and equaling the resulting expression to zero we obtain an equation determining the critical wavenumber α_x:

$$\frac{\alpha_*\frac{dF}{d\alpha}}{F}=\frac{4k_*(1-Ch2k_*-k_*^2ch2k_*-k_*^2+2k_*sh2k_*)}{\frac{1}{2}sh4k_*-sh2k_*-2k_*^2sh2k_*-2k_*ch2k_*+4k_*^3} \tag{3.13}$$

where $k_x=\alpha_x\eta_h$. Taking the properties of constructed function F into account, it can be easily shown that $\alpha_x\to\infty$ for $k_x\to\infty$. Retaining on the right-hand side of Equation (3.13) only the highest-order terms with respect to k_x, we obtain

$$\frac{\sqrt{\pi}}{2}\alpha_* = 8k_*^3 e^{-2k_*}, \frac{\sqrt{\pi}}{2\eta_h} = 8k_*^2 e^{-2k_*} \tag{3.14}$$

Thus, the dependence of the extremum wavenumber on the relative film thickness is determined from the relations

$$K_* = \frac{1}{2}\ln\eta_h, \alpha_* = \frac{1}{2\eta_h}\ln\eta_h \tag{3.15}$$

Expressions (3.14), (3.15) imply that most detrimental perturbations destroying the physical absorption process are those with wavelengths that are much greater than the interfacial diffusion layer thickness, but still much lower as compared to the film thickness. Performing a limiting transition $k_x\to\infty$, $\alpha_x\to 0$ in Equation (3.12), we arrive at an asymptotic expression for the minimum critical Marangoni number $Ma_x=4$.

For arbitrary values Le<<1 and sufficiently large values of η_h, the critical Marangoni numbers are obtained by numerical calculation. An analysis has revealed that the leveling of molecular diffusion coefficients of impurity and heat leads to a rather rapid growth of the critical wavenumbers and Marangoni numbers. (Table 5).

Table 5. Critical Marangoni numbers and wavenumbers for various Lewis numbers

Le	0,01	0,05	0,10	0,5	0,8
δ_x	0,30	0,50	0,60	0,90	1,05
Ma_x	6,2	8,8	11,1	35,6	110

For Le=1 we have $Ma_x=\infty$ and, hence, no instability due to the spreading of dissolution heat is observed if the diffusion coefficients of impurity and heat are equal. Moreover, the physical absorption process will be also stable for Le>>1. For the instability to develop, it is necessary that Le<<1 or æ>D_A.

Let us now turn to an analysis of the results obtained from the viewpoint of physical mechanisms. For the Marangoni numbers exceeding the critical level, Ma>Ma_x, the absorption process becomes unstable with respect to small perturbations.

There are two factors which are necessary for the onset of surface convection upon the physical absorption of gases: first, the heat of dissolution of an absorbed gas at the interface must be not too small and, second, the diffusion of heat in the liquid film must be sharper than the diffusion of matter, that is, the thermal diffusivity must exceed the diffusion coefficient. If the two quantities are equal, a complete similarity of the mass and heat transfer processes will result in the appearance of equal thermal and concentrational gradients in the system. Since the concentration of the absorbed gas at the interface is constant, the temperature and surface tension will also remain unchanged and the driving force of interfacial instability will vanish. On the other hand, in a system featuring a strong difference between the thermal diffusivity and diffusion coefficients a rapid transfer of the dissolution heat in the liquid will tend to create considerable temperature gradients over the liquid surface. These gradients will, in turn, favor the onset of instability due to the action of a positive feedback mechanism which will destroy, for supercritical Marangoni numbers, the initial absorption regime. The vertical temperature gradient appearing in the liquid film along its surface will be negative, because of a maximum concentration gradient and, hence, a maximum heat flux warming the film occurs in the beginning of interfacial diffusion layer. Downstream, the heat flux decreases with a corresponding drop in the liquid temperature. The negative temperature gradient along the film creates an additional motion in the surface layer, directed streamwise and accelerating the flow. As a result, the diffusional gas flux into the film grows, producing a corresponding increase in the heat flux and temperature, and the arising surface tension gradient accelerates the film flow still further; and so forth. Under appropriate external conditions, (large Marangoni numbers), the hydrodynamical, thermal and diffusion perturbations tend to grow. Eventually, this will lead to the establishment of a heat mass balance at a qualitatively new level characterized by a higher mass transfer coefficient.

It should be noted that the Marangoni instability appears due to the concentration gradient across the thickness of interfacial diffusion level, while the temperature gradient across this layer hinders the development of this instability. Because of this, the Marangoni numbers exhibit a minimum at Le=0 and increase with growing Le values. At Le=1, the effective mixing gradient is zero and the instability vanishes. The neutral disturbances arising upon the critical Marangoni numbers appear as surface waves moving downstream with the velocity equal to that of the film, and the period of these waves equals $2\pi/d_x$. The structure of these waves and their phase velocity provides some justification for the theoretical model of Ruckenstein and Berbente [51] which considered an increase in the mass transfer rate in a liquid film due to the development of vorticity appearing as the circulation cells move at a velocity equal to that of the film's surface. It should be noted that the above analysis of the heat-diffusion Marangoni instability was carried out in terms of two-dimensional perturbations. This limitation can be removed for the dimensionless complex values $\alpha Re/\eta h^2 << 1$, because the set of equations, with respect to perturbations, allows for the separation of variables in the case of three-dimensional perturbations such as the type of expi $\alpha_1(\bar{x}-\bar{\tau})e^{(i\alpha_2\bar{\tau})}, \alpha_1^2+\alpha_2^2=\alpha^2$. Therefore, it is expected that the above determined Marangoni numbers will also be the minimum ones for the class of three-dimensional perturbations.

Using the expression for the critical Marangoni number, it is possible to determine the critical time of phase contact which is necessary for the onset of instability

$$t_* = \frac{Ma_*^2}{4}\left(\frac{\mu\lambda}{\varepsilon \Delta H C_{Ai}\sqrt{\mathcal{D}_A}}\right)^2 \qquad (3.16)$$

This relation implies the critical time of phase contact to be directly proportional to the squared viscosity and heat conductivity and inversely proportional to the square of heat evolution value in the surface layer. This is quite clear from the physical point of view, because the heat of gas absorption by a liquid creates sufficiently large surface tension gradients in a low-heat conducting medium, these lead to a surface instability in the case of low viscosity.

The results of experiments [52] on carbon dioxide absorption by flowing down films of organic liquids, (methanol, toluene, isobutane, chlorobenzene), showed a 3-4-fold increase in the mass transfer rate as compared to the values calculated by the classical permeation theory. The surface tension of a liquid was found to decrease in the case of developed surface convection.

A decrease in the surface tension in the experiments could be attributed to the strong heating of the surface due to evolution of the absorption heat. Indeed, the surface convection sharply increases the diffusion flux of matter which, in turn, enhances in the same proportion the heat evolution in the surface layer. Low heat conductivity of organic liquids (e.g., 0.2 and 0.134 W/(mg deg) in methanol and toluene, respectively; cf. 0.6 W/(mg deg) in water) favors a sharp elevation of the temperature of liquid at the surface and a creation of intense destabilizing gradients of surface tension.

Quantitative calculation of the critical time of phase contact for the systems studied in Reference 52 yields t_x=11, 28, and 144 s for methanol, toluene, and water, respectively. Thus, water seems to be the most unfavorable absorbent for the manifestation of thermal Marangoni instability driven by the absorption heat. This is explained, according to Section 2, by relatively high viscosity and heat conductivity of water.

4. INSTABILITY IN A FLOWING DOWN FILM UPON CHEMISORPTION

Experimental investigations into mass transfer processes complicated by chemical reactions occurring in flowing down liquid films allowed for the establishment of several important facts [1-10] which have not yet received adequate theoretical explanation. Among these facts, the most interesting are the anomalous high values of the coefficient of mass transfer via the gas-liquid interface for the absorption of carbon dioxide by aqueous amine solutions, a lack of such enhancement of mass transfer for the carbon dioxide absorption by alkalis, and the independence of the coefficient of mass transfer of the diffusion coefficient of absorbed gas for sufficiently large concentrations of carbon dioxide in the gas phase.

According to the hypothesis proposed in Reference 4, 52, the above phenomena are driven by the concentrational-capillary convection of the Marangoni type appearing in a thin layer adjacent to the interface. It is important to note that the experiments [4, 52] were carried out in flowing down films of streams of liquid.

Apparently, the problem of creating a complete theory that accounts for the observed phenomena will be solved only in the future, because even the simplest case of a first-order reaction requires a study of three nonlinear processes, namely, the hydrodynamical and two diffusional (for the absorbed gas and the reaction product). At the first stage of investigation, aimed at the development of the nonlinear theory, it is necessary to determine the critical Marangoni numbers taking the processes of chemisorption into account.

Let a liquid chemisorbent film flow down a vertical wall, and a chemical reaction occurs between the absorbed gas and the chemically active component of the liquid.

Consider a laminar waveless flow of the liquid film featuring the processes of convective diffusion, heat transfer, and chemical transformation according to the scheme

$$c_A \xrightarrow{k} c_B + \Delta H,$$

where k is the rate constant of the first-order chemical reaction and ΔH is the reaction heat.

We shall assume that the surface tension depends linearly on the temperature T and the reaction product concentration c_B:

$$\sigma = \sigma_o + \frac{\varepsilon_1}{C_{Ai}} C_B + \frac{\varepsilon_2}{T_G}(T - T_G) \tag{4.1}$$

where ε_1 and ε_2 are the activity coefficients of the reaction product and temperature, respectively; c_{Ai} is the concentration of absorbed substance at the film surface; and δ_0 is the surface tension at a temperature of T_G.

Consider the case of $T_G(æ/\partial T)=\varepsilon_2<0$, while the sign of the derivative of surface tension with respect to concentration may be arbitrary.

A model problem will be treated using a set of equations of hydrodynamics and convective heat and mass exchange in a liquid phase, making an allowance for the evolution heat due to the chemical reaction:

$$s\left(\frac{\partial V_i}{\partial t}+V_j\frac{\partial V_i}{\partial X_j}\right)=sg_i-\frac{\partial p}{\partial X_i}+\mu\frac{\partial V_i}{\partial X_j^2};\frac{\partial V_j}{\partial X_j}=0$$

$$\frac{\partial C_A}{\partial t}+V_j\frac{\partial C_A}{\partial X_j}=D_A\frac{\partial^2 C_A}{\partial X_j^2}-kC_A$$

$$\frac{\partial C_B}{\partial t}+V_j\frac{\partial C_B}{\partial X_j}=D_B\frac{\partial^2 C_B}{\partial X_j^2}+kC_A \qquad (4.2)$$

$$\frac{\partial T}{\partial t}+V_j\frac{\partial T}{\partial X_j}=æ\frac{\partial^2 T_2}{\partial X_j^2}+\frac{k\Delta H}{scp}C_A$$

where c_A and c_B are the concentrations of the absorbed gas and reaction product, respectively; D_A and D_B are the molecular diffusion coefficients of substances participating in the reaction.

In the case of k=0, (i.e., in the absence of chemical reaction), assuming one-dimensional, stationary flow of a viscous liquid down a vertical wall, we obtain, (within a model of interfacial diffusion layer [15]), an equation of convective diffusion (3.3). A solution of this equation can be represented in the form of derivative $dc_A°/d\eta$ as given in Equation (3.4).

It should be noted that Equation (3.4) has been obtained assuming the velocity of a liquid within a thin interfacial diffusion layer, (with thickness δ), to be constant and equal to the surface velocity of the liquid film.

As a stationary state, we shall consider the regime of isothermal absorption characterized by the relations

$$\frac{dC_B}{d\eta}=0,\ \frac{dT}{d\eta}=0 \qquad (4.3)$$

and by Equation (3.4) for the distribution of concentration of the absorbed component.

The stationary approximation (3.4) for the gradient of absorbed gas concentration has a structure $cA°(Y,D_AX)$ which is typical for an interfacial layer, and for small diffusion coefficients the variable D_AX can be treated as a slowly varying one. This circumstance allows for the application of hypothesis of the parallelicity of flow, and below we neglect the derivatives along the flow in comparison with the derivatives along the norm in equations of heat and mass exchange. The application of this procedure to hydrodynamical interfacial layers has been justified to some extent in Reference 53.

The perturbed solution of the system of equations (4.2) can be obtained by substituting

$$p=p^0+p';c_A=\overset{\circ}{c}_A+c_A';c_B=c_B+c_B',T=T^0+T'$$

where the superscripts "0" refer to the stationary, and the primes, to the perturbed values of the quantities studied.

Let us consider the case where the heat exchange with absorbed gas at the interface obeys the Newton Law, and a non-volatile reaction product is formed. Moreover, it is assumed that the concentration and temperature perturbations are zero at a sufficiently large distance from the interface:

$$\begin{aligned} &C_A' = 0, \frac{\partial C_B'}{\partial Y} = 0, \lambda \frac{\partial T'}{\partial Y} = \alpha T' \text{ for } Y = 0; \\ &C_A' = C_B' = T' = 0 \text{ for } Y = \infty \end{aligned} \tag{4.4}$$

The velocity component that is normal to the interface obeys usual boundary conditions, with no allowance for the surface deformation and the surface viscosity and elasticity effects:

$$\begin{aligned} &V' = 0, \mu \frac{\partial^2 V'}{\partial Y^2} = \frac{\varepsilon_1}{C_{Ai}\delta^2} \Delta_1 C_B' + \frac{\varepsilon_2}{T_G \delta^2} \Delta_1 T' \text{ for } Y = 0; \\ &V' = \frac{\partial V'}{\partial Y} = 0 \text{ for } Y = h \end{aligned} \tag{4.5}$$

Equation (4.5) reflects the balance of tangential momentum with an allowance for the dependence (4.1).

By linearizing Equation (4.2) in the vicinity of a stationary solution of Equation (3.1) and taking Equations (4.3 and 4.4) into account, we obtain

$$\begin{aligned} &\Delta(\Delta V) = 0, \frac{\partial \overline{C}_A}{\partial \overline{\tau}} + \frac{\partial \overline{C}_A}{\partial \overline{x}} - \frac{2}{\sqrt{\pi}} V e^{-2^2} = \frac{1}{Pe_A} \Delta \overline{C}_A - \Lambda \overline{C}_A \\ &\frac{\partial \overline{C}_B}{\partial \overline{\tau}} + \frac{\partial \overline{C}_B}{\partial \overline{x}} = \frac{1}{Pe_B} \Delta \overline{C}_B + \Lambda \overline{C}_A, \frac{\partial \theta}{\partial \overline{\tau}} + \frac{\partial \theta}{\partial \overline{x}} = \frac{1}{Pe} \Delta \theta + \Lambda \overline{C}_A \\ &V = 0, \overline{C}_A = 1, \frac{\partial \overline{C}_B}{\partial \eta} = 0, \eta_h \frac{\partial \theta}{\partial \eta} = Bi\theta, \frac{\partial^2 V}{\partial \eta^2} = N_1 \Delta \overline{C}_B + N_2 \Delta_1 \theta \eta = 0; \\ &V = 0, \partial V / \partial \eta = 0 \text{ for } \eta = \eta_h; \overline{C}_A = \overline{C}_B = \theta = 0 \text{ for } \eta = \infty \end{aligned} \tag{4.6}$$

where

$$V = \frac{V'}{V_i};\ \overline{C}_A = \frac{C'_A}{C_{Ai}};\ \overline{C}_B = \frac{C'_B}{C_{Ai}};\ \theta = \frac{T\rho C_p}{C_{Ai}\Delta H};\ \overline{x} = \frac{X}{\delta};\ \eta = \frac{Y}{\delta};\ \overline{z} = \frac{z}{\delta}$$

$$\eta_h = \frac{h}{\delta};\ \overline{\tau} = \frac{tV_i}{\delta};\ Pe_A = \frac{V_i\delta}{D_A};\ Pe_B = \frac{V_i\delta}{D_B};\ Pe = \frac{V_i\delta}{æ}$$

$$N_1 = \frac{\varepsilon_1}{\mu V_i};\ N_2 = \frac{\varepsilon_2 C_{Ai}\Delta H}{T_G sc_p \mu V_i};\ \Lambda = \frac{k\delta}{V_i};\ B_i = \frac{\alpha_F h}{\lambda}$$

$$\varepsilon_1 = C_{Ai}\frac{\partial\sigma}{\partial C_B};\ \varepsilon_2 = T_G\frac{\partial\sigma}{\partial T};\ \Delta = \frac{\partial^2}{\partial x^2} + \frac{\partial^2}{\partial\eta^2} + \frac{\partial^2}{\partial z^2};\ \Delta_1 = \frac{\partial^2}{\partial x^2} + \frac{\partial^2}{\partial z^2}$$

V' is the oscillatory velocity component normal to the free surface of the film; U_i is the surface velocity of the film in a stationary state; c_A' and c_B' are the oscillatory values of concentrations of the absorbed gas and reaction product, respectively; c_{Ai} is the concentration of absorbed component on the film surface; T' is the oscillatory temperature value; ΔH is the reaction heat; X, Y, Z are the Cartesian coordinates; h is the liquid film thickness; and α_G is the coefficient of heat transfer from liquid to gas.

The solution of Equation (4.6) has the form of three-dimensional waves moving with a velocity equal to that of the film surface:

$$\left.\begin{aligned} V &= V(\eta) \\ \overline{C}_A &= \tilde{C}_A(\eta) \\ \overline{C}_B &= \tilde{C}_B(\eta) \\ \theta &= \tilde{\theta}(\eta) \end{aligned}\right\} \bullet \left\{ e^{i(\alpha_1\overline{x}+\alpha_2\overline{z})}e^{-i\alpha_i\tau} \right. \tag{4.7}$$

Substituting (4.7) into (4.6) we obtain a spectral problem for the neutral perturbations:

$$\frac{d^4\tilde{V}}{d\eta^4} - 2\alpha^2\frac{d^2\tilde{V}}{d\eta^2} + \alpha^4\tilde{V} = 0$$

$$\frac{1}{Pe_A}\left(\frac{d^2C_A}{d\eta^2} - \alpha^2 C_A\right) - \Lambda C_A = -\frac{2}{\sqrt{\pi}}\tilde{V}e^{-\eta^2}$$

$$\frac{1}{Pe_B}\left(\frac{d^2C_B}{d\eta^2} - \alpha^2 C_B\right) + \Lambda C_A = 0$$

$$\frac{1}{\mathrm{Pe}}\left(\frac{d^2\theta}{d\eta^2}-\alpha^2\theta\right)+\Lambda C_A=0$$

$$\tilde{V}=0, C_A=0, \frac{dC_B}{d\eta}=0, \eta_h\frac{d\theta}{d\eta}=\mathrm{Bi}\theta, \frac{d^2V}{d\eta^2}=-N_1\alpha^2C_B-N_2\alpha^2\theta \text{ for } \eta=0; \quad (4.8)$$

$$\tilde{V}=0, d\tilde{V}/d\eta=0 \text{ for } \eta=\eta_h;$$

$$C_A=C_B=\theta=0 \text{ for } \eta=\infty$$

The existence of eigenvalues of the spectral boundary problem represented by Equation (4.8) means that at critical Marangoni numbers a bifurcation occurs in the stationary motion and heat and mass transfer processes with a transition to processes of wave nature. This is accompanied by an essential change in the mass flux across the interface.

Let us now turn to an analysis of the linear problem (4.8). From the third and fourth equations of the set (4.8), it is easy to derive a relation between the temperature of a liquid and the concentration of the reaction product,

$$\frac{1}{Pe}\left(\frac{d^2\tilde{\theta}}{d\eta^2}-\alpha^2\tilde{\theta}\right)=\frac{1}{Pe_B}\left(\frac{d^2\tilde{C}_B}{d\eta^2}-\alpha^2\tilde{C}_B\right)$$

which, in turn, yields the dependence

$$\tilde{\theta}=\frac{Pe}{Pe_B}\tilde{C}_B+Ash\alpha\eta+Bch\alpha\eta$$

where A and B are the integration constants.

Using the boundary conditions for the temperature and concentration, we arrive at the expression

$$\tilde{\theta}=\frac{Pe}{Pe_B}\frac{\alpha\eta_h}{\eta_h\alpha+Bi}\tilde{C}_B, \eta=0 \quad (4.9)$$

Taking the value of temperature and concentration into account at the interface, we can derive, from the boundary condition of equality for the tangential forces and surface tension gradient

$$\frac{d^2\tilde{V}}{d\eta^2}=-N_1\alpha^2\left(1+\frac{W\alpha\eta_h}{\alpha\eta_h+Bi}\right)\tilde{C}_B; W=\frac{N_2Pe}{N_1Pe_B}; \eta=0 \quad (4.10)$$

where parameter W represents a relation between the surface tension gradients of the thermal and diffusional nature.

Obviously, the Marangoni instability in the problem considered will be determined by the competent processes of heat evolution in the liquid phase and diffusion of the reaction product towards the interface.

An equation for the hydrodynamical perturbations $\tilde{v}$ can be integrated exactly to yield

$$\tilde{V} = A_1 sh\alpha\eta + A_2 ch\alpha\eta + A_3 \eta s\alpha\eta + A_4 \eta ch\alpha\eta \tag{4.11}$$

where A_1 to A_4 are the integration constants.

In the investigation of mass transfer processes, it is usually assumed that the velocity of liquid within a thin interfacial diffusion layer is constant and equal to the surface velocity of the liquid film. The magnitude of the Nusselt diffusion number calculated under the above assumption agrees well with the experiment. In such a case, the mass flux across the interface is calculated using the principal expansion term of the parabolic profile of velocity with respect to coordinate normal to the flow.

Suggesting that the above procedure is also applicable to hydrodynamical perturbations, we shall describe these within the interfacial layer using the expansion of Equation (4.11) in the series with respect to and retaining only the principal expansion terms. Apparently, this will be justified provided that the thickness of interfacial diffusion layer is much less than the film thickness (η_h>>1), which condition is usually fulfilled. Using the boundary conditions for the normal velocity component and expanding the function and its second derivative in series, we obtain

$$\tilde{V} = Bz_1\eta; \frac{d^2\tilde{V}}{d\eta^2} = Bz_2 \text{ for } \eta = 0. \tag{4.12}$$

were B=const; z_1 and z_2 are determined by Equation (3.8).

Thus, we arrive at the system of equations

$$\frac{1}{Pe_A}\left(\frac{d^2\tilde{C}_A}{d\eta^2} - \alpha^2\tilde{C}_A\right) - \Lambda\tilde{C}_A = -\frac{2}{\sqrt{\pi}}Bz_1\eta e^{-\eta^2}; \frac{1}{Pe_B}\left(\frac{d^2\tilde{C}_B}{d\eta^2} - \alpha^2\tilde{C}_B\right) = -\Lambda\tilde{C}_A \tag{4.13}$$

Eliminating the function $\tilde{c}_A(\eta)$ from Equation (4.13), we obtain a single equation determining the perturbation of the reaction product concentration

$$\frac{d^4\tilde{C}_B}{d\eta^4}-(2\alpha^2+\Lambda Pe_A)\frac{d^2\tilde{C}_B}{d\eta^2}+(\alpha^4+\alpha^2\Lambda Pe_A)\tilde{C}_B=\frac{2}{\sqrt{\pi}}B\Lambda Pe_A Pe_B z,\eta e^{-\eta^2}$$

$$d\tilde{C}_B/d\eta=0,\ d^2C_B/d\eta^2=\alpha^2\tilde{C}_B,\ Bz_2=N_1\alpha^2\left(1+\frac{\alpha\eta_h w}{d\eta_h+Bi}\right)\tilde{C}_B \text{ for } \eta=0; \quad (4.14)$$

$$\tilde{C}_B=0,\ \frac{d^2\tilde{C}_B}{d\eta^2}=0 \text{ for } \eta=\infty$$

The linear equation (4.14) has a trivial solution $\tilde{c}_B=0$. Therefore, the existence of a nonzero solution will imply that, in addition to the unperturbed solution describing the process of physical absorption, the set of equations of hydrodynamics and mass exchange must have one more solution. In other words, a bifurcation of the mass transfer rate will occur at some critical values of the parameters due to the disturbing action of the chemical reaction. For supercritical Marangoni numbers, the system exhibits a transition to the alternative mechanism of mass transfer in the liquid film, whereby intense secondary motions arise at the free surface to play a determining role in the transfer process.

Let us construct the Green function G(η,ε) of the operator

$$\frac{d^4C_B}{d\eta^4}-(2\alpha^2+\Lambda Pe_A)\frac{d^2C_B}{d\eta^2}+\left(\alpha^4+\alpha^2 Pe_A\right)C_B=0$$

with the boundary conditions (4.14). The solution has a form

$$G(0,\xi)=\frac{1}{\alpha\Lambda Pe_A}(e^{-\alpha\xi}-e^{-\sqrt{\alpha^2+\Lambda Pe_A}\,\xi}) \quad (4.15)$$

To construct the Green function, it was necessary to know its value in a single point η=0. In order for the solution of Equation (4.14) to be nonzero, there must be a definite relation between the problem parameters. Assuming $\tilde{c}_B\neq 0, B\neq 0$, using the above expression for the Green function, and calculating the corresponding integrals, we obtain the dependence of a critical Marangoni number on the wavenumber α, hydrodynamical, and physicochemical characteristics of liquid and gas:

$$Ma=2\frac{sh2\alpha 2_h-2\alpha 2\eta_h}{sh^2\alpha\eta_h-\alpha^2\eta_h^2}\left[\sqrt{\alpha^2+\Lambda Pe_A}\exp\frac{\alpha^2+\Lambda Pe_A}{4}erfc\frac{\sqrt{\alpha^2+\Lambda Pe_A}}{4}-\alpha\exp\frac{\alpha^2}{4}erfc\frac{\alpha}{2}\right]^{-1}$$
$$\left(1+\frac{\alpha\eta_h w}{\alpha\eta_h+Bi}\right)^{-1} \quad (4.16)$$

where Ma is the Marangoni number for the given interfacial layer thickness, ΛPe_A is a dimensionless complex characterizing the ratio of times of diffusion and chemical reaction (dimensionless time of phase contact).

First, consider the case of W=0 (isothermal reaction). Equation (4.16) has the asymptotes

$$Ma = 2\left[\sqrt{\Lambda Pe_A}\exp\frac{\Lambda Pe_A}{4}\,erfc\frac{\sqrt{\Lambda Pe_A}}{2}\right]^{-1}(\alpha\eta_h)^{-1} \tag{4.17}$$

in the region of small wavenumbers $\alpha<<1$ (long-wave perturbations) and

$$Ma = \sqrt{\pi}\,\frac{\alpha^4+\alpha^2\Lambda Pe_A}{\Lambda Pe_A},\ \alpha>>1 \tag{4.18}$$

in the region of large wavenumbers (short-wave perturbations).

It should be noted that in the range of long-wave perturbations the critical Marangoni number depends on the relative thickness of liquid films, $\eta_h=h/\delta$, while no such dependence is observed in the short-wave perturbation range where the determining role belongs to the dimensionless time of phase contact, $\Lambda Pe_A=4kX/U_i$. Equations (4.16), (4.17) and (4.18) imply that, as the wavenumber tends to either zero or infinity, the Marangoni number indefinitely increases and, hence, expression (4.16) describes the curve of neutral stability for the considered mass transfer process complicated by a homogeneous chemical reaction.

In the case of large η_h values ($\eta_h>>1$) we may construct an asymptotic expression for the minimum critical Marangoni number:

$$Ma_* = 4\left[\sqrt{\Lambda Pe_A}\exp\frac{\Lambda Pe_A}{4}\,erfc\frac{\sqrt{\Lambda Pe_A}}{2}\right]^{-1}$$

and for small values of phase contact time, $\Lambda Pe_A<<1$, we have

$$Ma_* = \frac{4}{\sqrt{\pi\Lambda Pe_A}}$$

In these expressions we have used the asymptotic expression for the error function

$$erf\,u = \frac{2}{\sqrt{\pi}}\int_0^u e^{-u^2}du,\ erfc\,u = 1-erf\,u$$

$$erf\sqrt{u} = 1-\frac{1}{\sqrt{\pi}}e^{-u}\sum_{k=0}^{\infty}\frac{(-1)^k\Gamma(k+\frac{1}{2})}{u^{k+\frac{1}{2}}}$$

where Γ is the gamma function, $\Gamma(1/2)=\sqrt{\pi}$, and $\Gamma(3/2)=\sqrt{\pi}/2$.

Thus, the above consideration has revealed that, within the linear theory, the process of physical absorption at $Ma<2/\sqrt{\pi}$ will be stable with respect to infinitely small perturbations.

Applying the above procedure to the minimum wavenumber, we calculate that

$$\alpha_* = \frac{1}{2\eta_h} \ln \frac{4\eta_h}{Ma}$$

Thus, for the sides of a hypothetical cell in the case of large η_h we obtain

$$\frac{d}{h} = \frac{4}{3\pi\alpha_*\eta_h}$$

It is seen that as the diffusion layer thickness decreases, the cell size diminishes by a logarithmic law.

As the dimensionless time of phase contact decreases, the critical Marangoni number grows by the power law as $(\lambda Pe_A)^{-1/2}$. These results are quite reasonable from the physical point of view: the less the reaction rate constant, the more surface-inactive the reaction products must be in order to induce the onset of interfacial instability.

The general pattern of development of the instability appears as follows. The absorbed gas reacts with a liquid chemisorbent to yield surface-inactive products creating the gradients of surface tension and concentration. For the supercritical Marangoni numbers, these gradients favor the transition of free surface energy into the kinetic energy of convective motion of the liquid. As a result, the hydrodynamical structures similar to those that are typical for the Bénard heat convection are produced at the interface, and a new transfer mechanism arises, which is manifested by an anomalous increase in the mass flux across the interface.

Consider the reaction with considerable heat evolved or absorbed, (note that for an endothermal reaction $\Delta H<0, W>0$, and for an exothermal reaction $\Delta H>0$, $W<0$). Let the critical Marangoni number, now allowing for the reaction heat effect, be denoted by Ma_{x0}. Then, the critical Marangoni number for the process with heat evolution (or absorption) is

$$Ma_{*T} = Ma_{*0}\left(1+\frac{W\alpha\eta_h}{\alpha\eta_h+Bi}\right)^{-1}$$

For $Bi=0$ we have $\eta_h \to \infty$. Using the results of calculation of the critical wavenumbers and minimum Marangoni numbers carried out at $W=0$, we obtain:

$$Ma_{*T} = \frac{8}{\sqrt{\pi}}\left(\sqrt{\Lambda Pe_A}\exp\frac{\Lambda Pe_A}{4}\,erfc\frac{\sqrt{\Lambda Pe_A}}{2}\right)^{-1}(1+W)^{-1} \tag{4.19}$$

As the duration of chemical reaction grows, $\lambda Pe_A \to \infty$, Equation (4.19) gives an asymptotic formula

$$Ma_{*T} = \frac{4}{1+W},\ \Lambda Pe_A \to \infty \tag{4.20}$$

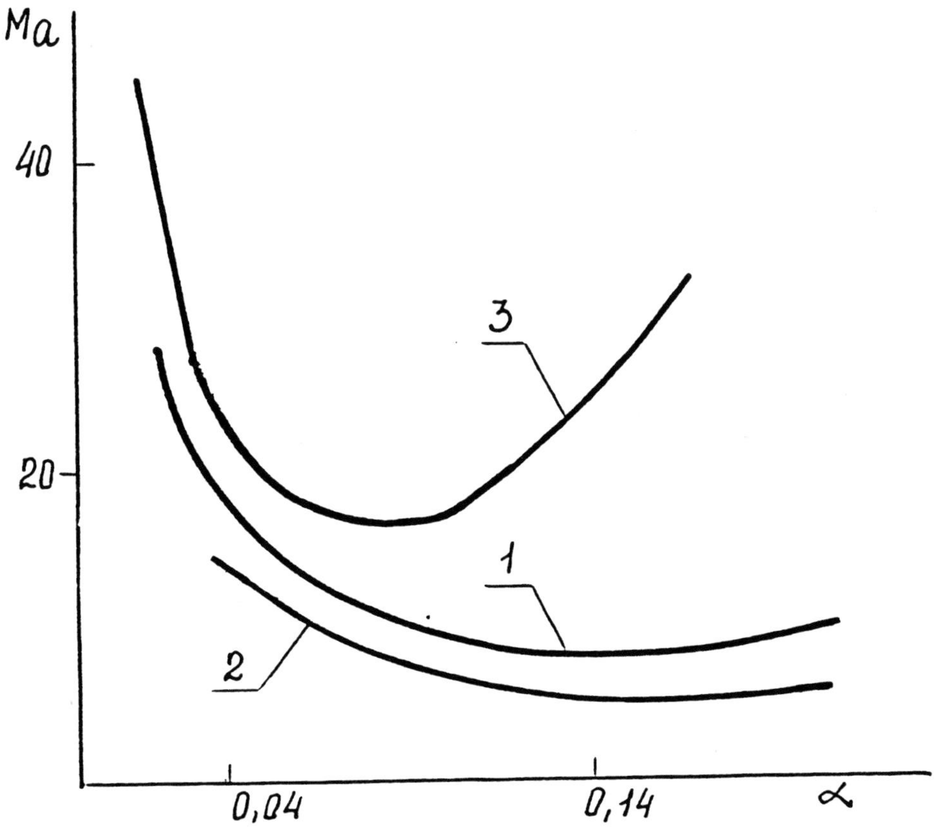

Figure 6. Critical Marangoni number vs. the wavenumber 55. (λPe_A=1; η_h=30; Bi=0.2); (1) W=0; (2) W=1; (3) W=-1.53.

If the reaction product does not affect the surface tension, $\varepsilon_1=0$, it is easy to evaluate the critical phase contact time from Equation (4.20):

$$t_* = \left(\frac{sc_p\mu \, Pe_B}{C_{Ai}\Delta H\varepsilon_2 Pe} \right)^2 D_B$$

For the typical values corresponding to the physical properties of water at cAi=1 mole/m^3, ΔH=200 kJ/mole, this formula yields t_x=0.4 s which is an achievable value for the real mass transfer conditions.

The instability of convective diffusion processes complicated by a non-isothermal chemical reaction arises as follows: The heat absorbed due to the chemical reaction produces a negative temperature gradient along the film which destabilizes the flow. The presence of an exothermal reaction results in a positive temperature gradient and increases the critical Marangoni number, eventually improving the stability of mass transfer.

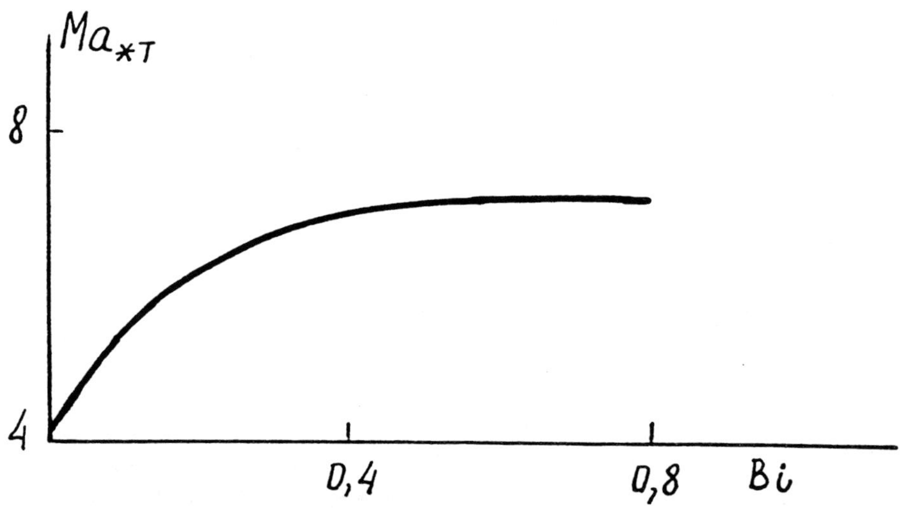

Figure 7. Minimum critical Marangoni number vs. the Biot number [55] for $\lambda Pe_A=1$; $\eta_h=100$; and W=1.

The interrelation of hydrodynamical, thermal, and diffusion processes leads to the appearance of positive feedbacks interacting according to the following scheme. Let a negative temperature gradient of fluctuation character arise along the film due to the absorbed reaction heat. This produces an increase in the surface tension gradient and a growth of surface velocity which, in turn, enhances the flux of the component transferred and stimulates the reaction. This further increases the temperature gradient, and so forth. In combination with the effect of a surface-inactive reaction product which also exhibits a positive acceleration, the processes of motion and heat and mass transfer will enhance each other, leading to the Marangoni instability.

In the case of an exothermal reaction, the appearing feedbacks are negative. The evolution of heat produces a decrease in the surface tension gradient and a drop in the surface velocity. This favors a reduction of the mass transfer and slows down the chemical reaction, with a resulting decrease in the heat evolution. In this case, the reaction heat evolution may suppress the destabilizing effect of surface-inactive products, provided that the coefficient in the linear temperature dependence of surface tension is greater than that in the concentrational dependence.

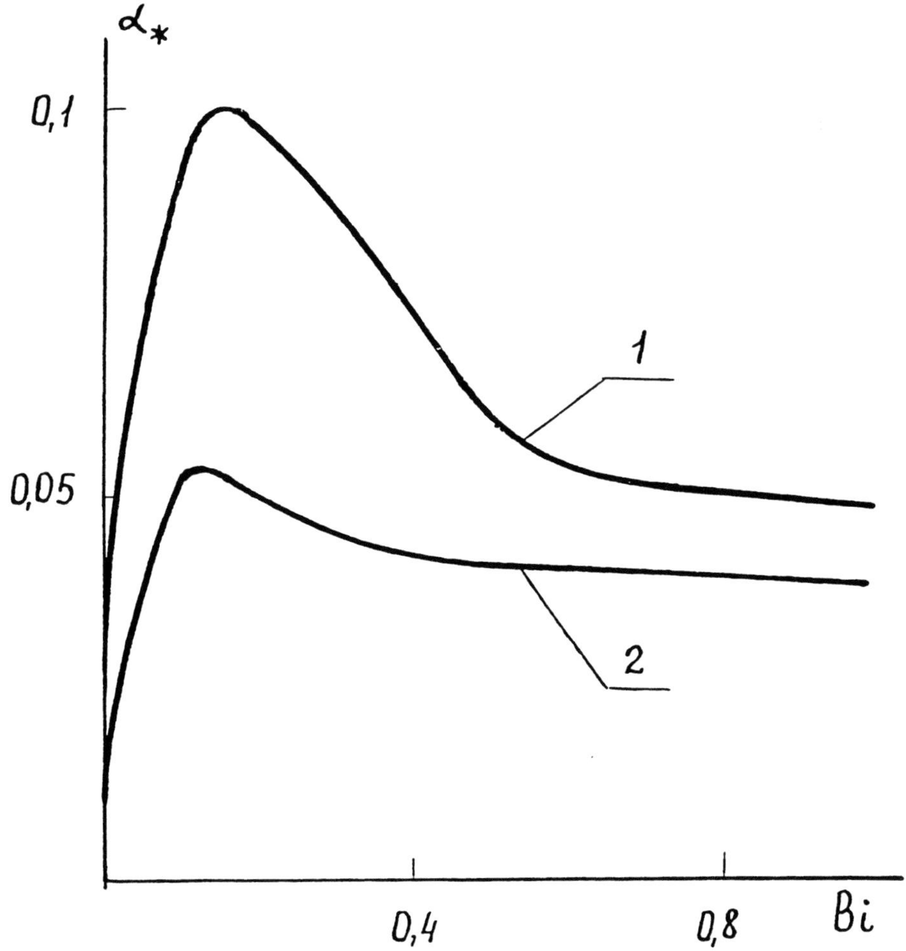

Figure 8a. Critical wavenumber (a) vs. the Biot number [55] for η_h=100; W=1; and λPe_A=1 (1) or 0.1 (2).

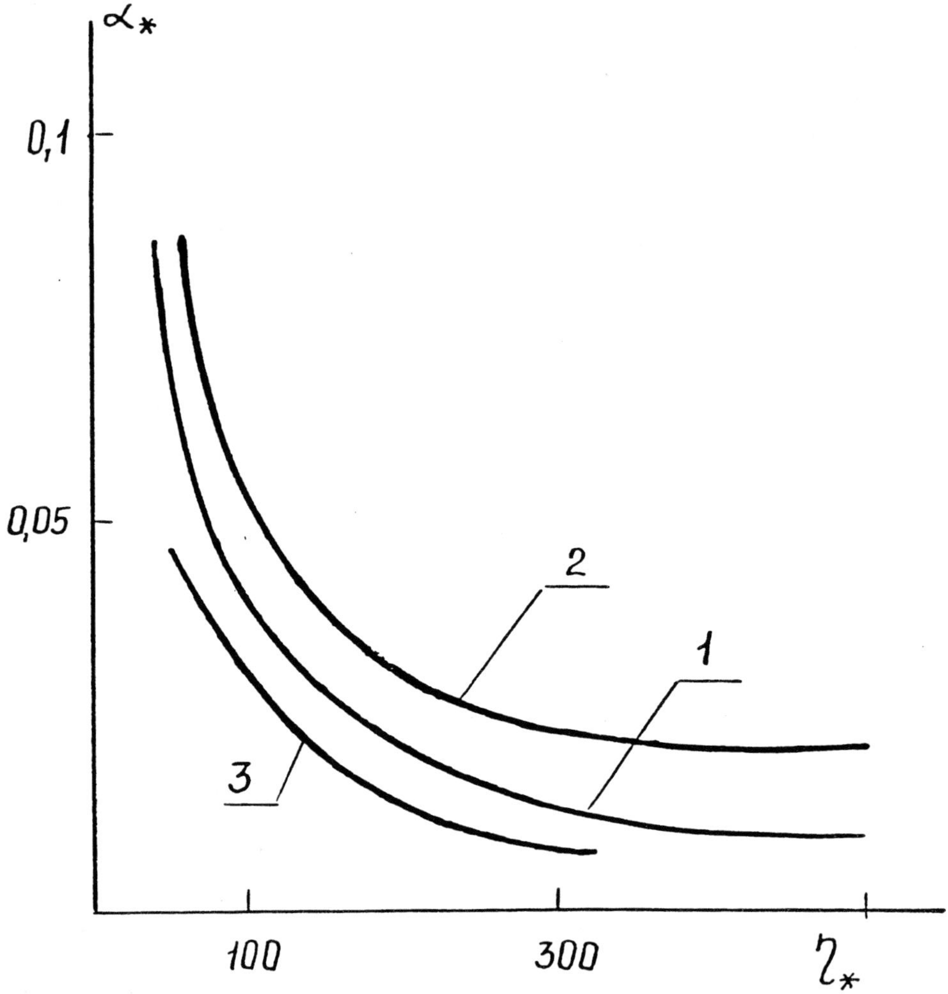

Figure 8b. Critical wavenumber (b) vs. the dimensionless film thickness for λPe_A=0.1; Bi=0.2; and W=0 (1), 1 (2), or -1.53 (3).

On the contrary, the absorption of heat due to chemical reaction may favor the development of Marangoni instability even in the case when the reaction products reduce the surface tension. Eventually the instability is determined by the competition of these processes.

The results of numerical calculations of the neutral stability curve, critical wavenumbers and minimum Marangoni numbers are presented in Figures 6 and 7. It is seen that the critical wavenumbers determining the size of periodic structures exhibit a sufficiently rapid growth (Figure 8) with increasing relative film thickness ($\eta_h \to \infty$). As the Biot number increases, the critical Marangoni number attains an asymptotic level (Figure 7), while the critical wavenumber exhibits an extremum (Figure 8) [54,55].

5. INSTABILITY DUE TO HORIZONTAL TEMPERATURE GRADIENT

In the preceding sections, we have considered the instabilities induced by the heat and mass transfer in the direction normal to the interface. This transfer was just the factor rendering the system's non-equilibrium. A non-isothermal Marangoni instability described in Section 1 was driven by the gradient directed normally to the interface. However, it is possible that an instability might be induced by the temperature gradient tangential to the surface of liquid.

Consider a model problem to illustrate this possibility. Let the motion of a viscous liquid occur in a flat tray whose width is much greater than the flow depth. The pressure along the free surface of the liquid is constant and equals the atmospheric pressure in an open tray. Let a specified amount of heat be supplied to the liquid from beneath, via the tray bottom, while the free surface is thermally insulated. Such a boundary condition is valid in the case when the ratio of heat conductivities of liquid and gas is large (for example, this ratio is about 30 for the air-water system, and even greater for the systems with liquid metals).

The set of equations of motion and heat exchange for the straight flow lines is

$$\mu \frac{d^2V}{dY^2} = -sg\sin\alpha; \; V\frac{\partial T}{\partial X} = æ\left(\frac{\partial^2 T}{\partial X^2} + \frac{\partial^2 T}{\partial Y^2}\right) \tag{5.1}$$

$$V = 0, \; \lambda \bullet \frac{\partial T}{\partial Y} = -q_w \text{ for } Y = 0;$$

$$\frac{\partial T}{\partial Y} = 0, \; \mu\frac{dV}{dY} = \frac{d\sigma}{dX} \text{ for } Y = h,$$

where α is the tray inclination angle with respect to the horizontal, and h is the depth of liquid in the tray.

The temperature dependence of surface tension is defined by Equation (1.6). The stationary state in the case of constant heat supply is ensured by the removal of the same amount of heat via convection, which results in the establishment of a temperature gradient along the free surface of liquid, $E = dT/dX$.

The solution of Equation (5.1) is represented in the form

$$T = T_0 + EX + T_1(Y)$$

where the function $T_1(Y)$ is determined from an ordinary differential equation

$$æ\frac{d^2T_1}{dY^2} = EV; \lambda\frac{dT_1}{dY} = -q_w, Y = 0; \frac{dT_1}{dY} = 0, Y = h \tag{5.2}$$

Integrating Equation (5.2) over the tray cross-section and taking the thermal boundary conditions into account, we obtain a relation between the unknown temperature gradient E, wetting density Q, and heat flux at the wall q_W.

$$q_w = sc_pQE, Q = \int_0^h VdY \tag{5.3}$$

The velocity distribution with an allowance for the appearance of surface tension gradients at the interface has the form

$$V = \frac{sg\sin\alpha}{\mu}(hY - \frac{Y^2}{2}) - \frac{\varepsilon E}{\mu}Y \tag{5.4}$$

It should be noted that in the case of heat supply (q_W>0, E>0) the force arising due to the tangential temperature gradient will retard the motion of the liquid film, (remember that the force is directed from lower to higher surface tensions due to spontaneous decrease of the Gibbs surface energy [56]).

By integrating Equation (5.4) over the film cross-section, we obtain a transcendental equation determining the dependence of the temperature gradient on the film's thickness:

$$\frac{E}{\mu}\left(\frac{\delta g\sin\alpha h^3}{3} - \frac{\varepsilon E h^2}{2}\right) = \frac{q_w}{sc_p} \tag{5.5}$$

For the dependence of friction stress at the wall on the film thickness and temperature gradient, we have

$$\tau_w = \mu\frac{dV}{dY} = sg\sin\alpha\cdot h - \varepsilon E, Y = 0. \tag{5.6}$$

Introducing the dimensionless quantities

$$a = \frac{3\varepsilon E}{2sgh\sin\alpha}, h = \sqrt[4]{\frac{2s^3C_pg^2\sin^2\alpha}{\partial\varepsilon\mu q_w}}h,$$

$$\tau_w = \frac{\tau_w}{\sqrt[4]{\frac{\partial s\varepsilon\mu q_w g^2\sin^2\alpha}{2C_p}}}, Q = \sqrt[4]{\frac{16s^5C_p^3g^2\mu\sin^2\alpha}{\partial q_w^3\varepsilon^3}}Q$$

and using Equation (5.5), (5.6) we obtain the parametric dependence of the friction stress on the flow rate of liquid:

$$\bar{\tau}_w = \frac{1-\frac{2}{3}a}{4\sqrt{a(1-a)}}, \; \overline{Q} = 4\sqrt{\frac{1-a}{a^3}} \tag{5.7}$$

An analysis of Equation (5.7) yields that $a \to 0$ we have $\overline{Q} \to \infty$, and $\tau_W \to \infty$ and for $a \to 1$ we have $\overline{Q} \to 0$ and $\tau_W \to \infty$. Thus, the dependence of friction stress on the flow rate is nonmonotonous because for both $\overline{Q} \to 0$ and $\overline{Q} \to \infty$ we have $d\tau_W / d\overline{Q} > 0$.

Using the known methods of investigation of the stability of stationary states, it is possible to show that the stationary regimes of motion and heat exchange are unstable in the range of "negative" friction. The stationary state is destroyed via the positive feedback mechanism which operates according to the following scheme. Let the liquid temperature grow and its gradient along the surface increase. Then the retarding action of surface tension gradients will slow down the film and, hence, the removal of heat from the heat-exchange region will decrease, resulting in a further elevation of the liquid temperature.

The described effect is most inherent in very thin liquid films with a strong temperature dependence of the surface tension, featuring an intense heat exchange with the ambient medium.

6. MASS TRANSFER AT SUBCRITICAL MARANGONI NUMBERS

On the basis of what has been considered above, the general pattern of the absorption process complicated by the chemical reaction resulting in the formation of surface-inactive substances appears as follows. The reaction products are accumulated at the interface and create a positive concentration gradient. For supercritical Marangoni numbers, the free surface energy of liquid transforms into the kinetic energy of its convective motion and the hydrodynamical structures, similar to the Benard cells produced upon the thermal convection, are developed at the interface. Thus, a new transfer mechanism arises and the absorption rate increases. However, it has already been noted that, even at subcritical Marangoni numbers, an increase in the longitudinal component of the film flow velocity may stimulate the mass exchange. It is important that the process occurs in stationary conditions.

Let us study the mass transfer at subcritical Marangoni numbers, i.e., under the conditions when the motion of liquid retains its stability and the interaction of the flow conditions and the mass transfer rate are mediated by the development of tangential interfacial stress due to surface tension gradients. The last circumstance allows for a considerable simplification of mathematical calculations.

By integrating the equations of motion, (with the origin of coordinates fixed at the free surface), over the film depth we obtain an expression for the velocity profile

$$V=\frac{1}{\mu}\left[\frac{sg}{2}(h^2-Y^2)+\frac{d\sigma}{dX}(h-Y)\right]$$

satisfying the boundary conditions

$$\mu\frac{\partial V}{\partial Y}=-\frac{d\sigma}{dX},\, Y=0;\, V=0,\, Y=h$$

Taking into account the fact that the problem is stationary, it is not difficult to derive the relations for the surface velocity and the rate of flow per unit length of wetted perimeter (called the wetting density)

$$V_i=\frac{1}{\mu}\left(\frac{sg}{2}h^2+\frac{d\sigma}{dX}h\right)$$
$$Q=\frac{1}{\mu}\left(\frac{sg}{3}h^3+\frac{d\sigma}{dX}\frac{h^2}{2}\right) \tag{6.1}$$

Let us assume, as is conventionally done for large Peclet numbers, that the velocity of liquid within the interfacial diffusion layer is constant and equals that at the surface. Moreover, let the molecular diffusion coefficients of the absorbed gas, D_A, and the reaction product, D_B, be equal, because this will not affect the essence of the problem considered. (The effect of D_A/D_B ratio on the mass transfer rate can be judged by the data presented in Reference 50).

The system of dimensionless equations describing the transfer process at subcritical Marangoni numbers, $Ma<Ma_x$, with an account for the above relations (6.1) has the form

$$\frac{\partial^2 C_A}{\partial \hat{y}^2}=\frac{\partial C_A}{\partial H}+C_A \quad \frac{\partial^2 C_B}{\partial \hat{y}^2}=\frac{\partial C_B}{\partial H}-C_A$$
$$\hat{h}^3+Ma\frac{H}{\hat{x}}\hat{h}^2\frac{dC_B}{dH}=1 \quad \hat{h}^2+\frac{4}{3}Ma\hat{h}\frac{dC_B}{dH}=\frac{\hat{x}}{H}$$
$$C_A=1,\frac{\partial C_B}{\partial y}=0,\hat{y}=0,H>0;C_A=0,C_B=0,\hat{y}>0,H>0;C_A=0,C_B=0,\hat{y}=\infty,H>0. \tag{6.2}$$

where c_A, c_B are the dimensionless concentrations of the absorbed gas and the reaction product, respectively, (rendered dimensionless by scaling to c_{Ai}, the surface concentration); $\hat{x}=kX\frac{2\upsilon}{g}(\frac{3\upsilon Q}{g})^{-2/3}$, $\hat{y}=Y\sqrt{k/D}$ are the dimensionless coordinates; $\hat{h} = h(\frac{3\upsilon Q}{g})^{-1/3}$ is the dimensionless film thickness; Ma=εk/(ρgQ) is the Marangoni number; and $H=kX/U_i$ is the dimensionless phase contact time.

The Marangoni number Ma entering in Equation (6.2) represents the ratio of the power developed by the chemical-reaction-driven surface force during the reaction time to the power of gravitation force maintaining the motion of viscous film (the powers related to the unit surface area of the film).

Furthermore, let us assume that

$$\frac{X}{V_i}\frac{dV_i}{dX} << 1 \tag{6.3}$$

The smallness of this quantity implies the smallness of the velocity component normal to the interface. This restriction is a usual proposition of the theory of convective interfacial diffusion layer.

Equation (6.2) represents a set of two linear equations with respect to unknown concentrations c_A, c_B and a set of two transcendental equations with respect to the phase contact time, H, and the dimensionless film thickness, $\hat{h}$. Therefore, the effects of hydrodynamical factors are accounted for by introducing the unknown time which depends on the surface tension gradient and is determined by jointly solving the equations of convective diffusion and hydrodynamics.

As it has been shown in Reference 10, the solution of the first two equations of system (6.2) with respect to c_A and c_B has the form

$$C_A = \frac{1}{2}\left[\exp(-\hat{y})erfc\left(\frac{\hat{y}}{2\sqrt{H}} - \sqrt{H}\right) + \exp(\hat{y})erfc\left(\frac{\hat{y}}{2\sqrt{H}} + \sqrt{H}\right)\right]$$

$$C_B(o,H) = W(H) - 1$$

$$W = \exp\left(-\frac{H}{2}\right)\left[(1+H)I_0\left(\frac{H}{2}\right)) + HI_1\left(\frac{H}{2}\right)\right]$$

where I_0, I_1 are the modified Bessel functions [57]. On the basis of this solution, we can determine the rate of absorption via the interface

$$j = -DC_{Ai}\frac{\partial C_A}{\partial Y}, Y = 0; \bar{j} = -\frac{\partial C_A}{\partial \hat{y}}, \hat{y} = 0$$

$$j = C_{Ai}\sqrt{kDj}; \bar{j} = erf\sqrt{H} + \frac{e^{-H}}{\sqrt{\pi H}}$$

and the length-averaged Sherwood number

$$Sh = \frac{qL}{D}, q = \frac{1}{C_{Ai}L}\int_0^L j(X)dX$$

Obviously,

$$\frac{Sh}{Sh_0} = \frac{\sqrt{\pi}}{2\sqrt{\hat{L}}}\int_0^{\hat{L}} \bar{j}d\hat{x} = \varphi(\hat{L}, Ma)$$

where $L = Lk\frac{2v}{g}(\frac{3vQ}{g})^{-2/3}$ is the dimensionless length of liquid film; and $Sh_0 = 2/(\sqrt{\pi})[D^{-1/2}(\frac{2v}{g})^{-1/2}(\frac{3vQ}{g})^{1/3}L^{1/2}]$ is the Sherwood number determined by the physical absorption process in the absence of chemical reactions.

Obviously, for the reaction rate constant tending to zero, $k \to 0$, we have $\hat{L} \to 0$ and $\varphi(\hat{L}, Ma) \to 1$. Thus, the solution of the problem formulated above will be determined if we represent the phase contact time H as a function of the longitudinal coordinate $\hat{x}$ and the Marangoni number Ma.

Let us calculate the gradient of the reaction product concentration:

$$\frac{dC_B}{dH} = \psi(H) = \frac{1}{4}\left[(1-H)I_0\left(\frac{H}{2}\right) + 6I_1\left(\frac{H}{2}\right) + HI_2\left(\frac{H}{2}\right)\right]\exp\left(-\frac{H}{2}\right)$$

Using the asymptotic expressions for the modified Bessel functions of mth order [57]

$$I_m(z) = (z/2)^m, z << 1; I_m(z) = (e^z/\sqrt{2\pi z})[1 + 0(1/z)], z >> 1,$$

we can note the following features of the function $\psi(H)$:

$$I_m(z) = \left(\frac{z}{2}\right)^m, z << 1; I_m(z) = e^z/\sqrt{2\pi z}[1 + 0(1/z)], z >> 1$$

$$\max_{0 \le H \le \infty} \psi(H) = \psi(0) = \frac{1}{2}; \lim \psi(H) = 0; \frac{d\psi}{dH} < 0; \frac{d\psi}{dH} = -\frac{1}{8}, H = 0$$

The fact that $\psi(H)$ is a decreasing function implies that the maximum gradient of the reaction product concentration occurs in the very beginning of the interfacial diffusion layer formation. Then the gradient decreases because the reaction and absorption rates attain a stabilized regime whose characteristics are given in detail in Reference 50.

Substituting the expressions for the gradient of concentration c_B into the two last equations of the system (6.2), we obtain nonlinear equations with respect to the film thickness $\hat{h}$ and the contact time H. Eliminating the Marangoni number from these equations, we arrive at a relation between the surface velocity and the film thickness:

$$\hat{h}^3 + 3\hat{v}\hat{h} - 4 = 0 \quad \hat{h} = \sqrt{2+\sqrt{4+\hat{v}}} + \sqrt{2-\sqrt{4+\hat{v}}}$$

where $\hat{v} = \hat{x}/H$ is the dimensionless velocity of the surface of liquid film.

Therefore, the phase contact time is an explicit function of the longitudinal coordinate $\hat{x}$ determined from the transcendental equation

$$\hat{h}^3 + Ma\frac{H}{\hat{x}}\hat{h}^2\psi - 1 = 0 \tag{6.4}$$

The nonlinear Equation (6.4) has no singularities and, hence, possesses a single solution determining the phase contact time as a function of the Marangoni number and the film length.

For Ma=0 we have an obvious solution $\hat{h} = 1, H = \hat{x}$, i.e., this case corresponds to a well known Danckwerts solution which takes no account of the surface tension gradient.

For small Marangoni numbers, an approximate solution of the last two equations of the system (6.2) can be easily constructed using the method of small parameter

$$\hat{h} = \sum_0^\infty \hat{h}_n Ma^n, \hat{v} = \sum_0^\infty \hat{v}_n Ma^n$$

and for the first approximation we obtain the exact expressions

$$\hat{h}_1 = -\psi(\hat{x})/3, \hat{v}_1 = 2\psi(\hat{x})/3$$

Thus, the expressions for the film thickness and surface velocity at small Marangoni numbers have the form

$$\begin{aligned} \hat{h} &= 1 - \frac{Ma}{12}\left[(2-\hat{x})I_0\left(\frac{\hat{x}}{2}\right) + 6I_1\left(\frac{\hat{x}}{2}\right) + \hat{x}I_2\left(\frac{\hat{x}}{2}\right)\right]\exp\left(-\frac{\hat{x}}{2}\right) \\ \hat{v} &= 1 - \frac{Ma}{6}\left[(2-\hat{x})I_0\left(\frac{\hat{x}}{2}\right) + 6I_1\left(\frac{\hat{x}}{2}\right) + \hat{x}I_2\left(\frac{\hat{x}}{2}\right)\right]\exp\left(-\frac{\hat{x}}{2}\right) \end{aligned} \tag{6.5}$$

Using the above mentioned properties of the ψ function, it is easy to derive asymptotic expressions for the surface velocity:

$$\hat{v}(\hat{x}) = 1 + \frac{Ma}{3}, \hat{x} << 1; \hat{v}(\hat{x}) = 1 + \frac{4Ma}{3\sqrt{\pi\hat{x}}}, \hat{x} >> 1 \tag{6.6}$$

As follows from the relations (6.5) and (6.6), at the beginning of the interfacial diffusion layer the surface velocity increases with the Marangoni number and then, as the chemical reaction stabilizes along the film length, the velocity of liquid drops as $\hat{x}^{-1/2}$.

Among the particular features to be noted is a rather slow asymptotic approach of the surface velocity to its value $\hat{v} = 1$ corresponding to the absence of chemical reactions. Equations (6.2) to (6.6) also imply that the surface velocity does not depend on the molecular diffusion coefficient of the absorbed gas and is determined mostly by the Marangoni number.

An automodel character of the motion of liquid in the surface layer with respect to the diffusion coefficient is explained by the fact that the rates of reaction product formation and inward diffusion are equally proportional to the gas absorption rate, so that the accumulation of the reaction product at the film surface is determined only by the hydrodynamical factors and the partial pressure of absorbed component in the gas phase.

Using the asymptotic expressions for the velocity of liquid, it is not difficult to calculate the absorption rate for a short-time phase contact

$$\bar{j}(\hat{x}) = \sqrt{\frac{1 + Ma/3}{\pi x}}$$

from which we can derive an expression for the Sherwood number in the case of small Marangoni numbers:

$$\frac{Sh}{Sh_0} = 1 + \frac{Ma}{6} \tag{6.7}$$

According to Equation (6.7), the presence of ions in solution results in a greater acceleration factor for the absorption rate, in agreement with the experimental data of Reference 2.

It should be noted that Equation (6.7) can be used for the determination of surface activity ε of the reaction product, once the factor Sh/Sh_0 is measured experimentally.

For the dimensionless time of phase contact, $H(\hat{x}, Ma)$, we can readily find the expression

$$H = \frac{\hat{x}}{1 - \frac{Ma}{3}}$$

and, hence, the gradient of the reaction product concentration for small H is given by the formula

$$\frac{dC_B}{dH} = \psi(H) = \psi(0) + H\frac{d\psi}{dH}(0) = \frac{1}{2} - \frac{H}{8} = \frac{1}{2} - \frac{\hat{x}}{8} + \frac{Ma\hat{x}}{24}$$

The last expression suggests that the surface tension gradient not only stimulates the mass transfer rate, but also increases the reaction product concentration at the film surface which is a direct consequence of the interrelation between hydrodynamical and mass transfer processes.

Thus, the problem considered features the development of mutually related physico-chemical and hydrodynamical factors enhancing each other according to the scheme:

surface velocity of liquid→absorption rate→accumulation rate of surface-inactive reaction products→surface tension gradients→surface velocity of liquid.

Then, the cycle is repeated, whereby the positive feedback is closed via the chemical reaction. For certain conditions, e.g., large Marangoni numbers, this scheme of development of the mass transfer processes leads, as has been shown above, to the flow bifurcations and the appearance of qualitatively new mass transfer regimes. However, investigations of these phenomena lies outside the scope of the present work.

The obtaining of analytical expression for the velocity of the liquid film surface (6.5) allows for the evaluation of error introduced by assuming the validity of inequality (6.3). Calculations show that the maximum error due to this condition does not exceed 11%; moreover, the error decreases at sufficiently small or large $\hat{x}$ values.

Let us now consider the range of large Marangoni numbers, Ma>>1. Such conditions have been provided, e.g., in the experiments with chemisorption carried out in a horizontal tray at small Reynolds numbers [2].

In such a case, the system of equations obtained from the last two equations (6.2) for the condition

$$\hat{h} << Ma\psi$$

will acquire the form

$$Ma\frac{H}{\hat{x}}\hat{h}^2\psi = 1, \frac{4}{3}Ma\frac{H}{\hat{x}}\hat{h}\psi = \frac{\hat{x}}{H},$$

whence

$$\hat{x} = H\sqrt[3]{\frac{16}{9}Ma\psi}$$

For a short-term phase contact (H<<1):

$$\hat{x} = H\sqrt[3]{\frac{8}{9}Ma},\ \hat{v} = \sqrt[3]{\frac{8}{9}Ma},\ \bar{j} = \sqrt{\left(\frac{8}{9}Ma\right)^{\frac{1}{3}}(\pi\hat{x})^{-1}}$$

$$\frac{sh}{sh_0} = \left(\frac{8}{9}Ma\right)^{\frac{1}{6}} \quad \frac{dc_B}{dx} = \frac{1}{2} - \frac{1}{8}\left(\frac{8}{9}Ma\right)^{-\frac{1}{3}}\hat{x}$$

In the range of large Marangoni numbers, the effects of enhancing the surface velocity of liquid film and the mass transfer rate are rather weakly dependent on the magnitude of this criterion. However, it must be taken into account that for $Ma > Ma_x$ the flow of a film with flat interface loses its stability and transforms into a more complex hydrodynamical regime.

In such a case, the coefficients of mass transfer may increase 5-6-fold and beyond and the qualitatively different mechanisms of transfer processes may arise. In these regimes, the heat of absorption, until it is dissipated in liquid, becomes a source for the self-organization of microflows occurring at the interface.

This work attempts the theoretical calculation of the critical Marangoni numbers determining the onset of a new regime.

Translated by Pavel P. Pozdeev

REFERENCES

1. Yu.V. Aksel'rod, V.V. Dil'man and Yu.V. Furmer, Teor. *Osnovy Khim. Tekhnol.*, 1971, Vol. 5, No. 5, p. 676.
2. V.V. Dil'man, Yu.V. Aksel'rod, and F.M. Khutoryanskii, *Teor. Osnovy Khim. Tekhnol.*, 1977, Vol. 11, No. 1, p. 11.
3. Yu.V. Furmer, Yu.V. Aksel'rod and V.V. Dil'man, *Zh. Priklad. Khim.*, 1973, Vol. 46, No. 7, p. 1507.
4. F.M. Khutoryanskii, Yu.V. Aksel'rod, V.V. Dil'man and Yu.V. Furmer, *Teor. Osnovy Khim. Tekhnol.*, 1975, Vol. 48, No. 1, p. 72.
5. Yu.V. Aksel'rod and V.V. Dil'man, *Teor. Osnovy Khim. Tekhnol*, 1980 Vol. 14, No. 6, p. 837.
6. V.V. Dil'man and L.M. Rabinovich, *Teor. Osnovy Khim. Tekhnol.*, 1979, Vol. 13, No. 6, p. 54.
7. G.Ya. Mironova, Yu.V. Aksel'rod, V.V. Dil'man et al., *Teor. Osnovy Khim. Tekhnol.*, 1979, Vol. 13, No. 5, p. 693.
8. Yu.V. Aksel'rod, A.P. Mitina, N.E. Legezin, et al., *Teor. Osnovy Khim. Tekhnol.*, 1988, Vol. 22, No. 3, p. 291.
9. Yu.V. Aksel'rod, *Gas-Liquid Chemisorption Processes* (Moscow: Khimiya, 1989), p. 240 (in Russian).
10. P.V. Danckwerts, *Chem. Eng. Sci.*, 1967, Vol. 11, No. 3, p. 472.

11. N.N. Kulov and V.A. Malyusov, *Teor. Osnovy Khim. Tekhnol.*, 1967, Vol. 1, No. 2, p. 313.
12. V.N. Ramm, *Absorption of Gases* (Moscow: Khimiya, 1976), p. 655 (in Russian).
13. N.N. Kulov, V.V. Maksimov, V.A. Malyusov and N.M. Zhavoronkov, *Teor. Osnovy Khim. Tekhnol*, 1983, Vol. 17, No. 3, p. 291.
14. V.V. Dil'man, *Teor. Osnovy Khim. Tekhnol.*, 1975, Vol. 9, No. 6, p. 844.
15. V.G. Levich, *Physico-Chemical Hydrodynamics* (Moscow: Fizmatgiz, 1959), p. 659 (in Russian).
16. P.V. Danckwerts, *Insights into Chemical Engineering*, (Oxford, Pergamon Press, 1981).
17. J.R.A. Pearson, *J. Fluid Mech.*, 1958, Vol. 4, p. 489.
18. L.E. Scriven and C.V. Sternling, *J. Fluid. Mech.*, 1963, Vol. 19, p. 321.
19. E. Ruckenstein and C.V. Berbente, *Chem. Eng. Sci.*, 1970, Vol. 25, p. 475.
20. Y.T. Chan and A.Z. Szeri, *Chem. Eng. Sci.*, 1974, Vol. 29, No. 11, p. 2219.
21. M. Hozawa, N. Komatsu, N. Imaishi, and K. Fujinawa, *J. Chem. Eng. Jap.*, 1984, Vol. 17, No. 2, p. 173.
22. M.G. Slin'ko, V.V. Dil'man, and L.M. Rabinovich, *Teor. Osnovy Khim. Tekhnol.*, 1983, Vol. 17, No. 1, p. 10.
23. V.V. Dil'man, Yu.V. Aksel'rod, and F.M. Khutoryanskii, *Teor. Osnovy Khim. Tekhnol*, 1977, Vol. 11, No. 1, p. 11.
24. V.S. Kryov, V.A. Malyusov, U. Nitshke, and V.A. Lotkhov, *Teor. Osnovy Khim. Tekhnol.*, 1985, Vol. 19, No. 1, p. 12.
25. M. Hennenberg, P.M. Bisch, M. Vignes-Adler and A. Sanfeld, in: *Lecture Notes in Physics, Vol. 105*, (Berlin: Springer, 1979), p. 229.
26. A. Sanfeld, A. Steinchen, M. Hennenberg, P.M. Bisch, D. Van Lamswerde-Gallez, W. Dalle-Vedeve, in: *Lecture Notes in Physics*, Vol. 105, (Berlin, Springer, 1979).
27. L.E. Scriven and C.V. Sternling, *Nature* (London), 1960, Vol. 187, p. 186.
28. J.C. Berg, A. Acrives, M. Boudart, *Adv. Chem. Eng.*, 1966, No. 1.
29. H. Sawistowski, *Ber. Bunsenges. Phys. Chem.* 1981, Vol. 85, No. 10, p. 9061.
30. Sternling, C.V. and L.E. Scriven, *AJChE J.*, 1959, Vol. 5, No. 4, p. 514.
31. Yu.A. Buevich, *Inzh.-Fiz. Zh.*, 1985, Vol. 48, No. 2, p. 230.
32. V.A. Malyusov, in: *Modern Problems of Heat and Mass Transfer in Chemical Technology: Papers Presented at Intern. School-Seminar* (May, 1986) (Minsk, ITMO im. A.V. Lykov, 1986) Sect. 4.1, p. 3 (in Russian).
33. A.P. Burdukov, N.S. Bufetov and A.R. Dorokhov, *Izv. SO AN SSSR*. Ser. Tekhn. Nauk, 1981, No. 3, Vyp. 1, p. 13.
34. V.S. Avduevskii, A.Yu. Ishlinskii and V.N. Polezhaev, *Vestnik AN SSSR*, 1987, No. 6, p. 3.
35. H.Wilke, W. Loser, *Crystal Res. Technol.*, 1983, Vol. 18, p. 825.
36. V.I. Polezhaev and A.I. Protsomolotov, *Mathematical Modeling. Preparation of Single Crystals and Semiconductor Structures.* (Moscow: Nauka, 1986), p. 66 (in Russian).
37. S.A. Nikitin, V.I. Polezhaev, and A.I. Fedyushkin, *Technological Experiments* (Sverdlovsk, 1983) p. 140 (in Russian).

38. V.S. Avduevskii and V.I. Polezhaev, *Abstr. of Papers Presented at 5th All-Union Meeting on Theoretical and Applied Mechanics* (Alma-Ata, AN Kaz. SSR, 1981), p. 9 (in Russian).
39. P.C. Reid, J.M. Prausnits and T.K. Sherwood, *The Properties of Gases and Liquids* (New York, 1978).
40. K.A. Smith, *J. Fluid Mech.*, 1966, Vol. 28, No. 2, p. 401.
41. M. Velarde and J. Castillo, in: *Hydrodynamics of Interfaces* (Moscow, Mir, 1984) (Russian translation).
42. H. Linde and E. Schwarz, *Z. Phys. Chem.*, 1961, Vol. 224, No. 5, p. 331.
43. H. Linde, K. Loeschke, *Chem. Ing. Tekhn.* , 1967, Vol. 39, No. 2, p. 65.
44. S.H. Davis and G.M. Homsy, *J. Fluid. Mech.*, 1980, Vol. 98, p. 527.
45. J.L. Castillo and M.G. Velarde, *J. Non-Equilib. Therm.* , 1980, Vol. 5, p. 111.
46. G.Z. Gershuni and E.M. Zhukhovitskii, *Convective Stability of Incompressible Liquids* (Moscow: Nauka, 1972), p. 392 (in Russian).
47. J. Reichenbach and H. Linde, *J. Coll. Interface Sci.*, 1981, Vol. 84, No. 2, p. 433.
48. V.V Dil'man, V.I. Naidenov and V.V. Olevskii, *Dokl. AN SSSR*, 1988, Vol. 298, No. 3, p. 676.
49. V.V. Dil'man and V.I. Naidenov, in: *Abstract of Papers Presented at 6th All-Union Meeting on Theoretical and Applied Mechanics* (Tashkent: Nauka, 1986), p. 251, (in Russian).
50. P.V. Danckwerts, *Gas-Liquid Reactions.*
51. E. Ruckenstein and C. Berbente, *Chem. Eng. Sci.*, 1970, Vol. 25, p. 475.
52. M. Hosawa, N. Komatsu, N. Jmaishi, *Fujinawa, J. Chem. Eng. Jap.*, 1984, Vol. 17, No. 2, p. 173.
53. C.H. Ling and W.C. Reynolds, *J. Fluid Mech.*, 1973, Vol. 9, p. 571.
54. V.V. Dil'man and V.I. Naidenov, *Teor. Osnovy Khim. Tekhnol.* 1986, Vol. 29, No. 3, p. 316.
55. V.V. Dil'man, V.I. Naidenov, and I.I. Khegai, *Teor. Osnovy Khim. Tekhnol.*, 1988, Vol. 22, No. 3, p. 362.
56. Yu.G. Frolov, *Treatise on Colloidal Chemistry* (Moscow: Khimiya, 1982) (in Russian).

Heat Transfer in Film Flow

E.G. Vorontsov

Symbols

$A_w^+ = \frac{\delta_c - \delta}{\bar{\delta}} res$	- relative wave layer thickness;
a	- temperature conductivity coefficient, M^2/s;
C_p	- heat content, kJ/(kg•K);
f_w	- wave frequency, Hz;
$f_w^+ = f_w(\frac{\nu}{g^2})^{1/3}$	- dimensionless wave frequency;
g	- acceleration due to gravity, m/s^2;
L_{exp}	- experimental portion length, m;
l	- length, m;
q	- heat flux density, kW/m^2;
r	- heat of phase transition, kJ/kg
S	- distributing device slot width; m; mm;
t	- temperature, K; °C;
u	- film velocity, m/s;
x	- axis of coordinates in flow direction, m;
x_h	- hydrodynamic stabilization portion length, m;
x_t	- thermal stabilization portion length, m;
y	- axis of coordinates normal to flow direction, m;
α	- heat transfer coefficient, $kW/(m^2•K)$
Γ_v	- volumetric liquid concentration, m^2/s;
δ	- local film thickness, m;
δ_c	- film thickness in wave crest, m;
δ_{res}	- residual film thickness (thickness in wave valley), m;
$\delta^+ = \delta\left(\frac{g}{\nu^2}\right)^{1/3}$	- dimensionless film thickness;
λ	- heat transfer coefficient, kW/(m•K)
λ_w	- wave length, m;

$\lambda_w^+ = \delta_w \left(\frac{g}{\nu^2} \right)^{1/3}$ - dimensionless wave length;

μ - dynamic viscosity, Pa•s;

ν - kinematic viscosity, M^2/s;

ρ - density, kg/m^3;

σ - liquid surface tension, N/m;

τ - tangient stress, N/m^2;

$Fr_{sl} = \frac{\overline{u}_{sl}^2}{g \bullet S}$ - Froud criterion for slot flow;

$K_f = \frac{\rho \cdot \sigma^3}{g \cdot \mu^4}$ - film number/Kapitsa's criterion

$K_u = \frac{q}{r \cdot \rho_v \overline{u} x}$ - boiling criterion;

$Nu^+ = \frac{\overline{\alpha}}{\lambda} \left(\frac{\nu^2}{g} \right)^{1/3}$ - modified Nusselt criterion;

$\Pr = \frac{\nu}{a}$ - Prandtle criterion;

$\mathrm{Re} = \frac{4\Gamma_v}{\nu} = \frac{4\overline{\delta} u_x}{\nu}$ - Reynolds number;

$\mathrm{Re}_{sl} = \frac{u_{sl} \cdot S}{\nu}$ - Reynolds number for slot flow.

Subscripts

w - wave
in - inlet
h - hydrodynamic
cr - critical
lam - laminar
bb - beginning of bubble boiling
exp - experimental
res - residual
v - vapor
f - film
wl - wall
t - thermal
turb - turbulent
st - steady-state
ph - phase
sl - slot
0^+ - dimensionless
$\overline{0}$ - mean (averaged)

Abstract—The simplified grapho-analytical method is proposed for the analysis of initial portions of falling films with allowance made for inlet conditions. The analysis of average statistical amplitude-frequency response characteristics of large waves is presented, on the basis of which the semi-empirical theory of convective transport in free-falling films of low-viscosity liquids is rectified. The generalized relationships are presented for the analysis of heat transfer in cases of heating, surface evaporation and bubble boiling in films and are compared with results of the known studies.

INTRODUCTION

The problems of saving resources and energy, intensification of technological processes and raising the quality of products have recently acquired an increasingly greater importance. These problems are crucial for such industries as chemical, pulp-and-paper, biochemical, food production, etc.

Conducting processes in these liquid layers is one of the most promising and universal methods of intensification of heat- and mass-transfer, possessing such important advantages as high intensity of transport processes, high specific surface areas of contact between the liquid and the wall or boundary gas or vapor, small treatment time, absence of hydrostatic pressure of the liquid in the apparatus, etc. [1-9]. The optimum operating conditions of film apparatuses can be achieved at low temperature heads, which permits utilization of the industrial low-potential heat wastes [2-4, 7-9]. This enables film flows to be utilized in such processes as the treatment of thermosensitive liquids, desalination, shower cooling, convective heating and cooling, stripping and distillation, absorption and rectification, crystallization and particularly in the processes conducted under the vacuum and those requiring a high quality of final products.

In commercial apparatuses, film processes are usually realized on tubular surfaces manufactured from rolled metal, the demand and cost of which have increased lately. The cost of tubes is often a determining factor in estimating the cost of equipment as a whole, thus necessitating the substitution of metal tubes with polymeric, ceramic or glass, or steel tubes with protective coatings, (enamel, polymers, etc.), whose surfaces are usually hydraulically smooth and hydrophobic. In modern desalination, distillation and other technological equipment, use is also made of thin-walled, (50 to 100 μm thick walls), polymeric tubes 25 to 35 mm in diameter which are highly elastic and sensitive to even minor pressure variations. Changes in wetting conditions and in the state and rigidity of the washed surface results in the transformation of a flow pattern and in changes of intensity of transport processes.

In technological processes heat transfer from the wall to the film is possible: in the absence of heat exchange near the outer surface of the film, (heating-cooling in film heat exchangers), in the presence of heat exchange on the outer surface of the film, (film evaporators, distillers, absorbers, etc.), in the case of bubble boiling in the film (boilers, strippers, etc.). However, in any heat transfer mechanism, the process intensity is determined by flow hydrodynamics and by the conditions on the boundary between the film and the wall or gas (vapor).

Out of the numerous patterns of film flows, (in the field of gravitational, centrifugal, vibrational and other forces, two-phase flows, etc.), the paper under review deals with the liquid films free-falling over vertical walls. This type of flow in industrial equipment is characterized by the haphazard three-dimensional wavy regime that is difficult to describe mathematically. Therefore, the principal qualitative investigation of these processes is the experiment which, in itself under conditions of the non-isothermal flow of the layer with the thickness constantly oscillating within 0.3 to 2.0 mm, is highly complex and labor-consuming, just as techniques for measuring the basic parameters which require the development of special sensors, transducers and devices [3, 5, 8, 10-12].

1. Inlet Hydrodynamic and Thermal Portions

In cases of free-falling thin liquid films, the film flow can be divided into the inlet and steady-state portions. When the liquid flow is formed into a film, within the inlet (initial) portions of the film flow velocity, temperature and concentration profiles are transformed from initial profiles determined by the inlet conditions to the steady-state (stabilized) profiles determined by the flow regime and by heat- and mass-transfer conditions [2, 8, 14-16]. In an engineering analysis, discrimination of these portions is reasonable in the cases where these lengths are commensurable with the length of the wetted surface. In this case, for instance, the mean coefficient of heat transfer from the wall to the film is averaged over the length

$$\overline{\alpha} = \frac{\alpha_{in} \cdot x_h + \alpha_{st}(l - x_h)}{l}, \tag{1}$$

where the heat-transfer coefficient α_{in} at the inlet portion may be either estimated from the analytical relationships [14-17] or from the known Nusselts equation [2]

The analysis of the local velocities and film thicknesses in the inlet portion is usually conducted by two methods. The first method is based on the solution of the integral equation of pulses [4]. It is less accurate, but is analytically simpler. The second method lies in the combined solution of the equations of motion and continuity [8, 15, 17]. This method was taken as a basis in the proposed simplified analysis of laminar film flow [8, 13] where the solution of the dimensionless non-linear equation of motion with respect to u^+ was found in the form of the

$$u^+ = \left(C_o y^+ - 0,5g^+ \cdot y^{+2}\right) + \sum_{i=1}^{6} fi(y^+)(x^+ + \varepsilon)^{-1}, \tag{2}$$

where

$$u^+ = \frac{Ux}{U_{sl}};\ y^+ = \left(\frac{\overline{U}_{sl}}{\nu S}\right)^{0,5} y;\ x^+ = \frac{x}{S};\ g^+ = \frac{gS}{U^2 sl};$$

$$S^+ = \left(\frac{\overline{U}_{sl}}{\nu S}\right)^{0,5} S;\ \delta_{in}^+ = \left(\frac{\overline{U}_{sl}}{\nu S}\right)^{0,5} \delta_{in};\ C_0 = \left(3\,\mathrm{Re}_{sl}^{0,5}\,Fr_{sl}^{-2}\right)^{\frac{1}{3}}$$

Here, f_i are the functions of the y coordinate [8, 13, 18], and ε is the constant required for obtaining the family of solutions for x=0. Further on, ε is selected within $0<\varepsilon<S^+$.

The distinguishing characteristic of function (2) as compared with the known functions [4, 6, 15] is the allowance for the initial velocity distribution in the flow (in this case at the exit from the slot distributor). The local film thickness δ_{in} in the inlet portion was determined by the graphoanalytical method [18]. The assumed model of the flow in the initial portion is well confirmed by the experimental results in the region of Re≤450, though it does not necessarily mean that the actual kinematic flow structure fully corresponds to the assumed limitations, particularly at $x^+<10$ [13].

The dimensionless length of the initial hydrodynamic portion x_h^+ for particular conditions (Re, S) was determined with the relative error of 1% from the variations of the velocity on the outside surface of the film.

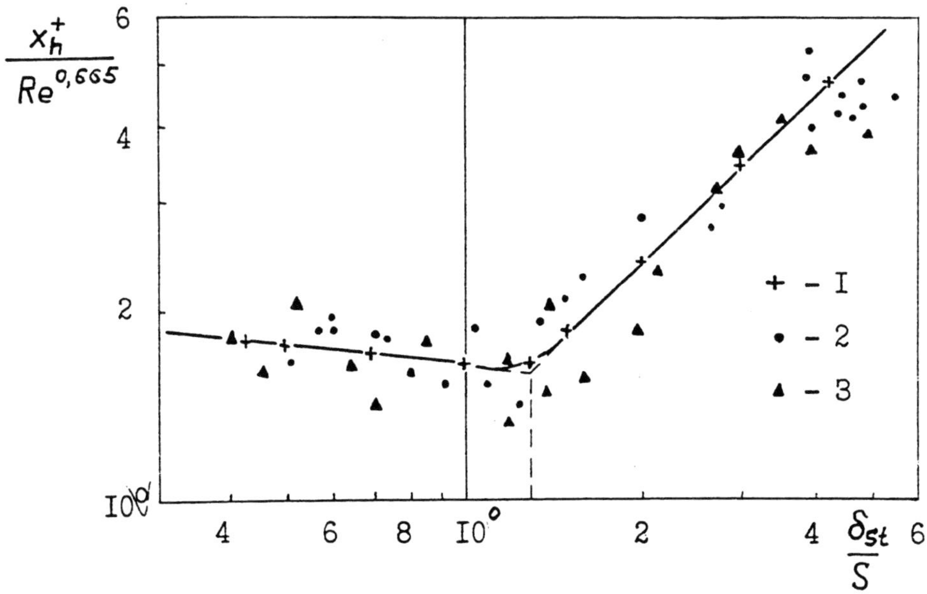

Figure 1. Generalization of data on dimensionless inlet portion length X_h^+: 1-analysis following proposed grapho-analytical method [19]; experimental data; 2-V.N. Miroshnichenko [20]; 3-author's

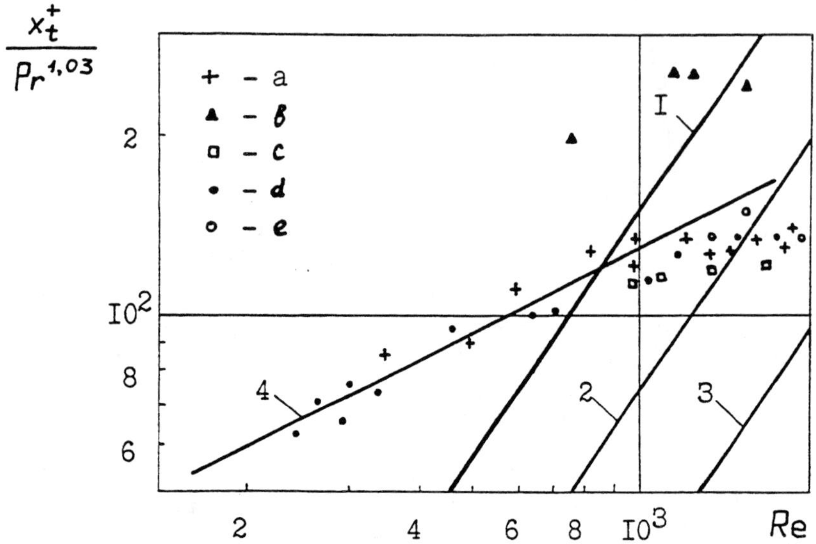

Figure 2. Dimensionless representation of data for thermal inlet portion length: 1,2,3 - analysis based on theoretical S.S. Kutateladze - V.Ye. Nakoryakov's equation [6] for water at S=0.5; 1.0 and 2.0 mm, respectively; 4 - from equation (6); experimental data: a,ε - Yu.M. Gandzuk [14] for alcohol solutions of hydrocholesterol 7 - DHC and water; β - Yu. Ye. Lukach et al. [2]; c - V. Wilke [21]; α - author's.

The equations for estimating have been obtained [19]:

$$0,3<\frac{\delta_{st}}{S}\leq 1,3 \qquad x_h^+ = 2,06\ \mathrm{Re}^{0,665}\left(\frac{\delta_{st}}{S}\right)^{-0,1} \tag{3}$$

$$1,3<\frac{\delta_{st}}{S}\leq 5,0 \qquad x_h^+ = 1,584\ \mathrm{Re}^{0,665}\left(\frac{\delta_{st}}{S}\right)^{-0,9}, \tag{4}$$

which show that for every Re number, (and for every steady-state film thickness δ_{st}) certain initial conditions (slot width S) exist at which x_h^+ shall be minimum. A comparison of the experimental and analytical data is presented in Figure 1.

This method was also used for a simplified analysis of the temperature profile and the length of the initial thermal portion x_t:

$$t^+ = \frac{t}{t_{sl}} = (B_1 - B_0 y^+) + \sum_{i=1}^{6} \varphi_i(y^+)(x^+ + \eta)^{-i} \quad (5)$$

$$x_t = 4,67\,\mathrm{Re}^{0,49}\,\mathrm{Pr}^{1,03}\; S = 2,06\,\mathrm{Re}^{0,82}\,\mathrm{Pr}^{1,03}\,Fr_{sl}^{-0,33}\delta_{st} \quad (6)$$

Here B_0, B_1 and η are constants (for simplicity it has been assumed that $\varepsilon=\eta$ [14, 16]). The experimental data agree with relationships (5) and (6) in the region of $Re \leq 450$, and for small path lengths $10 \leq x^+ \leq 80$ even at $Re \leq 760$ (Figure 2). At high liquid concentrations and high path lengths the mixing effect of waves becomes more pronounced and smoothes out the temperature variations throughout the film cross-section [2, 22].

2. Wave Film Flow

In the steady-state portion the film flow is accompanied by a well-developed process of wave formation. This process is essentially non-stationary, but with a number of simplifying assumptions, relationships can be obtained which are suitable for practical utilization. The developed wave flow is characterized by a simultaneous existence of wave disturbances of a different form and height which move with a different frequency and velocity. The detailed description and classification of the wave disturbances in films were given in [8, 1, 3, 5, 10]. The large "push" or "shock" waves transporting the bulk mass of the liquid and intensively mixing the flow are covered outside by fine capillary waves [1-3, 5, 8, 10, 23]. It is the large waves that affect the near-wall thermal boundary layer thereby deforming and destroying it and intensifying heat transport. This is convincingly confirmed by the oscillations of the measured local wall temperatures [2, 22, 23] and tangient stresses on the wall [6], which are always strictly synchronized with large-scale wave oscillations in the film flow. The small-scale waves may, to a certain extent, affect the heat- and mass-transport processes only on the outer surface of the film [5,6,8]. Therefore the study of the laws governing the behavior of the large waves and their effect on the transport processes in films is of great importance [5, 8, 24]. Given the random nature of the wave process, the most correct approach is to study the probabilistic determination of the basic wave characteristics through the spectral density of the wave process energy which provides both information on the contribution of all frequencies and determines the most probable harmonics [3, 5, 23, 24].

3. Experimental Conditions and Assessment of Experimental Errors

The experiments were conducted on tubes ranging from 21.0 to 249.0 mm in wetted diameter and from 0.21 to 4.0 m in the length of the portion under study and manufactured from various materials, (carbon and alloyed steel, copper, glass, carbon steel coated with enamel UES-200, 300 and with pentaplast with chromium oxide, as well as polymer-film tubes 50 to 100 μm in wall thickness from low-density polyethylene, polypropylene PN-4, polycarbonate 1 and polyamides). Heating was performed by steam, hot water and, in some cases, by electricity (q up to 75.5 kW/m^2). Taken as the operating fluids were water, aqueous solutions of NaCl, $NaNO_3$ and glycerine, hydrolysate M-2, (a product obtained from the processing of polymethylsiloxanes), alcohol solutions of 7-dehydrocholesterol (7-DHC) and others. The basic characteristics of these fluids ranged within the following limits: ρ=738 to 1143 kg/m3; $\nu \cdot 10^6$=0.295 to 10.4 m2/s; $\sigma \cdot 10^4$=177.3 to 798 N/m; Pr=1.75 to 10^9.7; $K_f=1 \cdot 10^6$ to $3.1 \cdot 10^{12}$. For creating the films use was made of slot (S=0.1 to 5.0 mm) and jet distributors. The Reynolds numbers varied up to Re=$1.95 \cdot 10^4$.

For determining integral and local characteristics of film flow, use was made of the power cutoff, tracks and electric resistance methods as well as the electrocontact and capacitive methods [8]. The maximum relative measurement errors were as follows: for $\bar{\delta}$ - ±4.11%; for δ_{res} and δ_c of large waves - ±2.0%; for λ_w - ±12%; for A^+_w - 13%; for U_w - ±19%; for U_{ph} - 22% and for α-±4.17%. The external wave structure of the film was also investigated by means of rapid photography with local introduction of dye in the flow.

4. Basic Characteristics of Wave Flow

For estimating the mean flow thickness $\bar{\delta}$ the equations were obtained in the form of

$$\bar{\delta}^+ = \delta\left(\frac{g}{\nu^2}\right)^{\frac{1}{3}} = C \cdot \mathrm{Re}_1^n \tag{7}$$

where for Re<Re_{cr} C=0.775 and n=0.355 for wetted surfaces from rolled steel (Re_{cr}=1600); and C=0.885 and n=0.325 for hydraulically-smooth wetted surfaces (Re_{cr}=1950). In the region where Re>Re_{cr} these parameters were equal to C=0.177 and 0.555 and n=0.202 and 0.52, respectively. The smaller values of $\bar{\delta}$ on the hydraulically-smooth wetted surfaces are accounted for by slower development of wave and turbulence phenomena in the film flow [5, 8].

The results of measurement of the thickness of the continuous layer δ_{res} of liquids, (in shallow wave valleys), were generalized by the following relationships for steel industrial tubes at x>0.485 m (S=0.5 mm):

$$\mathrm{Re} < 800 \quad \delta_{res}^+ = \delta_{res}\left(\frac{g}{\nu^2}\right)^{\frac{1}{3}} = 1,351\,\mathrm{Re}^{0,31}\left(\frac{x}{S}\right)^{-0,06} \tag{8}$$

$$\mathrm{Re} \geq 800 \quad \delta^{+}_{res} = \frac{\mathrm{Re}}{0,088\,\mathrm{Re} + 4,1}\left(\frac{x}{S}\right)^{-0,06} \tag{9}$$

and for smooth vitrous-enamel and polymer surfaces at x>0.655 mm

$$\mathrm{Re} < 1000 \quad \delta^{+}_{res} = 1,241\,\mathrm{Re}^{0,31}\left(\frac{x}{S}\right)^{-0,042} \tag{10}$$

$$\mathrm{Re} \geq 1000 \quad \delta^{+}_{res} = \frac{\mathrm{Re}}{0,08\,\mathrm{Re} + 14,6}\left(\frac{x}{S}\right)^{-0,042} \tag{11}$$

The value of δ_{res} is practically unaffected by the film number K_f.

In the case of the asymmetrical non-stationary wave spectrum $(\delta_c-\delta)>(\bar{\delta}-\delta_{res})$ the relative thickness of the wave layer A^{+}_{w} rapidly increases with an increase in the wetting density at x≤0.545 m and Re<2500 [2,25]. At small values of the path lengths A^{+}_{w} increases with increasing x, since at the inlet portion an intensive waves formation takes place. As the liquid concentration increases further, A^{+}_{w} grows much slower and then remains practically unchanged (Figure 3). The effect of the surface tension on the intensity of wave formation is appreciable mainly at small liquid concentrations (Re<250).

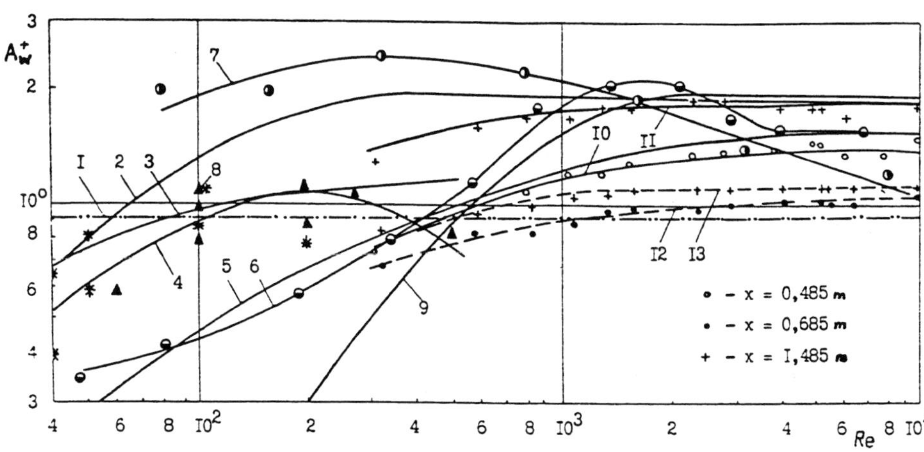

Figure 3. Relative wave layer thickness A^{+}_{w} as a function of Re-number according to the data of: 1 - P.L. Kapitsa [26]; 2 - G.I. Gimbutis [4]; 3 - L.P. Kholpanov et al. [27]; 4, 8 - I.A. Rogova et al. [7]; 5 - author [25]; 5 - K. Chu and A.E. Duckler [24]; 7 - G. Brauer [10]; 9 - B.G. Ganchev [3]; 10-13 - author: metal (10-11) and enameled (12-13) tubes.

The experimental results are well generalized by the following relationships:

- in the case of flows over rolled steel tubes

$$\mathrm{Re} \le 300 \quad A_w^+ = \frac{0,084\,\mathrm{Re}}{0,396\,\mathrm{Re} + 94,38}\left(\frac{x}{S}\right)^{0,43} Kf^{-0,04}, \tag{12}$$

$$300 < \mathrm{Re} < 1600 \quad A_w^+ = \frac{2,65\,\mathrm{Re}}{9,13\,\mathrm{Re} + 1165,6}\left(\frac{x}{S}\right)^{0,362} Kf^{-0,04}, \tag{13}$$

$$1600 \le \mathrm{Re} < 13000 \quad A_w^+ = 1,071\,\mathrm{Re}^{-0,001}\left(\frac{x}{S}\right)^{0,19} K_f^{-0,004}; \tag{14}$$

- in the case of flows over hydraulically-smooth tubes:

$$\mathrm{Re} \le 300 \quad A_w^+ = \frac{0,574\,\mathrm{Re}}{0,879\,\mathrm{Re} + 109,6}\left(\frac{x}{S}\right)^{0,28} K_f^{-0,07} \tag{15}$$

$$300 < \mathrm{Re} < 1800 \quad A_w^+ = \frac{5,425\,\mathrm{Re}}{4,091\,\mathrm{Re} + 633}\left(\frac{x}{S}\right)^{0,2} K_f^{-0,07} \tag{16}$$

$$1800 \le \mathrm{Re} < 13000 \quad A_w^+ = 2,312\,\mathrm{Re}^{0,028}\left(\frac{x}{S}\right)^{0,092} K_f^{-0,07}. \tag{17}$$

The analysis of spectral density of the wave process energy with the discrimination of the predominant ("carrying") frequency [23] has made it possible to experimentally determine the most probable wave frequencies. As the wetting density increases within the steady-state flow portion, the predominant frequency f_w increases and, as certain Reynolds number close in value to the critical one Re_{cr} is reached, stabilizes remaining further practically unchanged (Figure 4). The wave layer oscillation frequency decreases slightly with increasing liquid viscosity (curves 4, 5, 6), and the region of stabilization of f_w is shifted towards smaller Re numbers. For the steady-state flow portion, the following generalized dependencies have been obtained for the dimensionless predominant frequency of large waves:

- for flows over rolled steel tubes:

$$\mathrm{Re} < 1600 \; f_w^+ = 0,024\,\mathrm{Re}^{0,23} K_f^{-0,048} \tag{18}$$

$$\mathrm{Re} \ge 1600 \quad f_w^+ = 0,132 K_f^{-0,048}; \tag{19}$$

- for flows over glass, enameled or polymer surfaces

$$\mathrm{Re} < 1800 \quad f_w^+ = 0{,}027\,\mathrm{Re}^{0{,}186}\,K_f^{-0{,}062}, \tag{20}$$

$$\mathrm{Re} \geq 1800 \quad f_w^+ = 0{,}108 K_f^{-0{,}062} \tag{21}$$

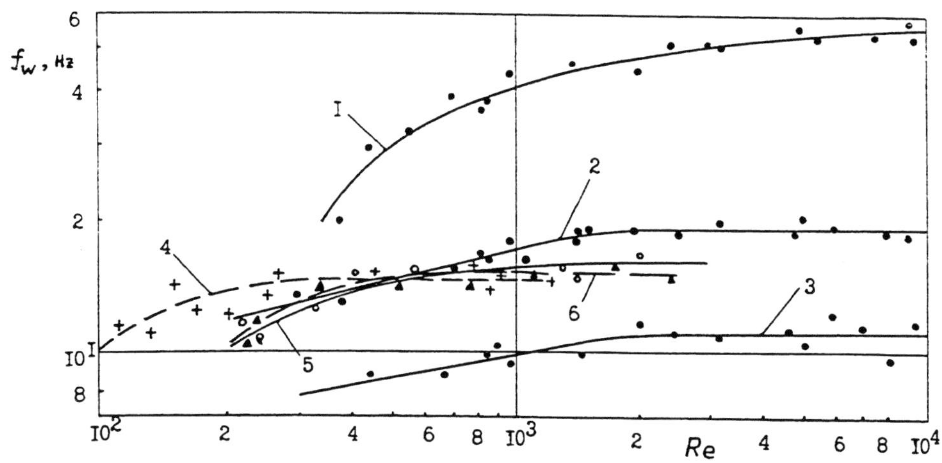

Figure 4. Prevailing wave frequency f_W as a function of Re-number of film flows: 1 - water (t=19.1°C, x=0.09 m); 2 - water (t=19.1°C, x=1.485 m); 3 - water (t=18.7°C enamelled tube); 4, 6-hydrolyzate M2 (t=18°C and 70°C); 5 - flycerine 50% (t=21°C).

Within the inlet portion, the predominant wave oscillation frequency is usually higher at a high liquid concentration than within the steady-state film flow, since it depends on the inlet conditions or on the conditions of transformation of the liquid flow is in the distributing device into the film flow, as well as on various random disturbances [8]. These high-frequency local oscillations die out as the film path length increases, but, at the same time, the regular wave disturbances begin to develop and then become predominant. The higher the liquid concentration, the higher the frequency of the forced random inlet oscillations, but at the same time the smaller their damping length.

Figure 5 illustrates comparison of the known data on predominant wave oscillation frequencies in the form of the function f_W=f(Re).

Within the region of very small Re numbers the theoretical values of the regular waves lengths λ_W decrease with increasing liquid concentration [1,6, 15], with the experimental values of λ_W being noticeably higher than the theoretical ones [3, 5, 8]. At considerable values of the film path lengths the wave length λ_W is normally determined either through averaging the distances between the fronts of multiple waves [5, 10] or by the pitch of the most probable harmonic wave [3, 23]. Within a large range of values of the Re numbers and physical properties of the wetting liquid the lengths of the large waves do not practically change and amount to 100 to 140 mm [5, 8, 23]. More exactly for the steady-state flow they can be calculated for R≤13000 from the following equations:

– for the flows over rolled steel tubes

$$\lambda_w^+ = 2521\,\mathrm{Re}^{0.03}\,K_f^{-0,024}, \tag{22}$$

– for the flows over glass, enameled and polymer surfaces

$$\lambda_w^+ = 51701\,\mathrm{Re}^{0,05}\,K_f^{-0,144} \tag{23}$$

The phase velocity of the large waves can be determined from the expression $u_{ph}=\lambda_w \bullet f_w$, where λ_w and f_w are determined from the above-mentioned equations (18-23).

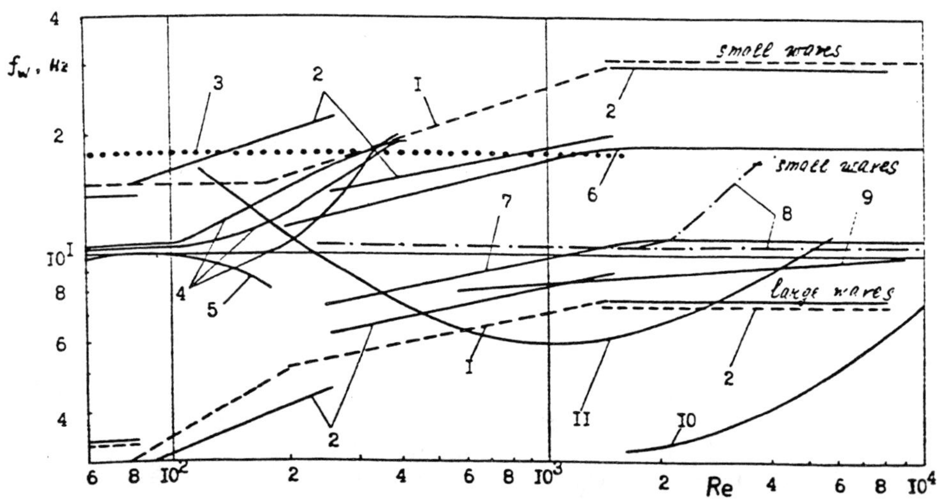

Figure 5. Variation of prevailing wave frequency with Re-number in water film flow based on data of: 1 - N.N. Kulov [5], 2 - V.V. Maksimov et al. [11]; 3 - B.G. Ganchev et al. [3]; 4 - I.A. Rogova et al. [7]; 5 - A.G. Sheinkman and V.M. Pakhaluev [29]; 10 - B.G. Ganchev [3]; 11 - R.P. Salazar and E. Marshall [31].

The most concentrated and useful information on the crosswise transport of the pulse, heat and mass at wave flow of a thin liquid film is carried by the mean statistical velocity u_w of transverse wave layer pulsations which generalizes the mean statistical amplitude-frequency response characteristics:

$$u_w = 2(\delta_c - \delta_{res})f_w = 2\bar{\delta}A^+_w f_w \qquad (24)$$

Irrespective of the state of the wetted surfaces, the transverse pulsations velocity in the case of the turbulent film flow is higher by the order of magnitude than that in the case of the laminar-wave flow. With all other conditions being equal, the values of u_W in the case of the film flows over rolled steel tubes are more than three times higher than those in the case of the film flows over smooth glass and enameled tubes, which is accounted for both by an increase in the velocity of the film falling over smooth surfaces and by the retarded development of the transverse wave pulsations in the viscous sublayer (Figure 6).

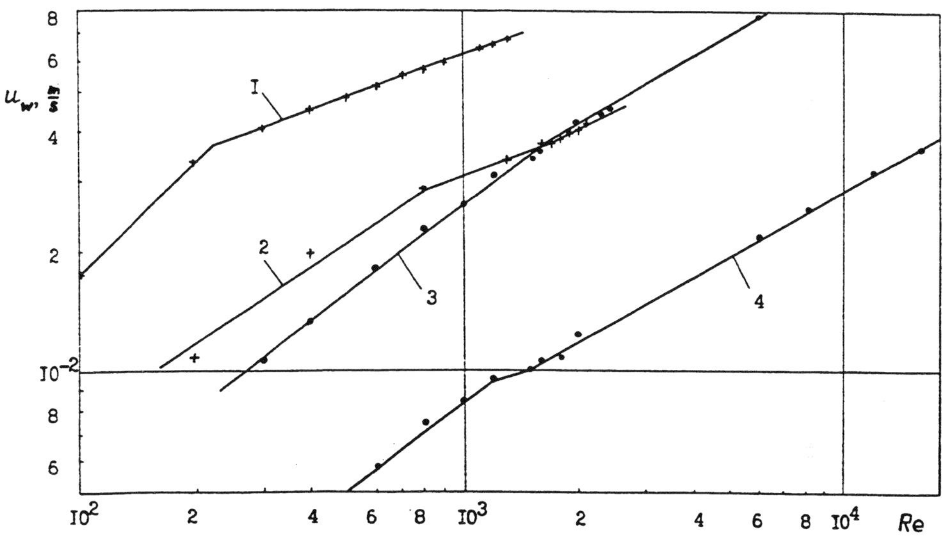

Figure 6. Transverse pulsations velocity u_W of wave layer as a function of Re-number for x=1.485 m in flow flows: 1,2 - hydrolyzate M-2 (t=18°C and 70°C); 3, 4 - water (t=19°C, steel and enameled tubes).

The data obtained for the wave characteristics of the large waves in a free-falling film flow make it possible to both assess the transport processes in the thin liquid layer [7] and to further clarify and expand the range of utilization of the semi-empirical wave and turbulent transport relationships [2, 4, 6, 8, 10, 17, 28].

It should be noted that the initial nonuniformity in the distribution of the thin liquid film around the wetted perimeter results in considerable changes in the flow character and transport processes [32].

5. BASIC LAWS GOVERNING CONVECTIVE HEAT TRANSPORT

The equations for the local and mean heat transfer coefficients for the laminar wave-free film flow proposed by V. Nusselt yield values lower than those obtained in experiments, since the presence of waves intensifies the heat transport. In the case of the laminar wave-free flow transport is caused only by the molecular heat transfer

$$q = q_{lam} = -\lambda \frac{dt}{dy} = -c_p \rho a \frac{dt}{dy} \tag{25}$$

whereas in the case of the laminar-wave flow, which takes place in practice, it is accompanied by the convective transport due to the mixing effect of waves, and then the total heat flux may be expressed as [4, 10]:

$$q = q_{lam} + q_w = -c_p \rho (a + a_w) \frac{dt}{dy} \tag{26}$$

Here a_w is the wave temperature conductivity coefficient allowing for the heat transport by the wave oscillations of the outer surface of the film. This coefficient is maximum in magnitude on the outer surface, decreases progressively in the direction towards the wall due to energy dissipation and is equal to zero on the wall

$$a_w = w_w \bullet l_{hd} \tag{27}$$

where w_d is the characteristic wave disturbance velocity and l_{hd} is the heat dissipation path length [4]. Assuming in the first approximation that $w_w \sim u_w$ and $l_{hd} \sim \delta_{res}$ and taking equation (24) into account we shall obtain

$$a_w = u_w \delta_{res} = 2\bar{\delta}\delta_{res} A_w^+ f_w \tag{28}$$

In the known theoretical analyses of the turbulent flow [2, 4, 6, 34] the heat flux in the film cross-section

$$q = -c_p \rho (a + a_t) \frac{dt}{dy} \tag{29}$$

is determined with the help of effective (turbulent) temperature conductivity coefficient a_t determined as function y from either R.G. Deissler's or T. Karman's equations. Using the model of the turbulent flow with a smooth outside surface of the film, some publications [6] deal with the two-layer flux and others [34]- with the three-layer flux. A considerable discrepancy is observed [3, 4] between the theoretical and experimental results obtained by different authors (Figure 7). Sometimes this discrepancy is as high as 50 to 500% [2]. Such a high discrepancy between the obtained theoretical results stems from the imperfection of the theoretical models of the flow of the film with a smooth outer surface, and between the experimental results, from different conditions under which the experiments were conducted by different authors.

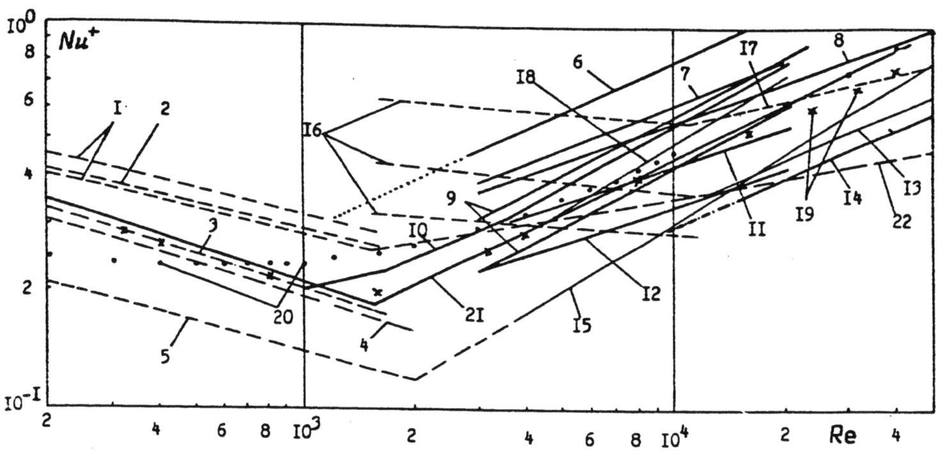

Figure 7. Comparison of results on convective heat transfer to film N_u^+=f(Re) at P_r=1.75 to 7.0 based on data of: 1-15 - G.M. Gebberd and U.L. Badger [2, 36]; 2 - P.L. Kapitsa [26, 2]; 3, 12, 13, 14 - K. Koziol and L. Broniarz [35]; 4 - V. Nusselt [2, 36]; 7 - G.S. Base and V.Ch. Moc-Adams [2, 36]; 9 - T.Sexauer [2]; 10 - V. Wilke [21]; 11 - L. Garvin and E.V. Knelli [2, 36]; 16 - G. Brauer [10]; 17, 20 - G.I. Gimbutis [4]; 18 - author [2]; 19 - O.A. Asbjornsen [2,8]; 21 - author; 22 - S.S. Kutateladze and V.V. Nakoryakov [6].

Using the approximate heat transfer theory for the turbulent film flow, allowing for wave phenomena that causes the deformation of free surface, we can write [4, 8, 10, 33] that

$$q = q_{lam} + q_w + q_{turb} = -c_p \rho (a + a_w + a_t) \frac{dt}{dy}, \tag{30}$$

where the turbulent temperature conductivity coefficient a_t can be estimated from the known expressions for the three-layer flow model [3, 4, 33, 34] with allowance made for the average statistical profile of the external surface of the film. Then

- for the laminar sub-layer

$$y^+ \le 5 \quad a_t = \nu (0{,}124 y^+)^4; \tag{31}$$

- for the transitory zone

$$5 < y^+ \le y^+_{res} \quad a_t = \nu(0,2y^+ - 1); \tag{32}$$

- for the turbulent layer

$$y^+ > y_{res} \quad a_t = \nu(0,4\, y^+ - 1), \tag{33}$$

where $y^+ = \left(\frac{\tau_{res}}{g}\right)^{0,5} \bullet \frac{y}{\nu}$; $y^+_{res} = \left(\frac{\tau_{res}}{g}\right)^{0,5} \bullet \frac{\delta_{res}}{\nu}$, and taking into account the fact that the total thickness of laminar and buffer sub-layer cannot be greater than δ_{res}, and from analysis also $y^+_{res} < 30$[33] δ_{res} is estimated from the formulae (8-11).

The analysis of convective heat transfer using relationships (26) and (30) allowing for the amplitude-frequency response characteristics of large waves are confirmed within the accuracy of the experiment by the generalized experimental data (with the exception of the region $1200 \le Re \le 2600$, where the analytical results are higher by a maximum of 35% than the experimental ones).

Numerous researches into convective heat transfer to heated films (Figure 8) have yielded the following generalized equations

$$Re \le Re_{cr} \quad Nu^+ = CRe^n\, Pr^{0,4} \tag{34}$$

For metal rolled surfaces $Re_{cr}=1600$; $C=0,77$; $n=-0,296$. For tubes with enamel or polymer protective coatings $Re_{cr}=1950$; $c=0,8$; $n=-0.31$.

$$Re > Re_{cr} \quad Nu^+ = C_1\, Re^m\, Pr^{0,4}, \tag{35}$$

where respectively $C_1=2.7 \cdot 10^{-3}$; $m=0,47$ and $C_1=2,35 \cdot 10^{-3}$; $m=0,46$.

Rectified equations for a thermally stable region (with the exception of the portion x_t) with allowance made for the length of run

$$Re \le Re_{cr} \text{ and } x\left(\frac{g}{\nu^2}\right)^{\frac{1}{3}} \ge C_2$$

$$Nu^+ = C_3\, Re^n\, Pr^{0,4} \left[x\left(\frac{g}{\nu^2}\right)^{\frac{1}{3}} \right]^k \tag{36}$$

For rolled metal tubes: $C_2=6120$; $C_3=0,127$; $k=0,17$. For tubes with enamel and polymer protective coatings: $C_2=8210$; $C_3=0,147$; $k=0,16$.

At Re>Re_{cr} and $x\cdot\left(\frac{g}{\nu^2}\right)^{\frac{1}{3}} \geq C_4$, respectively C_4=8210; C_3=4,7·10^{-3}; k=-5,3·10^{-2} and C_4=10300; C_3=4,7◊10-3; k=-8.1·10^{-2} for tubes with a smooth coating.

As shown in the studies conducted by V.L. Rakitsky,, the direction of the heat flux in relationships (35) and (36) may, whenever necessary, be taken into account in cooling the film by the complex $\left(\frac{Pr}{Pr_{wl}}\right)^{0.3}$ with the corresponding recalculation of coefficients C_1 and C_3.

In the laminar-wave region of the flow heat transfer is practically unaffected by the state of the wetted surface, whereas in the turbulent region on smooth polymer and vitrous enamel surfaces, with all other conditions being equal, it is lower by 20-25% than on rolled metal surfaces due to the displacement of the value Re_{cr} and retarded development of turbulence in the flow.

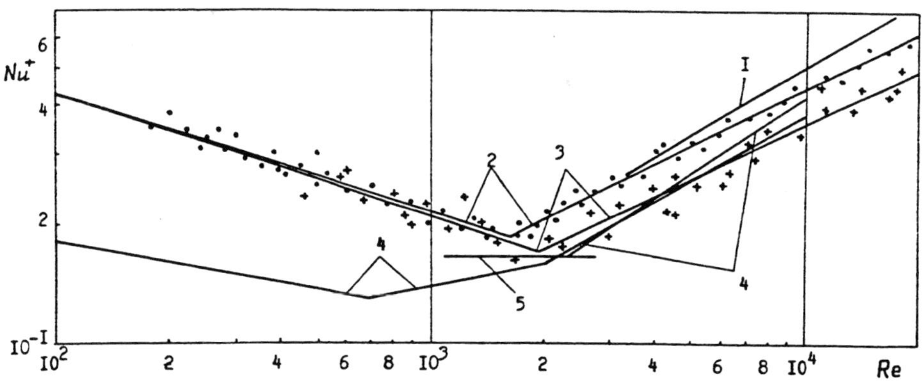

Figure 8. Experimental relationships Nu^+=f(Re) for P_r=3,6-7, 0 at film heating based on data of: 1 - author [2], metal tube L_{exp}=0,594 m; 2,3 - author, L_{exp}=1,885 (metal and enameled tubes, respectively; 4 - L.B. Radchenko [8], thick-walled tube from polypropylene; 5 - author, L_{exp}-1,5-1,9 m, elastic polymeric thin-walled tubes.

Equation (36) shows that if within the laminar-wave region the heat transfer coefficient α under thermally steady-state condition increases with an increasing length of run x because of well-developed wave phenomena, then within the turbulent region α decreases due to the equalizing of temperature throughout the film cross-section and drop of temperature head between the wall and the liquid. This is in good agreement with the data of [3, 37], though in these references an allowance for the relationship $\alpha=f(x)$ is more complex and formalized.

The narrow region of studies of heat transfer from thin-wall polymer tubes Re=1100-2800 (curve 5 in Figure 8) is accounted for by the hydrophoby and elasticity of a wetted surface and is limited on the bottom by the value of minimum spray density, and on top, by the disturbed continuity of the flow due to the emergence of "dry" spots and the partial breakaway of the fluid from the tube.

6. Heat Transfer Under Intensive Surface Evaporation and Bubble Boiling in Film

The following kinds of vapor formation are distinguished in free-falling films [2,8]: evaporation from free surface, (the transfer of heat by convection currents towards the outer surface of the film); suppressed boiling where vapor bubbles formed on heated wetted surface of the wall are condensed in cold liquid underheated to reach the saturation temperature; developed bubble boiling and film boiling related to the throw of the liquid film away from the surface of heating (V. Lyadenfrost's effect [2]). It is natural that the problems of determining the intensity of heat transfer shall be solved in different ways depending on the vapor formation mechanism [2, 3, 8, 38].

Heat transfer to the wall under intensive surface evaporation and under bubble boiling is of greatest practical interest. In any case, heat is transferred from the heating surface of the wall through the near-wall boundary layer, flow core and the boundary layer nearest the outer surface of the film, (i.e. the greatest percentage of the thermal resistance takes place in the boundary layers compared with the thermal resistance to convective transfer in the flow core is negligible) [2, 8, 39]. Thus, within the surface evaporation region, a smaller heat transfer coefficient should be expected than to the film being heated [8].

Some publications deal with processes of evaporation and boiling in a stationary thin layer of liquid [38] whose governing laws are erroneously transferred to the falling films [38]. The breaking diameter of the vapor bubble in the thin layer on the horizontal heating surface is on average by an order of magnitude greater than the film thickness [37-39]. The flowing film exerts pressure on the vapor bubble stuck to the wall. Under this pressure, and more so under wave-formation conditions, the bubble may break away at a considerably less diameter. The effect of the stuck and growing vapor bubble on hydrodynamics can therefore be considerable. Until its destruction, the separated bubble can move a certain distance both in the film core and on the surface of the film taking the heat of evaporation required by it from the film. In either of these cases, the initial thicknesses of boundary layers are reduced. The number of vapor-formation centers depends on the thermal load of the surface of the wall, its state and material [2, 8, 38]. Thus, in the case of bubble boiling, the film is jointly turbulized by waves and vapor bubbles.

In the case of boiling the film under small liquid concentration conditions, where a major part is played by the turbulizing effect of vapor bubbles, (the turbulizing effect of which is higher than in a large volume [2, 42]) the value of the heat transfer coefficient is higher than that in a large volume. As the thermal load increases, the liquid concentration has to be increased because of the danger of baring the surface of heating and, as a result, wave phenomena begin to increasingly exert a greater effect on the film turbulization [2, 37].

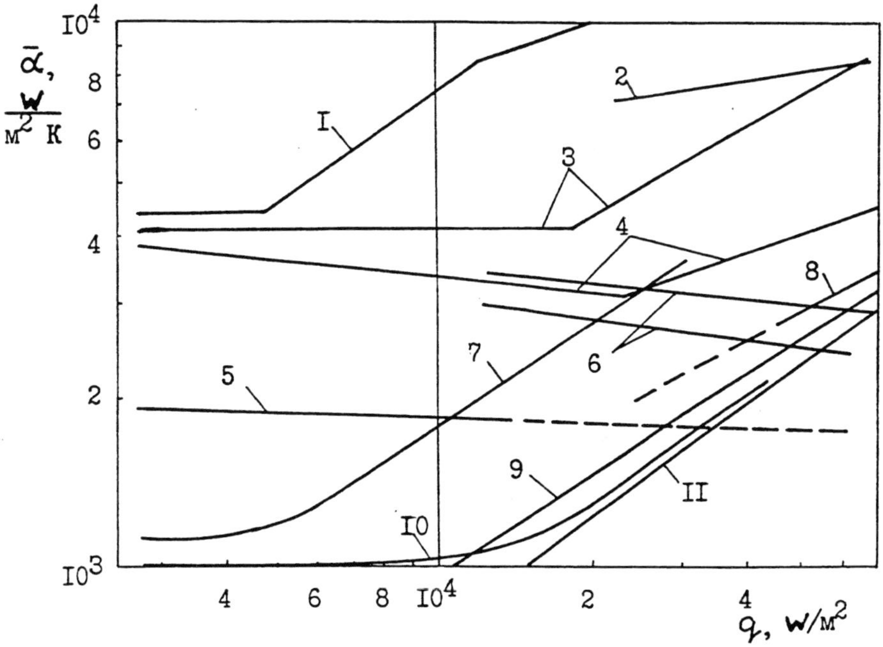

Figure 9. Function α=f(q) in evaporation and boiling of films (Re≈3200) based on experimental data of 1 - B. Haase [39], water; 2 - Yu.M. Tananayko [2], water; 3, 4 - author, water and 5-% solution of NaCl (metal tube); 5 - author; 2-% solution of NaCl (metal tube with polymeric coating); 6 - V.N. Slesarenko [40], 3.5% and 9.8% sea water; 7 - H. Struve [37], water; 8 - I.A. Kopchikov, et al. [2, 8]; 0 - U. Gropp [41], water; 10 - C. Elle [42], freon; 11 - water boiling in large volume [43].

The process of vaporization and boiling in a wetting film has been studied in detail on the basis of the equations of balance of transfer of pulse, heat and mass and with the help of dimensionalities analysis [39]. Out of the large number of the obtained criteria many prove to be of minor importance, and, as it was convincingly shown in the subsequent publications [3, 34, 40, 43, 44], these processes are described in the most simple and complete manner in the form of criterial relationship of the kind N_u^+=f (Re, Pr, K_u). We have already presented the critical review [2, 8, 44] of the known publications devoted to this problem.

Within the surface evaporation zone the heat transfer intensity is unaffected by the specific heat flux (Figure 9, curve 3). Intensive evaporation takes place here from the free surface of the liquid with a resultant increase in overheating in the layer [2]. As the vapor formation centers begin to emerge and the bubbles grow in number, the turbulization of the near-wall boundary layer grows in intensity thus improving heat transfer away from the wall and causing vapor formation to spread over the entire mass of the film. The heat transfer coefficient α grows in magnitude.

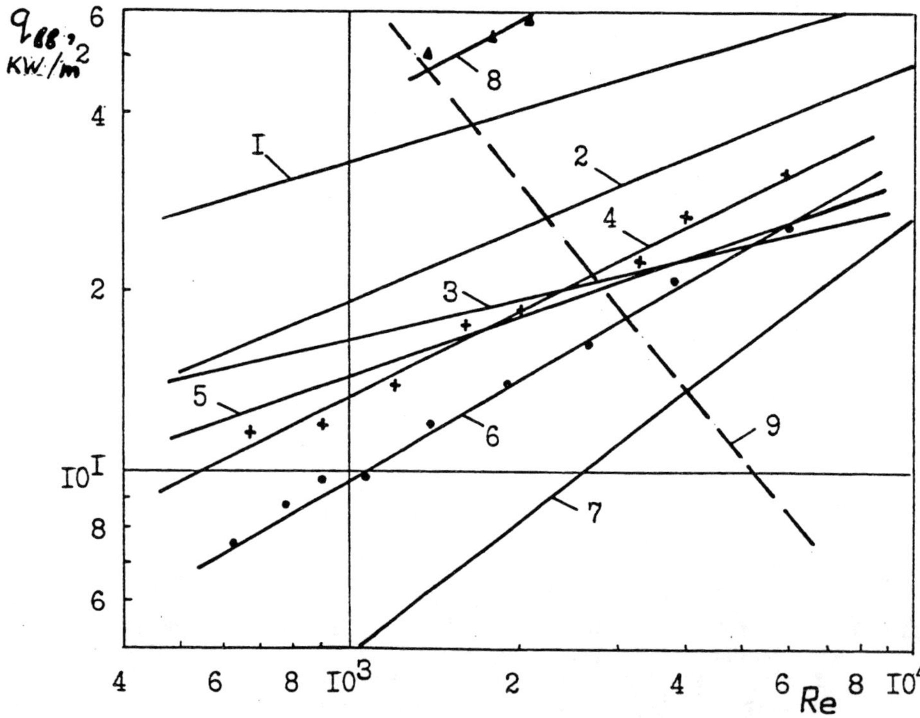

Figure 10. Heat flux q_{bb} at which bubble boiling begins as a function of the Re-number based on data of: 1,5 - Yu.M. Gandzyuk [44] (alcohol solution of 7-DHC and water); 2 - V.I. Leverash [2, 8]; 3 - V.A. Ardashev [44]; 4, 6 - author (2-5% solution of NaCl and water); 5 - B. Haase [39]; 8 - author (water, enameled and polymer tubes); 9 - V.A. Antonenko [38].

This conversion from the surface evaporation to the bubble boiling occurs at a certain fixed value of the specific heat flux q_{bb} which is determined in each particular case mainly by the liquid concentration (Figure 10).

When evaporating the solutions leaving deposits on the surface of heating (Figure 9, curve 4), α decreases to some degree with an increase in q within the surface evaporation region, since, probably, a certain effect of the thermal resistance of salt layer on the wall surface and related temperature head neglected in our experiments become felt here. On enameled and polymer-coated surfaces (curve 5) only the surface evaporation region has been determined.

Under surface evaporation and bubble boiling conditions at small values of q, the intensity of heat transfer to the film is considerably higher than that in a large volume (Figure 9, curve 11). Hence, the film apparatuses are effective under small specific thermal loads and small temperature heads and can be used to an advantage in multiple-effect evaporators and distillers and in the treatment of thermolabile products [2, 7, 9, 34, 37].

As the Reynolds number increases, the augmented turbulization and wave phenomena in the film intensifies the convective transport in the cross-section so much that the creation of the conditions on the heating surface necessary for the formation of bubbles requires considerable thermal loads. These conditions are naturally dependent on the properties of the liquid, and material and roughness of the surface. Thus, on wetting metal tubes with water film (Figure 10, curve 6) smaller values of q_{bb} were noted than in the case of wetting these tubes with solutions. On wetting vitrous-enamel and polymeric surfaces with films of aqueous solutions of NaCl and sea water an emergence of bubbles was observed only when a deposit was formed in the heating surface. When the deposit was removed the vapor formation centers on the wall disappeared again.

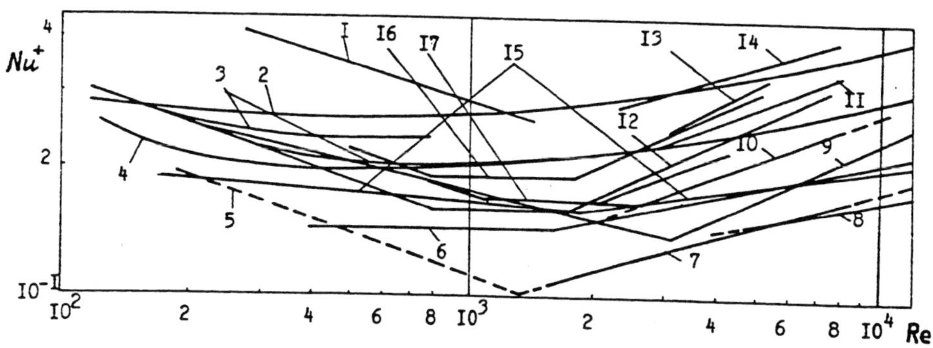

Figure 11. Comparison of generalized functions Nu+=f(Re) in the case of evaporation and boiling of water film based on the data of: 1, 8 - M. Serwinsky et al. [45]; 2 - I.V. Domanski [34]; 3 - W. Unterberg, D.K. Edwards [46]; 4 - A.E. Dukler [47]; 5 - V. Nusselt [2] at water heating (for comparison); 6 - R. Thom [37]; 7 - B. Haase [39]; 9 - R. Seban and A. Fagry [3]; 10 - I.R. Sinek and E.Ch. Jung [2,8]; 11 - C. Elle [42]; 12 - H. Struve [37]; 13 - V.I. Leverash [2,8]; 14 - Yu.M. Tananayko [2,8]; 15 - V.N. Slesarenko [48]; 16, 17 - author (metal tube and enameled and polymeric tubes).

In dependence on the flow regime and thermal condition three typical regimes of heat transfer are picked out (Figure 11, curve 16): laminar-wave, transient and turbulent. The appearance of transient regime here, (unlike non-isothermal convection flow), is connected with the increasing of heat transfer intensity due to liquid phase transition and its additional turbulization by separately arising steam generator centers [8, 37, 39, 42].

Generalization of the experimental data on intensive surface evaporation of films of water, weak solutions of NaCl and $NaNO_3$, sea water and alcohol solutions of 7-dehydrocholesterol in rolled metal tubes have yielded the following relationships:

$$\mathrm{Re} \leq 800 \quad Nu^{+} = 1.2\,\mathrm{Re}^{-0,31}\,\mathrm{Pr}^{0,4} \tag{37}$$

$$800 < \mathrm{Re} < 1800 \quad Nu^{+} = 0,152\,\mathrm{Pr}^{0,4} \tag{38}$$

$$\mathrm{Re} \geq 1800 \quad Nu^{+} = 5.21 \cdot 10^{-3}\,\mathrm{Re}^{0,45}\,\mathrm{Pr}^{0,4} \tag{39}$$

In wetting clean polymeric and vitrous-enamel surfaces

$$\mathrm{Re} \leq 1100 \quad Nu^{+} 1,138\,\mathrm{Re}^{-0,32}\,\mathrm{Pr}^{0,4}; \tag{40}$$

$$1100 < \mathrm{Re} < 2100 \quad Nu^{+} = 0,12\,\mathrm{Pr}^{0,4}; \tag{41}$$

$$\mathrm{Re} \geq 2100 \quad Nu^{+} = 3,59 \cdot 10^{-3}\,\mathrm{Re}^{0,46}\,\mathrm{Pr}^{0,4}. \tag{42}$$

For well-developed bubble boiling of the above-mentioned liquids in the films ($K_u > K'_u$) on the metal surface

$$\mathrm{Re} \leq 800 \quad Nu^{+} = 1,2\,\mathrm{Re}^{-0,31}\,\mathrm{Pr}^{0,4}\left(\frac{Ku}{K'_u}\right)^{0,55}; \tag{43}$$

$$800 < \mathrm{Re} < 1800 \quad Nu^{+} = 0,512\,\mathrm{Pr}^{0,4}\left(\frac{Ku}{K'_u}\right)^{0,55}; \tag{44}$$

$$\mathrm{Re} \geq 1800 \quad Nu^{+} = 5,21 \cdot 10^{-3}\,\mathrm{Re}^{0,45}\,\mathrm{Pr}^{0,4}\left(\frac{Ku}{K'_u}\right)^{0,55}, \tag{45}$$

where $K'_u = \dfrac{q_{bb}}{r\delta \bar{u}_x}$.

Comparison of experimental relationships N_u^+=f(Re) used in the known publications (Figure 11) reveals large differences in the obtained results (up to 2.6-fold). This is accounted for by the fact that many of the experimenters used short tubes where a considerable effect of the initial conditions of liquid distribution and heat transfer is felt, that often underheated liquid was delivered, etc.

The obtained results have constituted the basis for the analysis and design of film apparatuses where the processes of convective heat transfer, boiling, evaporation, distillation and film cooling of equipment and polymeric materials, (in the course of their production), are used, for working out recommendations for their optimum operation and for further improvement of film equipment parameters.

REFERENCES

1. A.A. Andreevsky "Wave flow of thin layers of viscous liquid." In: *Temperature conditions and hydraulics of steam generators,* L. Nauka, 1978, pp. 181-230.
2. Ye.G. Vorontsov, Yu.M. Nananayaka *Heat transfer in fluid films,* Kiev: Tekhnika, 1972, p. 196.
3. B.G. Ganchev, *Cooling of nuclear reactor elements by falling films.* M.: Energoatomizdat, 1987, p. 192.
4. G. Gimbutis, *Heat transfer in free-falling liquid films.* Vilnius: Mosklas, 1988, p. 233.
5. N.N. Kulov, *Hydrodynamics and mass transfer is downward two-phase film-dispersion currents: Thesis of doctor's dissertation.* M. 10NKh Acad. of Sc. of the USSR, 1984, p. 45.
6. S.S. Kutateladze, V.Ye. Nakoryakov, *Heat- and mass-transfer and waves in gas-liquid systems.* Novosibirsk: Nauka, 1984, p. 301.
7. *Film heat- and mass-transfer equipment.* V.M. Olevsky, V.R. Ruchinsky, A.M. Kalashnikov, V.I. Chernyshev. M. i Khimiya, 1988, p. 240.
8. Yu.M. Tananayko, Ye.G. Vorontsov, *Methods of analysis and research of film processes.* Kiev: Tekhnika, 1975, p. 312.
9. V.B. Kogan, M.A. Kharisov, *Equipment for mixtures separation under vacuum,* L.: Khimiya, 1976, p. 376.
10. H. Brauer, *Stromung und Warmeubergang bei Rieselfilmen.* VDI-Forschungsheft 457. Dusseldorf: VDI-Verlag, 1956, s. 40.
11. V.V. Maksimov, N.N. Kulov, V.A. Kalyusov, "Studies of wave characteristics of falling liquid film." In: *Non-linear wave processes in two phase media.* Novosibirsk: Institute of thermal physics. SO USSR Acad. So., 1977, pp. 153-158.
12. *Velocity profiles in falling liquid films.* N.N. Kulov, M.Yu. Muravyev, V.A. Malyusov, N.M. Zhavoronkov. Teor. osnovy khim, tekhnol. 1982, v. 16, No. 4, pp. 499-509.
13. Ye.G. Vorontsov, Yu.M. Gandzyuk, *Studies of velocity field in inlet film flow portion.* J. Inzh. fiz. 1986, v. 50, No. 6, pp. 913-917.

14. Ye.G. Vorontsov, Yu.M. Gandzyuk, *Studies of thermal inlet portion of film flow*. J. Inzh. fizika, 1986, v. 51, No. 6, pp. 892-896.
15. L.P. Kholpanov, *Heat- and mass-transfer and hydrodynamics of fluid film flow*. Teor. osnovy khim. tekhnol., 1987, v. 21, No. 1, pp. 86-94.
16. E.G. Vorontsov, O.M. Yakhno, *Formation of thermal initial section at non-stabilize film flow*. CHISA 84, Praha, 1984, E 1.7, p. 12.
17. L.P. Kholpanov, V.A. Malyusov, N.M. Zhavoronkov, *Hydrodynamics and mass transfer in turbulent liquid film flow involving the inlet portion*. Collection Czechoslovak Chem. Commun., 1981, V. 46, pp. 467-477.
18. Ye.G. Vorontsov, O.M. Yakhno, *Local thickness of wetting film at inlet portion*. Zhurn. prikladn. mekhaniki i tekhn. fiziki, 1974, No. 4, pp. 64-69.
19. Ye.G. Vorontsov, O.M. Yakhno, *Length of inlet portion of laminar flow of wetting films*, Gidromekhanika Ac. Sc. Ukr. SSR, Kiev, 1978, v. 38, pp. 48-52.
20. V.N. Miroshnichenko *On determining the length of inlet portion in laminar flow of liquid film. Proceedings of Nikolaev Ship-Building Institute*, 1977, issue 124, pp. 79-83.
21. W. Wilke, *Warmeubergang an Rieselfilme*. VDI-Forschungsheft 490, Dusseldorf: VDI-Verlag, 1962, 36 S.
22. Ye.G. Vorontsov, *Some peculiarities of temperature distribution in cross-section of wetting film*. J. Inzh. fiz., 1968, V. 15, No. 3, pp. 540-542.
23. V.L. Rakitsky, Ye.G. Vorontsov, *Application of random functions theory for disclosing latent periodicities in studying wave flow of liquid films*. Kiev Polytechnical institute, Kiev, 1989, p. 53., Dep. V. UkrNIINTI 16.06.89, No. 1765, UK-89.
24. K.J. Chu, A.E. Dukler, "Statistical characteristics of thin, wavy films. Part 2. Studies of the substrate and its wave structure." In: *AIChE Journal*, 1984, V. 20, No. 4, pp. 695-706.
25. Ye.G. Vorontsov, "Peculiar features of film flow with varying film run length." In: *J. Inzh. fiz.*, 1969, V. 16, No. 1, pp. 140-144.
26. P.L. Kapitsa, *Wave flow of thin layers of viscous liquid. Zhurn. eksperim. i tekhn. fiziki*, 1948, V. 18, iss. 1, pp. 3-28.
27. "On analysis of wave characteristics of falling liquid film." L.P. Kholpanov, V.Ya. Shkadov, V.A. Malyusov, N.M. Zhavoronkov. *Teor. osnovy khim. tekhnol.*, 1971, V. 5, No. 4, pp. 559-563.
28. V.V. Dilman, V.I. Naydenov, "On diffusion and thermal Marangoni instability in falling liquid film." In: *Teor. osnovy khim. teknol.*, 1989, V. 23, No. 5, pp. 575-582.
29. A.G. Sheinkman, V.M. Pakhaluev, "Peculiar features of gravity wave flow of fluid films on vertical and inclined plate." In: *Hydrodynamics and heat transfer*. Sverdlovsk Publ. of Ural Research Center of Acad. Sc. of the USSR, 1974, pp. 40-44.
30. D. Braun, *Die effektive Diffusion in welligen Rieselfilm*. Dissertation, Aachen, 1969, 75 S.
31. R.P. Salazar, E. Marschall "Statistical properties of the thickness of a falling liquid film" In: Acta Mechanica, 1978, V. 29, pp. 239-255.

32. Ye.G. Vorontsov, O.M. Yakhno *"Specific features of gravity flow of ununiformly-distributed thin liquid layers."* In: *Inzh. fiz. zhurn.*, 1974, V. 26, No. 2, pp. 272-279.
33. Ye.G. Vorontsov, O.M. Yakhno, *"Analysis of temperature field in wetting films,"* In: Inzh. fiz. zhurn., 1975, V. 28, No. 4, pp. 733-734.
34. I.V. Domansky, *Hydrodynamics and heat transfer in gas-liquid chemical-engineering apparatuses. Thes. of dissertation (Mr. Sc) L.: LTI named after Lensovet, 1976, p. 42.*
35. *K. Koziol, L. Broniarz, "Nowicka-Kubiszewska T. Pasywna turbulizacja prezeplywu i jej wplyw na intensyfikacje procesow wymiany ciepla i masy." In* Zeszyty naukowe politechniki Poznanskiej, 1980, No. 15, S. 309-333.
36. G.D. Fulford, "The flow of liquids in this film." In: *Advances in Chemical Engineering.* New York-London: Academic Press, 1964, V. 5, pp. 151-236.
37. H. Struve, *Der Warmeubergang an einem verdampfenden Rieselfilm,* VDI-Forschungsheft 534. Dusseldorf: VDI-Verlag, 1969, 36 S.
38. V.I. Tolyninsky *Heat transfer in boiling.* Kiev: Naukova dumka, 1980, p. 316.
39. V. Slesarenko *This film desalination plants and their role in water supply.* In: Desalination, 1987, V. 65, pp. 63-73.
41. U. Gropp, *Warme- und Stoffubergang bei der Oberflachenverdampfung und beim Blasensieden eines binaren Kaltemittelgemisches am Rieselfilm.* Dissertation, Karlsruhe, 1985, 167 S.
42. C. Elle, *Der Warmeubergang bei der Rieselfilmverdampfung des Kaltemittel-Ol-Gemisches RII-51 KM33.* Dissertation, Dresden, 1970, 150 S.
43. A.M. Kutepov, L.S. Sterman, N.G. Styushin, *Hydrodynamics and heat transfer under evaporation conditions.* M. Vysshaya Shkola, 1986, p. 448.
44. M. Serwinski, Z. Mikolajczyk *Wnikanie ciepla podczas wrzenia cieczy w wyparce blonkowej ze splyem grawitaxyjnym.* Inzynieria chem. i proc., 1983, T. 4, No. 4, S. 757-774.
46. W. Unterberg, D.K. Edwards, *Evaporation from falling saline water films in laminar transitional flow,* Sympos. on evaporation F. F.N.M., AIChE (Houston), 1965, p. 84.
47. A.E. Dukler, "Fluid mechanics and heat transfer in vertical falling-film systems." In: Chem. Engng. Progr. (Sympos. Ser. heat transport), 1960, V. 56, No. 30, pp. 1-10.
48. V. Slesarenko, "Characteristic features of desalination plants with thin film heat exchangers," In: Desalination, 1989, V. 74, pp. 305-316.

Rectification Using Film Type Equipment

V.M. Olevsky, A.V. Shafranovsky

Abstract - This chapter deals with the results of studies carried out at the State Institute of Nitrogen Industry (GIAP) and is aimed at the development of a new type of film rectification equipment.

A comparative evaluation of various methods of effecting mass transfer on rectification in a slot channel with one wetted wall is made and the results of experimental studies are presented.

Notation

d_c	internal diameter of the film type column, m;
d_r	diameter of the coaxial cylindrical rotor, m;
$d_e=2\Delta$	equivalent diameter of the vapor (gas) flow in the slot channel with one wetted wall, m;
G	molar flow rate of the vapor (gas) phase, kmol/h;
g	free fall acceleration, m/s^2;
h_{tg}	total height of a transfer unit calculated
h_g	phase height of a transfer unit for the vapor phase, m;
h_l	phase height of a transfer unit for the liquid film, m;
K_{oy}	mass transfer coefficient calculated for the gas phase, kmol/m^{2}h;
L	molar flow rate of the liquid phase, kmol/h;
l	path length of the gas flow in the slot channel, m;
$\overline{M}$	average molecular weight of the phase;
m	phase equilibrium constant;
r_c	radius of the column, m;
r_r	radius of the cylindrical rotor, m;
$r_m=\dfrac{r_c+r_r}{2}$	average radius of the annular gap, m;
$U=\dfrac{\pi d_r n}{60}$	peripheral velocity of the surface of the cylindrical rotor, m/s
U_{cr}	critical value of the peripheral velocity of the rotor determined from formula (11), m/s;

$U_{Ta_{cr}}$ — peripheral velocity of the rotor corresponding to Ta_{cr}, m/s;
W_g — average velocity of the gas phase, m/s;
$W_{eg}=\sqrt{W_g^2+0.1915U^2}$ — average velocity of the gas phase in the annular gap of the rotor column, m/s;
X — mole fraction of the volatile component in the liquid;
Y — mole fraction of the volatile component in the vapor;
β_x, β_y — individual mass transfer coefficients in the liquid and gas phase, respectively, kmol/m^{2}h;
β_1, β_g — same, m/s;
Γ — liquid feed rate per wetted perimeter unit, m^2/s;
Δ — gap width in the slot channel, m;
δ — average liquid film thickness, m;
$\nu=(\nu_1^2/g)^{1/3}$ — reduced liquid film thickness, m;
μ — dynamic viscosity of the phase; N/m.s;
ν — kinematic viscosity of the phase, m^2/s;
$\xi_g=\beta_y/\beta_g=3600\rho_g/M_g$, kmol.s/m^{3}h;
$\xi_1=\beta_x/\beta_1=3600\,\rho_1/M_1$, kmol.s/m^{3}h;
ρ — phase density, kg/m3;
$\omega=2\pi n/60$ — angular rotor velocity, s^{-1};
$Re_g=\dfrac{w_g d_e}{\nu_g}$ — Reynolds number for the gas phase;
$Re_l=\dfrac{4\Gamma}{\nu_1}$ — Reynolds number for the liquid film;
$Re_c=\dfrac{Ud_e}{\nu_g}$ — centrifugal Reynolds number;
$Re_e=\dfrac{w_e d_e}{\nu_g}=\sqrt{Re_g^2+0.1915\,Re_c^2}$ — effective Reynolds number;
$Ta=\dfrac{\omega r_m^{1/2}\Delta^{3/2}}{\nu_g}$ — Taylor number for the gas flow;
Ta_{cr} — critical Taylor number;
$Sh_g=\beta_g d_e/D_g$ — Sherwood number for the gas phase;
$Sc=\dfrac{\nu}{D}$ — Schmidt number;
$\dfrac{\beta_1\nu}{D_1}$ — dimensionless individual mass transfer coefficient for the liquid film

Indices

x,1	liquid phase parameters;
y,g	gas phase parameters;
c	column;
r	rotor;
cr	critical value;
*	equilibrium value;
e	equivalent.

Abbreviations

NTU_{tg}	total number of transfer units

1. COMPARATIVE EVALUATION OF THE EFFICIENCY OF FILM AND NON-FILM MASS TRANSFER CONTACTORS IN RECTIFICATION

Low-pressure rectification and distillation processes find increasingly extensive application in chemical engineering. This is due primarily to the fact that many substances encountered in industrial organic synthesis are highly temperature-sensitive, which rules out rectification processes under atmospheric pressure. In some cases, the use of a vacuum in the rectification and distillation of mixtures noticeably raises the relative volatilities of the components which is of particular importance in the rectification of closely-boiling substances. Finally, vacuum rectification makes it possible to strongly reduce the treatment temperature of the mixtures of high-boiling products and to replace high-temperature heating by using electric devices or special heat carriers by more economic and safe heating with high-pressure steam. The lowering of the process temperature also decreases the corrosion rate of the structural materials contacting with the working medium and allows for the use of polymeric materials.

An important factor in determining the quality of operating vacuum rectification equipment is the pressure drop between the top and bottom of the column, i.e. its hydraulic resistance. The quest for maximum reduction of the pressure drop, at least by removing the hydrostatic column of the liquid in the path of the vapor flow, has led to the development of film type equipment.

As film type equipment was studied and improved, other promising fields of application were gradually found. Much attention was paid to such specific features of film type equipment as high vapor phase throughput, insignificant spray entrainment and low liquid holding capacity.

Since 1973, in view of the drastic growth of prices of primary energy, new factors have come into action; these factors can greatly raise the proportion of film type rectification equipment. Thus, a renewed interest in heat pumps has arisen (secondary vapor recompression) and multiple-effect rectification (or evaporation) units with a view to utilize the heat of condensation of secondary vapor in heating the bottom of the rectification column. As it is well known, when utilizing the heat of condensation of secondary vapor, particularly in the case of heating the same bottom portion of the column where the vapor is being formed, it is very important that the rectification column, where secondary vapor is purified prior to its condensation, should feature the minimum temperature drop between the top and bottom. Low hydraulic resistance of film type equipment enables a more economical utilization of heat in rectification processes with the recovery of the heat of condensation of secondary vapor. This makes the use of film type equipment economically sound at pressures approaching and even exceeding the atmospheric pressure.

Because of the development of more stringent environmental standards, problems removing noxious impurities from the gases discharged by chemical enterprises become more acute. Absorption is an efficient method for treating large gas flows. The employment of film type mass transfer devices with insignificant hydraulic resistance in absorption allows for power-saving gas treatment under atmospheric pressure with the use of standard fans instead of high-head blowers.

In improving the design of regular packing, it is advisable to take the path of turbulizing the vapors and the liquid film, reducing the longitudinal and transverse nonuniformities in the distribution of the contacting phases, rationally using mass transfer end effects, maintaining a stable redistribution of the flows between the elements of the packing, increasing its specific area, and imparting a thin layer shape to the gas phase flow.

All these principles formed a basis for the design of a corrugated coiled band packing [1] made of thin metallic or plastic bands 25-200 mm wide with inclined corrugations rolled on a bush to form a double-start Archimedes' spiral of a certain diameter. The band is corrugated by passing it between two engaged helical gear wheels. The geometrical surface, S, of one coil with inclined corrugations is calculated from the formula [2, p. 16]

$$S = 2.22(D^2 - d^2)\sqrt{(h^2 + 4b^2)/b^2} \tag{1}$$

where D and d are the diameter of the coil and the central bush, respectively; h is the height of the coil; b is the distance between the coil turns for the cross-section of the corrugation having the shape of an isosceles rectangular triangle.

An important feature of this packing is the mounting/ dismounting easiness. Even if a certain skew occurs in the coils of the mounting, the flow of the wetting liquid from the center to the wall is virtually excluded. The packing with inclined corrugations is characterized by a complete absence of the fall-through of the liquid and of the stagnation zones since the corrugations of adjacent bands are oriented cross-wise and their generatrixes are in virtually one-point contact.

It is advisable to mount the packing coils on a central rod with a bottom rest alternately with radially directed spacer plates to form separate blocks 2-3 m high which will make it possible to utilize the end effects in full measure. The small height of the packing in each block strongly reduces the mechanical load on the coils which in this case can be made of thinner bands. The weight of the coiled packing made of a thin band is 1/2 to 1/3 that of densely packed rolls, fractionation precision being the same.

It is evident that the well-known Sulzer regular packing used in the form of packs of corrugated mesh made of wire 0.15-0.25 mm thick [3] is more efficient, due to its capillary properties, than the corrugated band packing. However, in real chemical production conditions all advantages of corrugated mesh packings may be reduced to zero for the following reasons:

- the liquid undergoing separation may contain suspended particles clogging the mesh;
- as a result of partial thermal decomposition, the liquid may leave a deposit on the working surface;
- the processed substance, (vapor, liquid), is not infrequently corrosive to the packing material but the allowance for corrosion cannot be made since an increase in the wire thickness will strongly deteriorate the capillary properties of the packing;
- because of a large total area of the wires forming the mesh, the catalytic effect manifests itself; selfignition of an organic substance, (or its decomposition products), retained by the mesh is sometimes possible on contact of the surface of the mesh with atmospheric oxygen.
- studies of hydraulic and mass transfer characteristics were carried out [4] on such subjects as corrugated coiled packing, plane-parallel glass fiber packing, and contact devices of traditional types: bubble cap trays, large-hole sieve trays, slotted perforation trays, cocurrent flow valve trays, and on a combined contact device.

All these mass transfer contact devices were studied under strictly identical separation conditions on a model rectification rig including a column 420 mm in diameter intended for the atmospheric rectification of a methanol-water mixture with an infinite reflux ratio. The tray spacing was constant (600 mm). In analyzing the efficiency of the contact devices of the tray and packing types, the common characteristic, the height of a transfer unit, was employed in order to facilitate comparison.

The tested devices are briefly characterized below.

CORRUGATED COILED PACKING

Coil height 50 mm, the angle between a corrugation and the vertical 20°, equivalent diameter 9.9 mm, corrugation height 10 mm. Total bed height 2610 mm.

Plane-Parallel Glass Fiber Packing

The packing comprised a squirrel cage frame, (height 460 mm, external diameter 400 mm), made of tubes with glass fabric pulled over the vertical tubes. The spacing between the neighboring fabric planes was 8 mm.

Large-hole sieve trays

Free cross-sectional area 11.7%, weir height 25 mm. Diameter of holes 12 mm.

Slotted perforation trays

Free cross-sectional area 17.8%, tray thickness 3 mm, slot width 4.5 mm.

Cocurrent-flow valve trays

Each tray contained ten round disk valves 40 mm in diameter ensuring a cocurrent flow of the vapor and liquid.

Combined contact device

The device consisted of a 150 mm high spider supporting a corrugated coiled packing (d_e=9.9 mm, l=50 mm) and mounted on a slotted perforation tray, the total height of the packing bed on each tray being 350 mm.

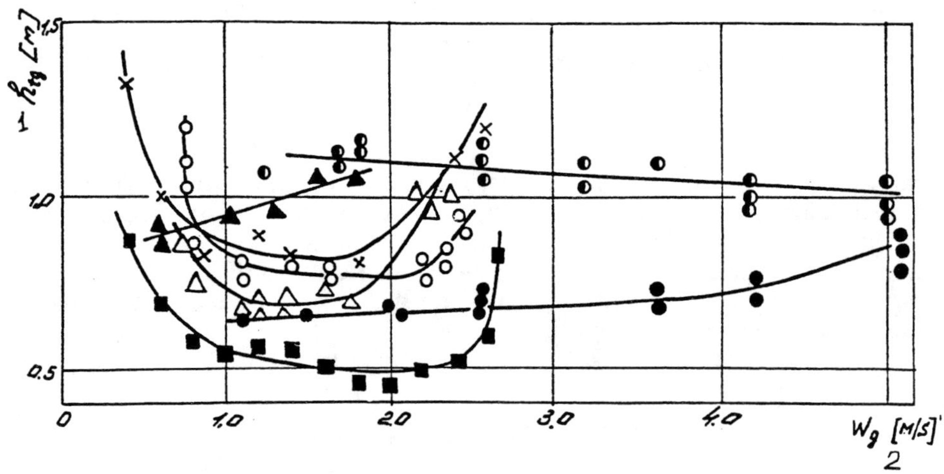

Figure 1. Dependence of the total height of a transfer unit, h_{tg}, on vapor velocity, W_g; o-sieve tray; -cocurrent-flow valve tray; ∘-coiled packing with inclined corrugation; -plane-parallel glass fiber packing; Δ-bubble cap tray; n-combined contact device; ×-perforated tray; 1-h_{tg},m; 2-W_g,m/s.

The generalized data on the effect of the vapor velocity in the column on mass transfer efficiency for all contact devices in question are presented in Figure 1; specific hydraulic loss (ΔP/NTU) values are given in Figure 2.

In the case of glass fiber plane-parallel packing (PPP) the experiments were run in the 4.2-19.5 m^3/m^2h feed rate range at linear vapor velocities, (full cross-section area base), Of 0.8-5.0 m/s. The efficiency of glass fiber PPP in the whole load range studied remained constant (h_{tg}≈1 m). The independence of the column efficiency of the load is peculiar to PPP, as confirmed by the data obtained for metallic packings in rectification of various organic mixtures in a broad range of loads and residual pressures [5, 6].

The fact that a reduction in the feed rate to 4.2 m^3/m^2 did not lead to the efficiency drop (as was the case with a metallic PPP) is of interest. This phenomenon can be explained by a better wettability of the glass fiber as a result of action of capillary forces and by the presence of larger amounts of the retained liquid.

The experiments revealed a very broad range of operation stability of the corrugated coiled packing (W_g=1-5 m/s) and a high separation efficiency (h_{tg}=0.6-0.7 m).

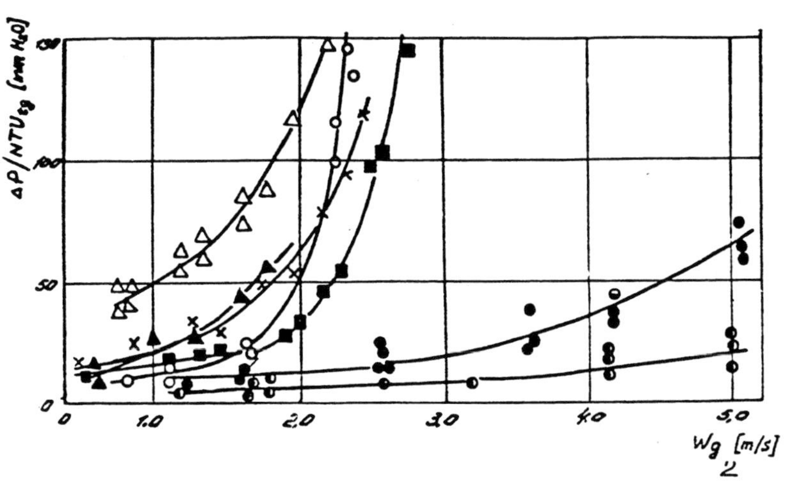

Figure 2. Dependence of specific hydraulic resistance, $\Delta P/NTU_{tg}$, on vapor velocity, Wγ: ○-sieve tray; -cocurrent-flow valve tray; •-coiled packing with inclined corrugation; -plane-parallel glass fiber packing; Δ-bubble cap tray; n-combined contact device; ×-perforated tray; 1-$\Delta P/NTU_{tg}$, mm H_2O; 2-W_g, m/s.

The data displayed in Figures 1 and 2 made it possible to compare the efficiencies of the contact devices mentioned above and, taking the specific hydraulic resistance as a criterion, to arrive at the following conclusions:

- the most promising design, from the standpoint of its use in mass transfer processes under the atmospheric and lower pressure, (particularly at high feed rates), is a regular coiled packing with inclined corrugations featuring high operation stability, low specific hydraulic resistance, and a comparatively high efficiency; plane-parallel glass fiber packing features a very low specific hydraulic resistance; with mechanical strength ensured, the packing can find commercial use;
- a combined contact device possesses the highest separation efficiency and holds much promise for such units where hydraulic resistance is not a limiting factor;
- bubble cap trays are known to have many benefits in those processes where the fall-through between the trays cannot be tolerated and in the columns with high ratios of liquid and vapor weight loads;
- perforated, cocurrent-flow valve, and large-hold seize trays can ensure satisfactory operation of a mass transfer column, (slightly more intense as compared with bubble cap trays), for medium feed rates and superficial vapor velocities.

2. Mass Transfer in Corrugated Band Coils

Kinetic feature studies of mass transfer in the liquid phase in the presence of a coiled corrugated packing dealt with the desorption of difficulty soluble CO_2 from water by a nitrogen flow [4]. The experimental data were generalized by the following dimensionless equation

$$\frac{h_l}{\nu} = A\,\mathrm{Re}_l^{e}\left(\frac{l_f}{\nu}\right)^{b}\mathrm{Sc}_l^{0.5} \tag{2}$$

where l_f is the height of the coil.

The exponent at Sc_l in Equation (2) was taken to be 0.5 without any additional studies. The unknown coefficient A and the exponents a and b in Equation (2) were determined by computer processing of the data obtained in all of the variables: $8.5 \le Re_l \le 357$ and $510 \le \left(\frac{l_f}{\nu}\right) \le 2040$.

Table 1 lists the values of the coefficient and exponents.

Table 1. Constants in Equation (2).

Constants	$8.5 \le Re_l < 28$	$28 \le Re_l \le 337$
A	20.02±2.32	7.264±0.082
a	0±0.164	0.360±0.015
b	0.367±0.119	0.321±0.023

Gas phase mass transfer under physical absorption conditions was studied in an ammonia/air-water system. With a view to develop a more justified method of calculating mass transfer efficiency in columns with coiled corrugated packing, the operation of the packing was studied under conditions of physical absorption and rectification; the experimental data were compared taking the thermal effect in rectification into account. The following equation for the determination of the height of a transfer unit in the gas phase was derived [4];

$$h_g = 10.7 d_e \,\mathrm{Re}_g^{0.12}\, Sc_g^{0.67} \left(\frac{l_f}{d_e} \right)^{0.42}; \tag{3}$$

the Equation was verified in the following ranges:

$$\mathrm{Re}_g = 260 - 4000;\ d_e = 3.6 - 15.0\ mm;\ \frac{l_f}{d_e} = 3.5 - 28.$$

3. New Types of Regular Packing

The corrugated band coil and plane-parallel glass fiber types of packing described in the previous paragraphs possess low specific hydraulic resistance and high vapor phase throughput; however, they ensure a comparatively high mass transfer efficiency only at high feed rates. It happens very often in vacuum rectification, particularly at low reflux ratios, that the amount of the feed is minor and the use of Sulzer type mesh packing, even if efficient in these conditions, is not always possible for the reasons explained above.

The following engineering approaches to the development of new types of regular band packings retaining their operation efficiency at lowered feed rates were taken at GIAP:

- the assembling of short sections into a sprayed packing with vapor redistribution in going from one section to another;
- the arrangement of the flow of the liquid film on the packing with a view to have the least possible total spraying perimeter by directing the whole flow along the narrow flanged strips forming the stack;
- the arrangement of a cross-flow scheme of contact of the vapor phase with the liquid film;
- the interferences of rotary motion with the interaction between the liquid and vapor flows.

These approaches were implemented in the design of rotor and non-rotor film type equipment developed at GIAP.

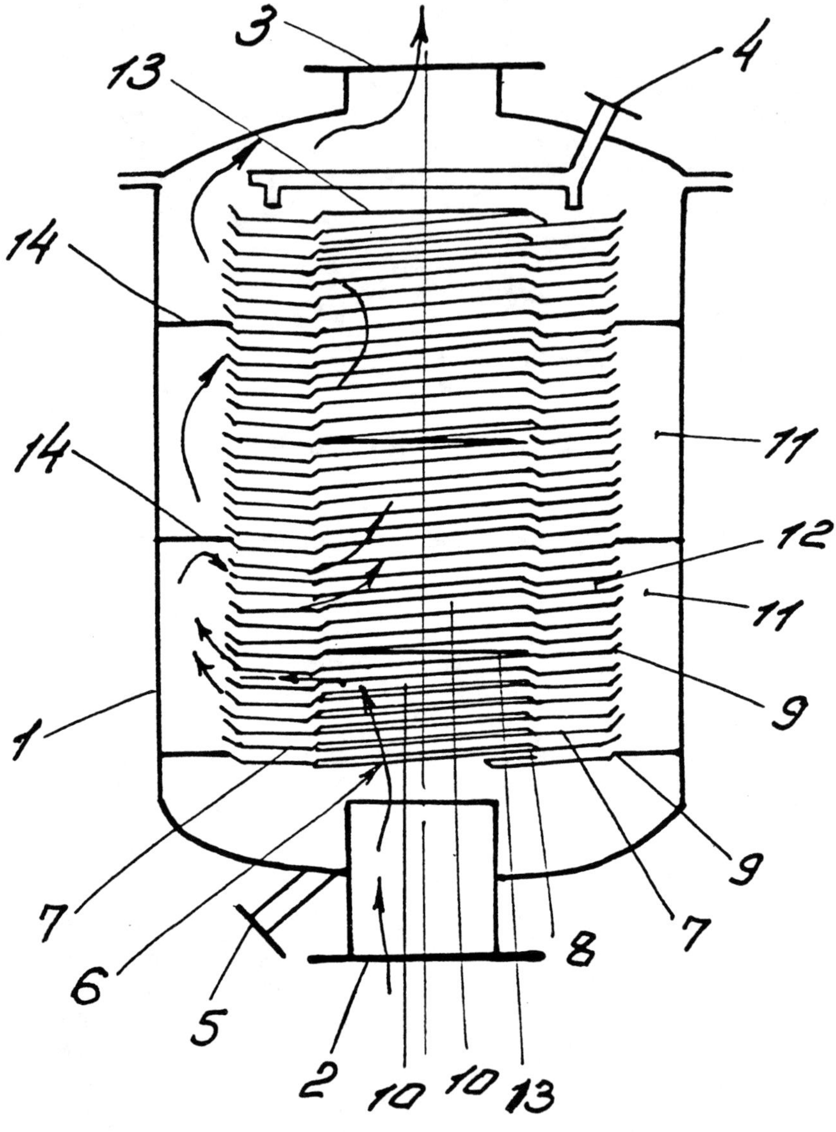

Figure 3. Film type rectification column with a stationary regular packing in the form of a coil; 1- shell; 2 - vapor inlet; 3 - vapor outlet; 4- reflux inlet; 5-liquid outlet; 6-wetted packing in the form of a coil; 7-flanged plates; 8-inner flanges of plates; 9-outer flanges of plates; 10-inner vapor channels; 11-outer vapor channels; 12-gaps between flanged plates; 13-inner transverse battles; 14-outer transverse baffles.

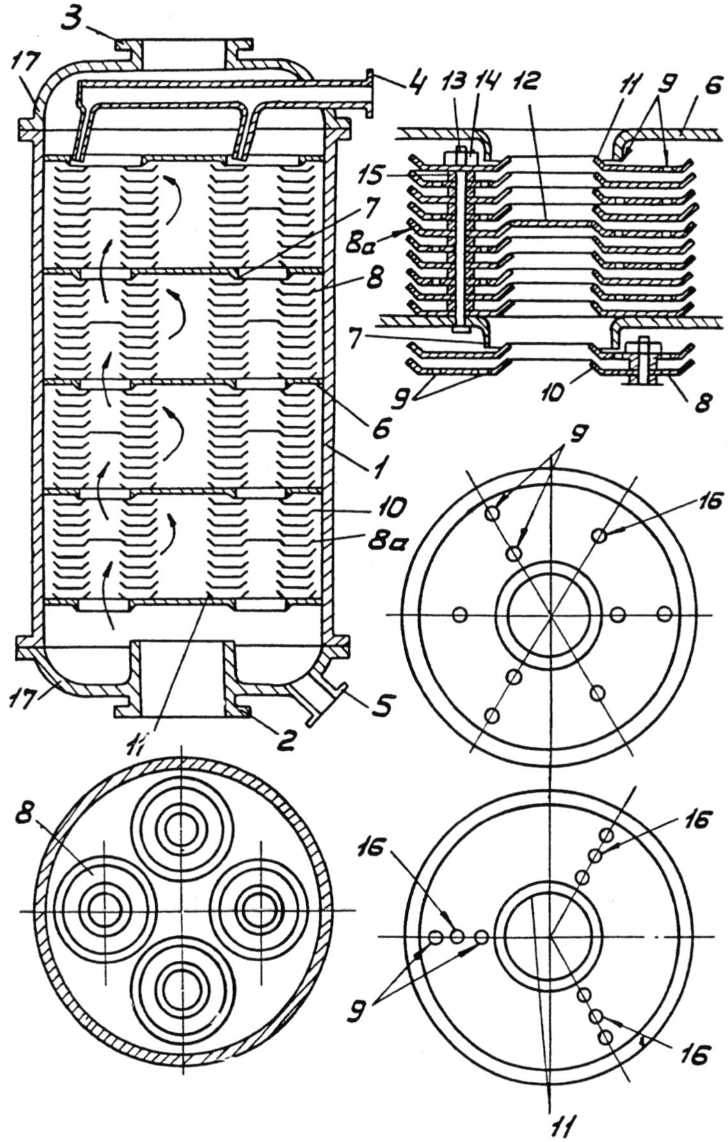

Figure 4. Film type rectification column with a stationary packing in the form of stacks of horizontal annular plates with flanged edges: 1-shell; 2-vapor inlet; 3-vapor outlet; 4-reflux inlet; 5-liquid outlet; 6-transverse load-carrying baffles; 7-drainage holes in baffles; 8-horizontal annular plates; 8a-blank annular plates; 9-drainage holes in plates; 10-flanged edges of plates; 11-top and bottom annular plates of the stack; 12-blank baffle of the stack of plates; 13-assembling bolts; 14-nuts; 15-spacers; 16-assembling holes (for bolts); 17-column lid or bottom.

A schematic diagram of a film type column with a stationary regular packing embodying these principles (except the last one) is given in Figure 3 [7-10].

The vertical cylindrical shell 1 of the column contains a helix assembled from inclined plates with flanges 9 and transverse baffles: outer 14 and inner 13. The column has gas inlet 2, gas outlet 3, liquid distributor 4 and liquid outlet 5. Plates 7 are made of a thin-sheet material.

The liquid is fed onto the helix from distributor 4 in such a manner as to wet the upper plates 7 of the stack. Due to the small width of the channels limited by flanges 9, the bottom of the channels will be fully wetted even in the case of pointwise liquid supply. The liquid film flows along the upper plane of plates 7 following a descending helical trajectory. The vapor, on its way from the bottom to the top of the column, follows a broken bath bouncing from transverse baffles 13 and 14 and multiply crossing the stack of plates 7. This design ensures a cross-motion of the liquid film and the vapor so that for the vapor flow the slot between plates 7 is a short wetted section.

Figure 4 presents a film type column with a stationary regular packing consisting of horizontal plates.

Mounted in a cylindrical shell 1 are transverse load-carrying baffles 6 supporting stacks of horizontal annular plates with flanged edges 10. The middle plate of the stack is joined to a partition (12). Plates 8 in the stacks are tied together and with load-carrying baffle 6 with the help of long narrow bolts 13 and spacers 15. Holes 16 in annular plates 8 are designed for passing bolts 13. Holes 9 serve for draining the liquid onto the underlying annular plate 8. Holes 9 in the underlying plate 8 are shifted by 60° with respect to holes 9 of the upper plate 8; thus, the liquid enters each plate 8 in the points which are equidistant from holes 9 of the same plate which ensures an ordered flow of the liquid film on the upper surface of the plate towards holes 9.

The vapor in the column follows a tortuous trajectory (indicated by arrows) multiply crossing the stacks of plates 8. The liquid is fed to the stacks from above and flows from one plate 8 to another covering them with a film. From the stacks the liquid flows onto the transverse load-carrying baffle 6 and further, through drainage holes 7, onto the underlying stacks of plates 8.

The photograph (Figure 5) shows a disassembled rectification column of the described design manufactured by a specialized chemical engineering plant [11-13]. Its diameter is 0.8 m; the column consists of three sections 2 m high. Figure 6 shows the column packing consisting of 4 stacks of plates fitted in the cross-section of the shell. The thickness of the sheet used in manufacturing the annular plates with flanged edges is 0.3-0.5 mm.

The column is intended for vacuum rectification of caprolactam or similar substances under residual pressures of about 1-10 Torr.

Such columns were first used in caprolactam purification. The feedstock was a caprolactam-oligomer concentrate obtained on a continuous polymerization and fiber formation unit after vacuum depolymerization of polycaproamide. Caprolactam recovery and purification are required for its recycle into the polymerization step in polycaproamide production. Feedstock distillation in standard bulk evaporators did not ensure the desired purification of recovered caprolactam.

Figure 5. Commercial rectification column (800 mm in diameter) comprising 3 sections 2000 mm each with regular packing consisting of horizontal annular plates. Principal view.

Presently, columns with a regular packing consisting of horizontal annular plates find expanding application in polycaproamide fiber production. These rectification columns possess a specific feature which is not peculiar to other types of equipment: packing rearrangement can give rise to an increase in the vapor throughput of the column with a slight concurrent reduction in the separation efficiency and vice versa. This property of the packing was revealed when carrying out preparatory and commissioning operations at one of the plants with the participation of GIAP experts. These operations were aimed at raising the capacity of a periodic rectification unit to 4 t/day, the quality parameters of the distillate satisfying the plant specifications. In addition to the still, the unit also included the column shown in Figure 5.

As is the case with many other monomers, stringent requirements are imposed on the purity of caprolactam: the allowable level of impurities, (water not included), in commercial caprolactam should not exceed 0.05%. Because of the low concentration of impurities allowed for caprolactam, serious difficulties in quantitative analysis appear. For this reason the quality of the purified product is evaluated using indirect indicators, the basic ones being the permanganate number, color, concentration of volatile bases, acidity, crystallization temperature, etc.

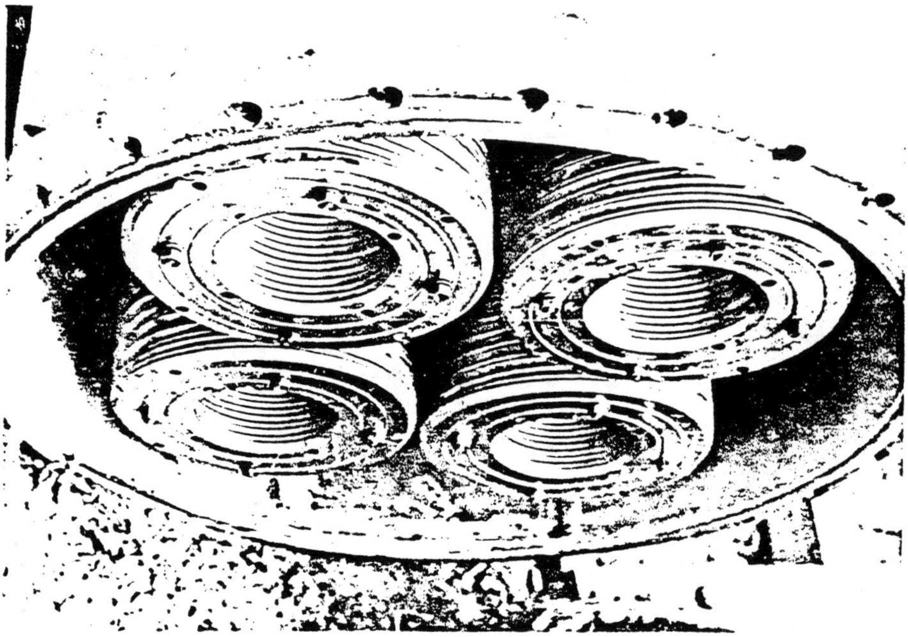

Figure 6. Commercial rectification column (800 mm in diameter). Principal view of plate stacks.

The column shown in Figure 5 and consisting of 3 sections was tested in three versions:

- as designed;
- completely free from packing;
- the height of the stacks in the second section was twice, and in the highest section three times as large as the designed height [13].

The last version was implemented through the removal of some transverse load-carrying baffles out of the column (Figure 4, item 6).

The test results are displayed in Table 2.

The growth of the column's capacity after a complete removal of the packing is explained by a decrease in the hydraulic resistance, lowering the residual pressure in the still and of the boiling point of the liquid in it, and consequently, by the growth of the temperature drop between the condensing vapor and the bottom liquid (i.e. the driving force of the heat transfer process in the still).

Table 2. Caprolactam Purification Efficiency as a Function of the Packing Layout

Column design	Capacity, t/day	Permanganate number, s	Color, ml	Highly volatile compounds, meq/kg	Acidity	Crystallization temperature, °C
As designed	1.5	7530	0.87	0.5	0.0008	68.9
No packing (hollow)	6-7	4300	1.56	0.8	0.0018	68.7
Increased height of stacks	4-4.5	4000	1.50	0.45	0.0009	68.8
Specified parameters	–	–	not more 1.50	not more 0.60	not more 0.0015	not less 68.8

On increasing the height of stacks, the total number of annular plates remained unchanged (2800 plates); the hydraulic resistance of the column decreased due to the growth of the free area for the flow of the vapor through the stack and to the reduction of the total number of crossings of the stacks by the vapor flow moving from the column's inlet to the outlet.

Table 2 gives the average values for the qualitative characteristics of recovered caprolactam, since they undergo changes in the course of rectification. Thus, at the beginning of caprolactam distillation its permanganate number is 9-12 thousands and at the end it drops to 2-3 thousand which is caused by the carryover of heavy fractions. In consequence, the reflux ratio has to be increased at the end of distillation which lowers the capacity of the column.

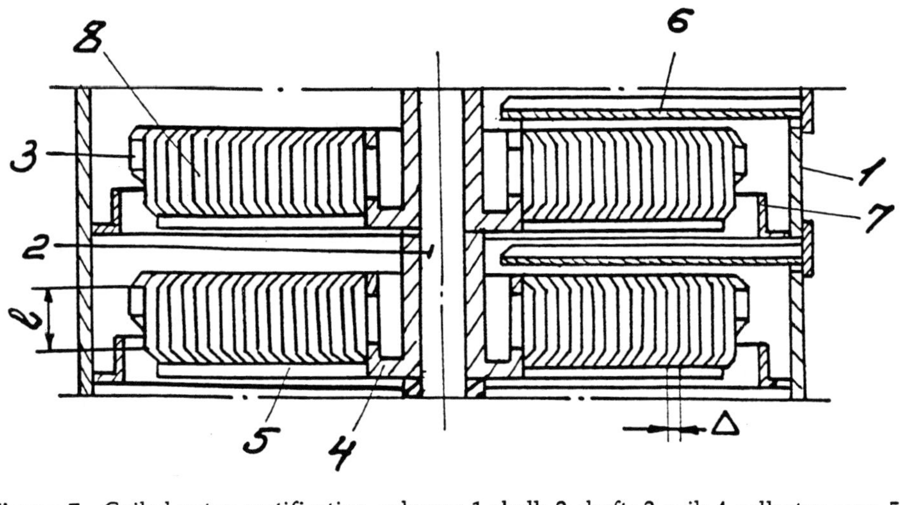

Figure 7. Coiled rotor rectification column: 1-shell; 2-shaft; 3-coil; 4-collector cup; 5-radial rib; 6-duct; 7-annular pocket.

Figure 8. Film type rotor rectification columns 1400 mm in diameter at a polycaproamide fiber plant.

Thus, the obtained recovered caprolactam is directed, together with primary caprolactam, to the production of polycaproamide cord fiber.

A unit featuring the interference of rotary motion with the interaction between liquid and vapor flows is shown in schematic form in Figure 7 [14,15].

A vertical cylindrical shell 1 contains a vertical shaft 2 supporting coils of flanged bands 3. Placed in the center of the coils are collector cups 4. Radial ribs 5 join rigidly coiled bands 3 together and to cup 4. Radial ducts 6 are intended for the transfer of the liquid to cups 4. Annular pockets 7 attached to shell 1 collect the liquid shedded from the rotating coils.

The liquid is moved by centrifugal force in the form of a film on the inner surface of flanged coiled bands 3 facing shaft 2, from the center of the coil to its periphery.

Slots 8 between coiled bands 3 ensure the permeability of the coil to the vapor flow in the axial direction. The flow is separated into thin layers making contact with the liquid film in a cross-flow pattern.

The length of the trajectory (l) of the vapor flow makes contact with the liquid film and is equal to the width of the coiled band, i.e., very small. In the space between the coils containing only ducts 6, the transverse velocity and concentration profiles formed in the vapor flow on its contact with the liquid film on the preceding coil collapse. Thus, within each coil the contact of the vapor with the liquid film occurs in a short section where there is insufficient time for a complete formation of the velocity and concentration profiles of the transferred substance in the vapor flow. This is known to favor the improvement of mass transfer in the vapor phase.

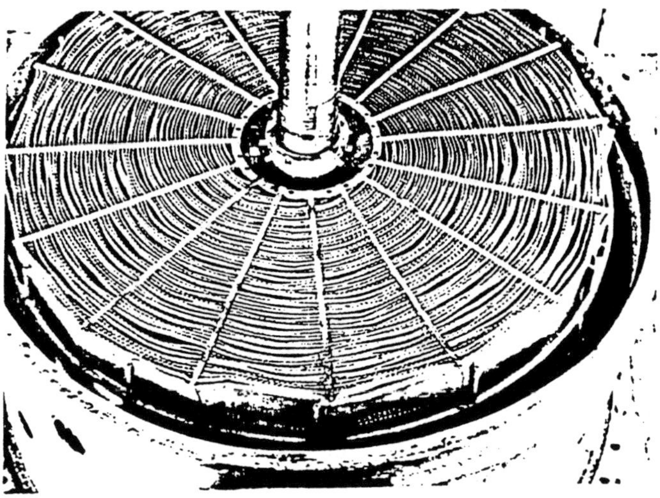

Figure 9. Film type rotor rectification column 1400 mm in diameter with removed lid.

Fail safe and full wetting of the packing with low amounts of liquid even in large-diameter columns is effected by the forced movement of the whole mass of the liquid along a small number of narrow bands 3 forming a coil with a large total area. Thus, the linear liquid feed rate in the column is artificially raised by diminishing the wetter perimeter which in this case is equal to the width of the coiled band multiplied by the number of bands in the coil and hence is very small.

The photograph (Figure 8) displays rotor film columns, (1400 mm in diameter), used as part of a caprolactam recovery unit at a synthetic fiber plant. A geared motor rotating the shaft is installed on the cover of the column. Figure 9 shows one of the columns with a removed cover; one can see the upper coil made of flanged bands fixed around the central collector cups with the help of radial ribs.

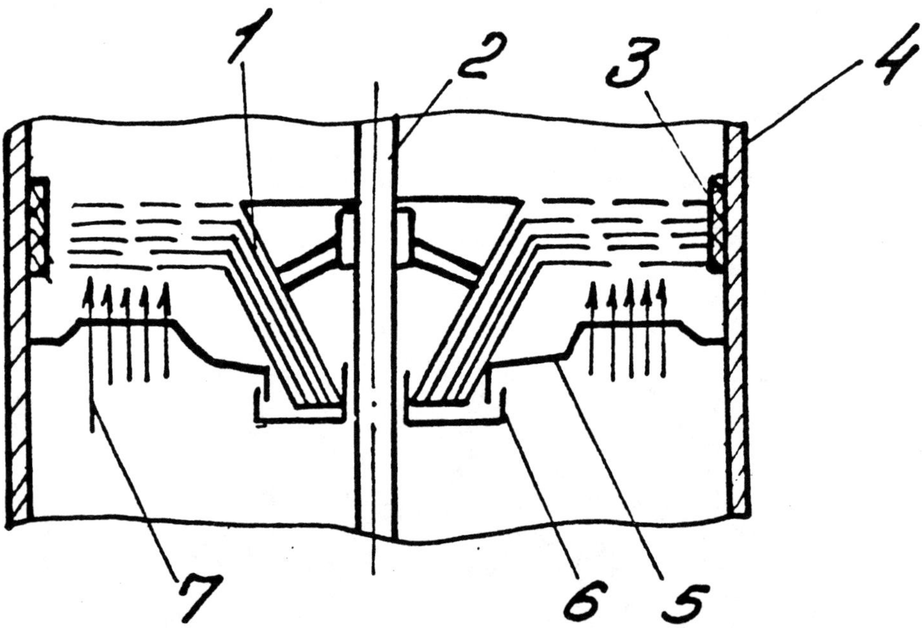

Figure 10. Kirschbaum-Stor rectification column: 1 - stack of cones; 2 - shaft; 3 - demister; 4 - column shell; 5 - collector tray; 6 - weir; 7 - vapor flow.

The new rotor columns have the following features: shell diameter 1400 mm, vapor inlet diameter 800 mm, working section height 3230 mm, number of steps 20, step spacing 160 mm, gap between bands 10m, band thickness 0.3-0.5 mm, working band width 68 mm, overall dimensions of the column 7100x2050x2050 mm, rotor speed 129 rpm, drive power 5.5 kW, working pressure 3-12 Torr, working temperature up to 140°C, working medium caprolactam.

The column has top and bottom end seals and external bearing assemblies.

The prototype of the GIAP design commercial rotor film column was made by revamping one of Kirschbaum-Stor columns (1.4 m in diameter) with eight separation steps used in the production of caprolactam. Kirschbaum-Stor type columns [16] have received wide acceptance in the industry of many countries including the USSR. A column of this type comprises a cylindrical shell and a coaxial shaft carrying spraying devices. The shaft is mounted in bearings installed in the top and bottom covers of the column. The shaft holes in the covers are sealed with double packing.

Each step of the column (Figure 10) contains [17] a stack of concentric truncated cones 1 fixed on a rotating shaft 2, demisters 3 attached to the inner surface of shell 4, and a collector tray 5 with weir 6 and vapor risers 7. Under the action of the centrifugal force, the liquid climbs up the cones to the upper edge and is sprayed in the form of parallel curtains of droplets making contact with the vapor.

The revamping project envisions the use of all ancillary equipment of the production unit; the shaft of the old rotor was also used. For practical reasons, no changes in the shell were made. After the removal of trays 5 (Figure 10), an insert 1.3 m in diameter was placed in the shell of the column (1.4 m in diameter); the gap between the insert and shell was thoroughly packed with graphitized asbestos. Further on, the insert served as the shell of the rectification column.

The rotor speed in the Kirschbaum-Stor column was 520 rpm. After drive overhaul, the rotor speed was first reduced to 90 rpm and subsequently raised to 120 rpm. The stacks of cones 1 were removed from shaft 2 of the column (Figure 10) and replaced by twelve identical steps in the form of coiled flanged bands 3 (Figure 7).

The reflux was fed to the second step from above so that the top step acted as a separator. The column operated in a continuous regime, and the feedstock, (molten caprolactam containing high-boiling impurities, such as resins, oligomers, etc.), was introduced into the seventh step from above. The direction of rotation of the shaft coincided with the direction of winding the coils in the step.

The number of bands in each coiled step was 24. The height of the coil in a step (band width) was 100 mm, the gap between the two neighboring bands being 12 mm.

Given below are the performance data in caprolactam rectification which was collected simultaneously for the Kirschbaum-Stor and GIAP rotor columns.

It should be noted that the GIAP design rotor column, under identical conditions of operation, featured a load which was 2.1 times, (2.42 times, taking account of the reduction of the cross-section diameter from 1.4 to 1.3 m), higher than that of the Kirschbaum-Stor column, hydraulic resistance values being identical.

Characteristics	Columns	
	GIAP	Kirschbaum-Stor
Column diameter, m	1.3	1.4
Working part height, m	2.3	2.8
Number of operating steps	11	8
Shaft speed, rpm	120	520
Step spacing, m	0.19	0.30
Residual pressure in the column, Torr	3-4	3-4
Temperature of caprolactam vapor (top of the column), °C	128-132	128-132
Feed rate, l/h	1720	850
Distillate (purified caprolactam) take-off, l/h	1680	800
Reflux ratio	0.3	0.3
Bottom product take-off, l/h	40	50
Same, % of feed	2.3	5.9
Installed drive power, kW	4.5	20

Several years of operation of the GIAP design rotor column showed that the resulting caprolactam compared favorably with the product from the Kirschbaum-Stor column. Thus, the extinction coefficient of the aqueous solution did not exceed 0.01 as compared with 0.04 in the other case. Rectification in the new column led to the stabilization of other characteristics featuring a small time scatter which is very important since the variation of properties between batches complicates the caprolactam processing at chemical fiber plants, causing high spoilage and scrapping.

The use of the GIAP rotor column made it possible to greatly raise the recovery of caprolactam by reducing the amount of the bottom product take-off. In this case, the increase in the percentage of high-boiling impurities in the bottom product did not affect the quality of the distillate, whereas attempts to lower the take-off of the bottom product from the Kirschbaum-Stor column inevitably led to the deterioration of the quality of the distillate. Comparing these facts, one arrives at the conclusion that the GIAP rotor rectification column possesses a higher separating power than the Kirschbaum-Stor rotor column with greater vapor capacity.

The reduction of the rotor speed from 520 to 120 rpm with a decrease in power consumption to 1.15 kW is operationally advantageous with the installed drive power being 4.5 kW.

The usefulness of rotor type rectification columns with contacting steps made of flanged coiled bands is due to their ability to process large fluxes of vapor at very low feed rates and pressure drops.

This design of equipment is included in new projects for ammonium nitrate and carbamide production at stages of purification of the secondary steam which is formed in the vacuum evaporation of aqueous solutions and melts of fertilizers.

In the GIAP rotor columns, (Figure 7), the vapor flow follows a rectilinear trajectory with rotation while in the GIAP, film columns with a stationary regular packing in the form of flanged plates (Figures 3, 4) its trajectory is a broken line. Therefore, rotor columns are preferable in the cases where the highest possible vapor capacity should be ensured for the given column diameter and pressure drop. On the other hand, film columns with stationary regular packing, having no design limitations on height are more suitable for separation processes that require an elevated number of theoretical plates.

4. EXPERIMENTAL STUDIES OF MASS TRANSFER ON RECTIFICATION IN A SLOT CHANNEL WITH ONE WETTED WALL

The importance of film type equipment is not confined to its use in the separation of mixtures under production and laboratory conditions. Because of the stability of the phase contact area and the geometric simplicity of the shape of the flows in contact, mass transfer devices with a wetted surface are often used as instruments suitable for studying the effect of various factors on mass transfer in gas-liquid and vapor-liquid systems.

Film type equipment is often employed in laboratory studies of mass transfer kinetics in rectification and absorption processes and in situations where it is needed to minimize the effect of variations in the interphase contact specific area on the experimental results.

In order to develop a method for calculating the efficiency of mass transfer in rectification in GIAP film type equipment with flanged sheet packing wetted from one side as a function of the geometric dimensions of packing and the character of motion of the contacting phases, it was necessary to carry out experimental studies on the idealized model of a contacting element where the character of motion of the liquid and vapor phases and the kinetics of mass transfer coincide in the main with those peculiar to commercial equipment.

Such a model can be exemplified by a contacting element made in the form of an inclined wetted duct [18] (Figure 11). Bottom 1 and cover 3 of the duct have inclined flanges 2. The liquid film flows by gravity along bottom 1 of the duct. The vapor passes through slot 4 normally to flanges 2, from one flange to the other above the liquid film. Thus, the duct directs the liquid and vapor in cross-current, as is the case with the wetted elements of the sheet type packing in GIAP film equipment. The duct is nothing but an inclined slot channel with one wetted wall whose geometric parameters affect mass transfer efficiency.

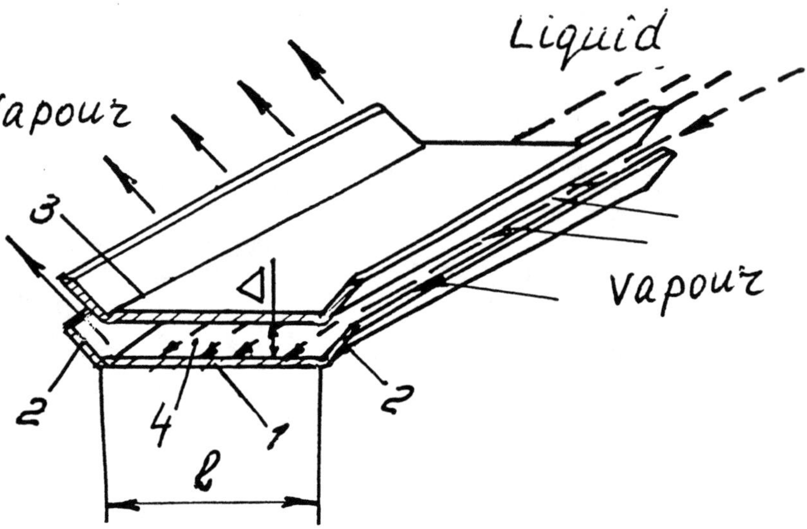

Figure 11. Model of a mass transfer contacting element with a cross-current of the gas and liquid film: 1 - bottom; 2 - inclined flanges; 3 - cover; 4 - slot.

Mass transfer in such a channel was studied on an experimental unit specially designed and mounted for atmospheric rectification experiments on model mixtures [18]. The unit included a still, a condenser, and a rectification section shown in Figure 12 in three projections. The section comprised a load-carrying frame 1 and two wings attached to it; the wings served for the supply (2) and withdrawal (3) of the vapor through inlets 4 and 5, respectively. Frame 1 had separable bottom 6 and lid 7 whose configuration enabled the formation of a slot channel with the wetted bottom wall. The conditions of vapor entry and withdrawal were the same as in the case of the duct shown in Figure 11. Because of the replaceability of bottom 6 and lid 7, a channel with any dimensions could be made, i.e. its height (Δ) or length (l) in the vapor flow direction could be altered.

Pipe 8 delivered the liquid, (heated to a temperature close to the boiling point), to bottom 6 of the rectification section; lower pipe 9 withdrew the liquid from the section. The section was slightly inclined so that the liquid film could flow freely from inlet 8 to outlet 9.

In order to keep watch on the state of the film surface in the course of rectification, the section was fitted with sight glasses 10 placed in pressure-tight sealants on both sides of frame 1.

The rectification section was mainly intended for studying mass transfer kinetics on the cross-flow of the liquid and vapor.

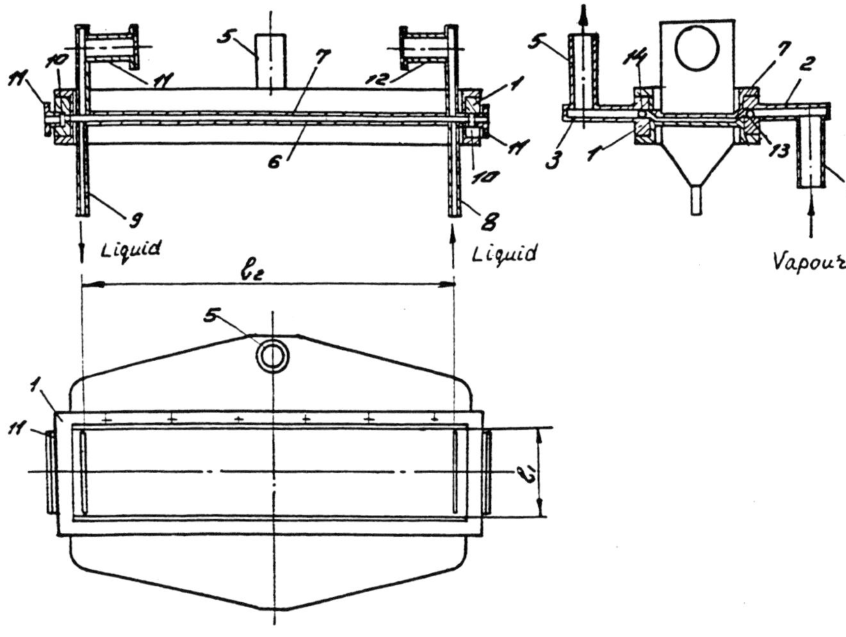

Figure 12. Rectification section: 1 - load-carrying frame; 2 - wing for the supply of the vapor to the working channel; 3 - wing for the withdrawal of the vapor from the working channel; 4 - vapor inlet for cross-current; 5 - vapor outlet for cross-current; 6 - bottom; 7 - lid; 8 - liquid inlet; 9 - liquid outlet; 10 - sight glasses; 11, 12 - vapor inlet and outlet for cocurrent and countercurrent conditions; 13, 14 - twisted metal mesh.

Afterwards, lid 7 was equipped with vapor tubes 11 and 12 in order to study mass transfer under cocurrent and countercurrent flow conditions. In this case, elastic tubes wrapped in a Teflon type film (not shown in Figure 12) were clamped between the beveled edges of lid 7 and bottom 6 in order to segregate the inner space of the slot channel and wings 2 and 3. Vapor pipes 4 and 5 were blanked off. The distance between tubes 8 and 9 (l_2) was taken as the path of the vapor flow moves counter- or cocurrent with respect to the liquid film.

In the case of cross-flow of the vapor and liquid, the width of the descending film (l_1) was taken as the length of the vapor flow. For the vapor flow to enter the slot with a uniform velocity throughout its length from tube 8 to tube 9, wire coils wrapped in a fine metal mesh (13, 14) offering small additional resistance to the vapor flow were installed in wings 2 and 3.

The rectification section was provided with external electric heating and had an asbestos sheet and cord insulation preventing vapor condensation on the walls and the formation of additional uncontrolled mass transfer zones.

It is well known that, in the general case, the intensity of mass transfer in rectification depends on the hydrodynamic state of both the vapor and liquid phases. The principle of additivity of phase diffusion resistance is illustrated by

$$\frac{1}{K_{oy}} = \frac{1}{\beta_y} + \frac{m}{\beta_x} \tag{4}$$

or

$$h_{tg} = h_g + m\frac{G}{L}h_l \tag{5}$$

where m is the phase equilibrium constant included in the relationship

$$Y^* = mX + const \tag{6}$$

where X is the concentration of the highly volatile component in the liquid phase;
Y* is the concentration of the highly volatile component in the vapor phase in thermodynamic equilibrium with the liquid with concentration X.

The physical meaning of the values $\frac{m}{\beta_x}$ and $\frac{1}{\beta_y}$ is the resistance of each phase to mass transfer. The total resistance to mass transfer is expressed by $\frac{1}{K_{oy}}$.

The data available in the literature [19] indicate that, on rectification in film type equipment, the resistance to mass transfer from the vapor phase ($1/K_{oy}$) prevails greatly over the contribution of the liquid phase. Thus, the paramount task is to investigate vapor phase mass transfer in a slot channel. One should also remember that the force interaction between the vapor flow and the liquid film at vapor velocities of up to 80-85% of the flooding limit is insignificant, making it possible to analyze mass transfer kinetics in each phase independently [19].

In order to harmonize the experimental data on film mass transfer for rectification in a slot channel with one wetted wall and the individual mass transfer or heat transfer data in a similar channel, steps were taken to exclude even a minor perturbing effect of the liquid phase on the overall mass transfer coefficient determined on rectification. To this end, such binary mixtures whose phase equilibrium constant m, (see Equation (6)), was low were chosen as model systems for the experiment. Thus, for an acetone-butanol mixture in the 60-100 mol.% acetone concentration range m=0.09 which means that in Equation (4) the m/β_x value becomes negligibly small as compared to $1/\beta_y$, i.e. $1/\beta_y >> m/\beta_x$.

Since m/β_x is negligibly small as compared to $1/\beta_y$, Equation (4) may be written as $1/K_{oy}=1/\beta_y$; hence $\beta_y=K_{oy}$, i.e. the individual mass transfer coefficient in the vapor phase on rectification of the acetone-butanol mixture in a film type column is numerically equal to the overall mass transfer coefficient, K_{oy}, which can be easily determined under experimental conditions.

The individual mass transfer coefficient in the liquid film, β_x, is known to depend on the flow rate of the liquid phase, i.e. on Re_l. Therefore, the absence of the dependence of $1/K_{oy}$ or $NTU_{tg}=l/h_{tg}$ on Re_l at a constant flow rate of the vapor phase is an indirect indicator of the prevailing role of the vapor phase in the resistance to mass transfer, according to Equation (4) or (5).

These types of experiments were run with acetone-butanol (m=0.09), methanol-butenol (m=0.15), hexane-benzene (m=0.8), acetone-benzene (m=0.64) mixtures on ducts with width l=60 and 122 mm, lid-to-bottom gap Δ=10 mm, and a 3° angle of inclination to the horizontal. Countercurrent, cocurrent, and cross-current schemes of motion of the vapor flow and the liquid film were employed [18].

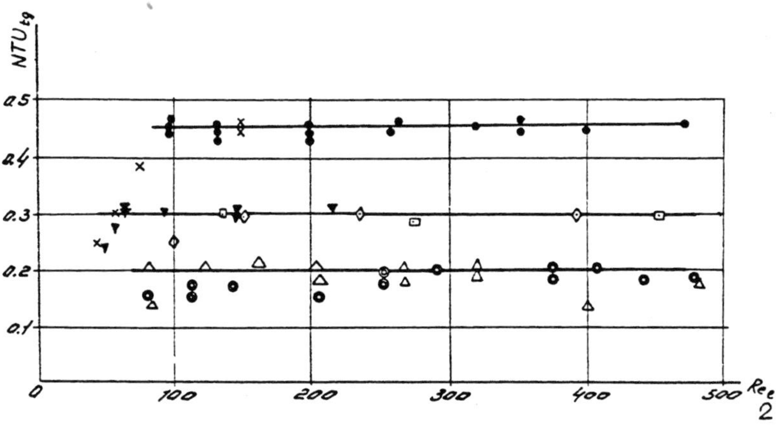

Figure 13. Dependence of NTU_{tg} on Re_l at Re_g=const in rectification of binary mixtures in a duct (slope angle 3°, gap Δ=0.01 m). Symbols: 1 - vapor flow path length, b - vapor flow width, ◎ - acetone-benzene, cocurrent flow, Re_g=11000, l=0.565 m, b=0.122 m; Δ - hexane-benzene, cocurrent flow, Re_g=11000-11500, l=0.565 m, b=0.122 m; t - methanol-butanol, cross current flow, Re_g=550, l=0.122 m, b=0.565 m; q - acetone-butanol, cross-current flow, Re_g=1540, l=0.122 m, b=0.565 m; l - acetone-butanol, countercurrent flow, Re_g=6500, l=0.565 m, b=0.06 m; u - acetone-butanol, cocurrent flow, Re_g=7000, l=0.565 m, b=0.06 m; 5 - acetone-butanol, cocurrent flow, Re_g=6700, l=0.565 m, b=0.122 m. 1 - NTU_{tg}; 2 - Re_l.

The experimental results are displayed in Figure 13. With Re_l changing by a factor of 4-5 NTU_{tg} (the number of transfer units) proved to be independent of Re_l with the exception of low feed rates. Strong reduction in NTU_{tg} at low feed rates (Re_l<60) can be explained by the appearance of ruptures or stagnation zones in the film.

Thus, the experiment supports the suggestion that $1/\beta_y >> m/\beta_x$, i.e. $\beta_y = K_{oy}$ is also the case where m is close to unity (hexane-benzene, acetone-benzene mixtures).

However, also chosen for further experiments aimed at elucidating the effect of geometric parameters of the wetted slot channel and other factors on the intensity of mass transfer in the vapor phase was the acetone-butanol mixture whose phase equilibrium constant m in the working concentration range had the lowest value. Among the factors mentioned above were the duct width, gap height, duct inclination angle, flow rate of the vapor phase and mutual direction of the flows of the phases in contact. Mass transfer in the vapor phase on rectification was studied under the conditions of laminar flow of the vapor phase. This choice was due to the fact that the GIAP design film rectification equipment with packing made of sheets with flanged edges, (one-wall wetting), are intended for vacuum rectification where the working load of the column is deliberately limited in order to reduce its hydraulic resistance without bringing it to packed flooding. Because of the comparatively small gaps which are usually designed for the vapor (Δ=5-12 mm), the Reynolds number of the vapor flow does not exceed 2300 i.e. the flow is laminar.

The mass transfer process in the vapor phase on rectification in a slot channel with one wetted wall can be regarded as similar to the heat transfer process in a flat channel where one wall is heated and the other wall is heat-insulated, the distribution of velocities and temperatures at the inlet are uniform and the boundary conditions are of the first kind (for a specified distribution of the wall temperature, i.e. the temperature at the gas/wall interface along the gas flow).

Based on the analogy between the heat and mass transfer processes in a flat channel and on the experimental heat transfer data [20] for a flat channel, preliminary conclusions about the character of the mass transfer process in the duct can be made.

In mass transfer experiments with a cross-current motion of phases, the parameters were changed in the following ranges: gap $\Delta=(2.5\text{-}12)\times10^{-3}$ m, duct width l=0.03-0.122 m, inclination to the horizontal 0-20°, Reynolds number of the vapor phase Re_g=100-2000. The duct length was the same in all tests (0.565 m). However, taking into account the fact that mass transfer in the duct under countercurrent, cocurrent, and cross-current conditions was modeled by changing the direction of the vapor flow, the channel length, (along the vapor flow), can be regarded as changing from 0.03 to 0.565 m.

The geometric parameters of the slot channel in heat-mass transfer are usually included in the form of a simplex l/d_e which in our experiments ranged between 1.25 and 80 (here $d_e=2\Delta$) [20].

The relative length of the initial hydrodynamic section of a flat tube (l_{it}) where the effect of the initial velocity distribution on the flow vanishes is, according to the theory [20]

$$\frac{l_{it}}{d_e} = 0.113\,\mathrm{Re}_g\,. \tag{7}$$

In isothermal flow, (which is the case in the rectification under discussion), the initial section was regarded as the interval where the value of the axial velocity differed no more than by 1% from the axial velocity of stabilized flow.

Since, under the experimental conditions Re_g was changed in the 100-2000 range, the relative length of the initial hydrodynamic section varied between 1.13 and 22.6, respectively. In the cross-current case the simplex l/d_e changed from 1.25 to 17. Thus, in the case where the vapor flow was directed from one flange to the other end and the liquid-flowed in the form of a film down the duct grade, (cross-current flow), mass exchange took place in the initial hydrodynamic section.

In the case of a cocurrent or countercurrent flow in the duct (l/d_e=16-80), with the vapor moving along the duct, the initial hydrodynamic section extended over the whole length of the duct only in the experiments with the maximum values of the gap (Δ) and Re_g. With minimum gaps and vapor mass velocities, the initial hydrodynamic section constituted an insignificant proportion of the whole length of the duct. Thus, at $\Delta=2.5\times10^{-3}$ m and Re_g=100 the length of the section was only 1.5% of the whole path of the vapor flow.

The experimental data are presented in Figure 14 in the form of a graph with complex $\frac{1}{\mathrm{Re}_g}\bullet\frac{l}{d_e}$ laid off as absissa and Shg as ordinate.

This method of data processing is usually used in studying heat transfer kinetics in flat and round channels [20]. The solid sloping line in Figure 14 refers to mass transfer in the initial (short) section of hydrodynamic stabilization according to the equation

$$\frac{Sh_g}{Sc_g^{1/3}} = 0.664\left(\frac{1}{\mathrm{Re}_g}\bullet\frac{l}{d_e}\right)^{-1/2} \tag{8}$$

obtained by analogy with the corresponding equation for heat transfer in a short section [20]. In plotting the sloping line from Equation (8), it was assumed that Sc_g=1. The solid horizontal line depicts mass transfer in a long section where the velocity profiles differ slightly from parabolic and Sh_g=4.86 for all values of Sc_g. Thus, the solid lines represent two limiting cases (very short vs. very long section) while the intermediate situations (medium length sections) should be visualized as a smooth line touching both straight lines.

The experimental points obtained for countercurrent, cocurrent, and cross-current conditions mostly on the acetone-butanol mixture are marked in Figure 14 (m=0.09); in this case the diffusion resistance of the liquid phase could be neglected. Some series of experiments were run on other mixtures: methanol-butanol (m=0.15) and n-heptane-ethyl benzene.

It was expected that, with the value of the complex $\frac{1}{Re_g} \bullet \frac{l}{d_e}$ decreasing, the experimental points referring to mass transfer in the initial section, i.e. in narrow ducts at large Re_g values (however, not exceeding 2300), would be located closer to the solid sloping line and at large values of the complex $\frac{1}{Re_g} \bullet \frac{l}{d_e}$ would fall on the solid horizontal line (Sh_g=4.86).

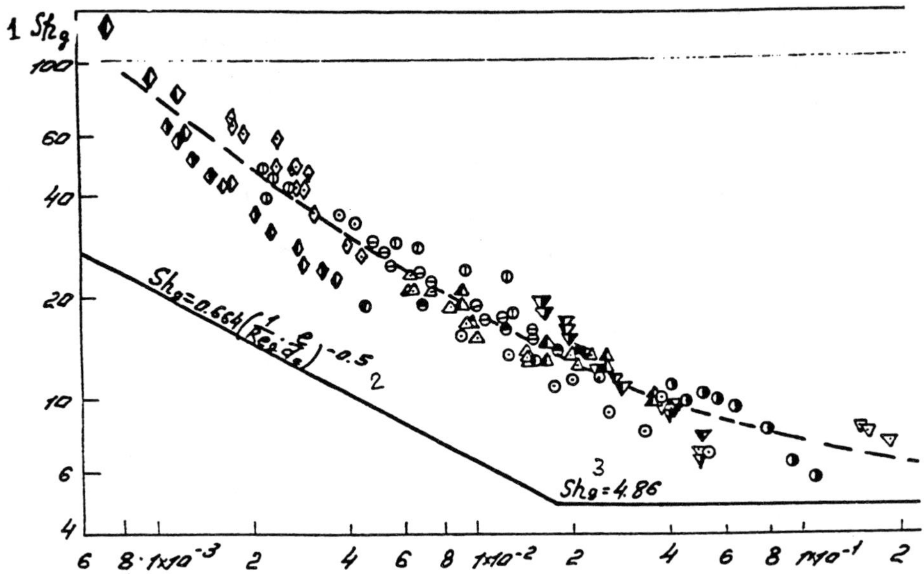

Figure 14. Dependence of Sh_g on the complex $\frac{1}{Re_g} \bullet \frac{l}{d_e}$ in rectification in a flat slot channel with one wetted wall under cross-current, cocurrent, and countercurrent flow conditions. Symbols: same as in Figure 18. 1 - Sh_g; 2 - Sh_g=0.664 ($\frac{1}{Re_g} \bullet \frac{l}{d_e}$)-0.5; 3 - Sh_g=4.86; 4 - $\frac{1}{Re_g} \bullet \frac{l}{d_e}$.

However, in Figure 14, the experimental points at lower values of the complex deviate more and more upwards from the sloped line which seems to contradict the theory and practice of mass transfer on a short flat plate. This fact in itself deserves a positive appreciation since it is indicative of a high intensity of mass transfer exceeding 3-5 times the expected theoretical values; however, explanations are needed.

It seems that secondary flows or Gertler eddies intensifying mass transfer arise in the vapor phase. The generation of eddies is due to the fact that the vapor flow enters the working slot and leaves it at an angle of 45° owing to the effect of the inclined edges of the duct (Figure 11,12). In other words, the vapor inside the slot follows a curved trajectory. The less the value of the complex $\frac{1}{Re_g} \bullet \frac{l}{d_e}$ the higher Re_g and the linear velocity of the vapor flow will be

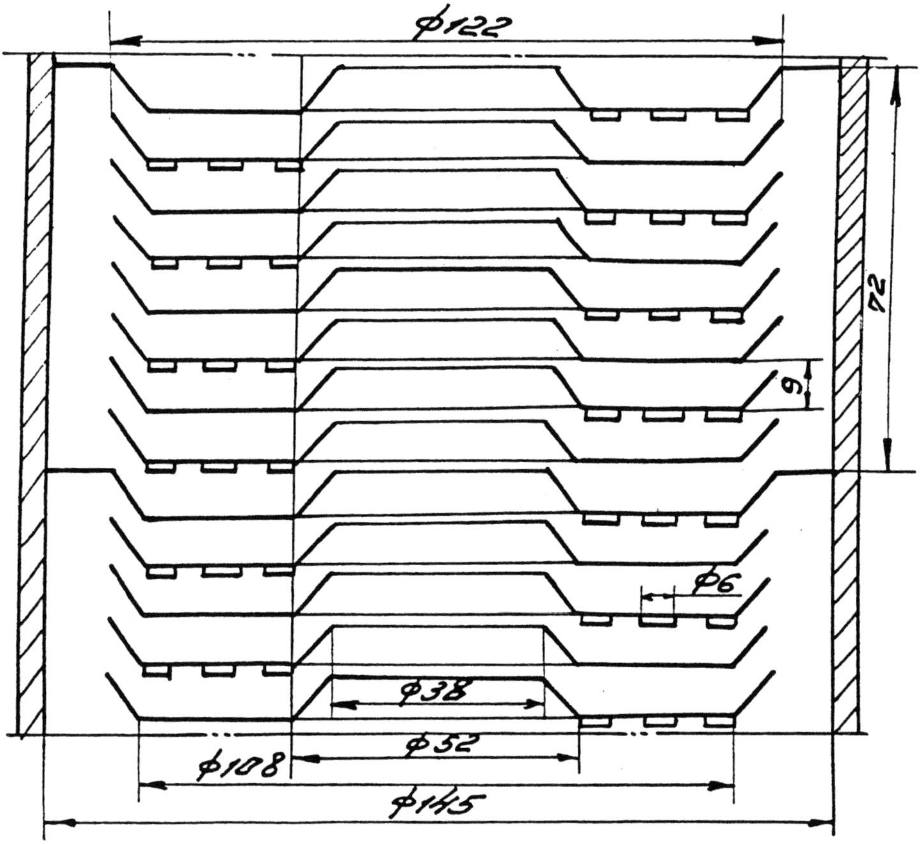

Figure 15. Design and dimensions of regular packing made of horizontal annular flanged plates and tested on a laboratory vacuum unit.

which will entail a more vigorous development of secondary motions in the vapor phase and a more intense mass transfer. At small linear velocities, the bending of the flow does not disrupt its stability and, in the limit, at large values of the complex $\frac{1}{\mathrm{Re}_g} \bullet \frac{l}{d_e}$ mass transfer in the duct is virtually indistinguishable from mass transfer in a long channel for which the expression $Sh_g=4.86$ is valid.

All atmospheric rectification data obtained for different mixtures using various patterns of contact between the vapor flow and the liquid film in a wetted channel with dimensions altered in a broad range fall on the graph in Figure 14 with a certain scatter along the general curve.

In order to resolve the problem of the applicability of the data obtained on a wetted slot channel to the calculations of GIAP film type equipment with a regular wetted packing (Figures 3-6), a large-scale vacuum rectification unit was fabricated and installed for the purpose of studying mass transfer in a film type column of a similar design.

The inner structure of the rectification section is shown in Figure 15. The shell of the section houses the stacks of wetted plates of annular shape with inner and outer flanges. The plate spacing (i.e. the gap) was 9 mm, the width of a plate in its wetted part was 28 mm. The annular plates were stamped out of 0.5 mm sheet. The bottom of each annular plate had 6 mm holes for liquid drainage. In order to prevent the wetting of the lower surface of the plate's bottom, the holes were flanged down. Each plate had three rows of holes, three holes in a row (Figure 15). The rows of holes were located on each plate at an angle of 120° to each other. The holes in the neighboring plates were shifted by 60° to each other.

The vapors pass through the gaps between the plates making contact with the liquid film covering the bottom of each plate. The vapor flow moving in a certain direction, (towards the axis or the shell of the column), is divided, on crossing the stack, into four thin parallel layers 9 mm thick and 28 mm wide.

Rectification under a residual pressure of 3-5 Torr in the top portion of the rectification section was run on a m-cresol-diethylene glycol binary mixture used as a model system. The Reynolds number in the experiments was changed in the 100-2000 range; thus, the motion of the vapor phase was laminar.

Rectification under atmospheric pressure and a residual pressure of 50 Torr was performed on an n-octane-ethyl benzene mixture.

The experimental data were processed, as in the single channel case, in $\frac{1}{\mathrm{Re}_g} \bullet \frac{l}{d_e}$ vs. Sh_g coordinates. It follows from Figure 14 that the value of the residual pressure in itself had virtually no influence on the efficiency of mass transfer in the vapor phase in the range studied: it will suffice to compare the results of runs on a n-octane-ethyl benzene mixture at 760 and 50 Torr.

These data and the results of rectification experiments on a m-cresol-diethylene glycol mixture at a residual pressure of 3-5 Torr are in good mutual agreement and correlate well with the experimental data obtained on a single slot channel using other binary mixtures.

For the complex $\frac{1}{\mathrm{Re}_g} \bullet \frac{l}{d_e}$ changing in the 0.001-0.030 range, as is the case with rectification in GIAP design commercial columns, the experimental data on mass transfer efficiency can be described by the equation [21]

$$\frac{Sh_g}{Sc_g^{1/3}} = 0.78\left(\frac{1}{\mathrm{Re}_g} \bullet \frac{l}{d_e}\right)^{-0.7} \tag{9}$$

neglecting the resistance to mass transfer due to the liquid phase.

5. Studies of the Effect of Mechanical Phase Mixing on Mass Transfer in a Slot Channel with One Wetted Wall

It is convenient to study mass transfer in artificially mixed thin layers taking account of all factors affecting the rectification process on an example of such a model which makes it possible to simplify the analysis of the mass transfer process and possess the distinctive features of rotor film equipment. This model is a tubular film column with a coaxial cylindrical rotor forming an annular gap with the wetted inner surface of the column.

A rectification unit intended for experiments with this model was fabricated and installed [22].

The schematic drawing of the unit is shown in Figure 16.

The main parts of the unit: still I, column II, and condenser III are made of stainless steel. Column II was installed in a strictly vertical position. Rotor 5 of the column was made in the form of a hollow smooth cylinder having in its top part an annular pocket with distribution pipes 6. Column II was equipped with electric compensation heater 3 ensuring the adiabatic conditions of rectification. All parts of the unit where heat losses into the environment were possible were fitted with asbestos sheet and cord insulation.

Operation of the unit. Vapor from still I entered column II. The vapor phase was intensely mixed by rotating rotor 5 in the annular gap of the column between its inner surface covered with the liquid film and the rotor. The vapors enriched in the highly volatile component passes from column II into condenser III. The condensate formed on the surface of cooling coil 9 and the inner surface of jacket 10 ran down into annular pocket 8. The condensate left pocket 8 through glass hydraulic seal 14, passed calibrated glass flow gauge 15, and entered metal hydraulic seal 17 equipped with an electric heater and thermometer 16. In hydraulic seal 17 the liquid was heated to a temperature 2-3°C below its boiling point. The reflux was fed through pipe 18 into the annular pocket rotating together with rotor 5. The reflux passing through radial tubes 6 was uniformly distributed throughout the upper part of the inner

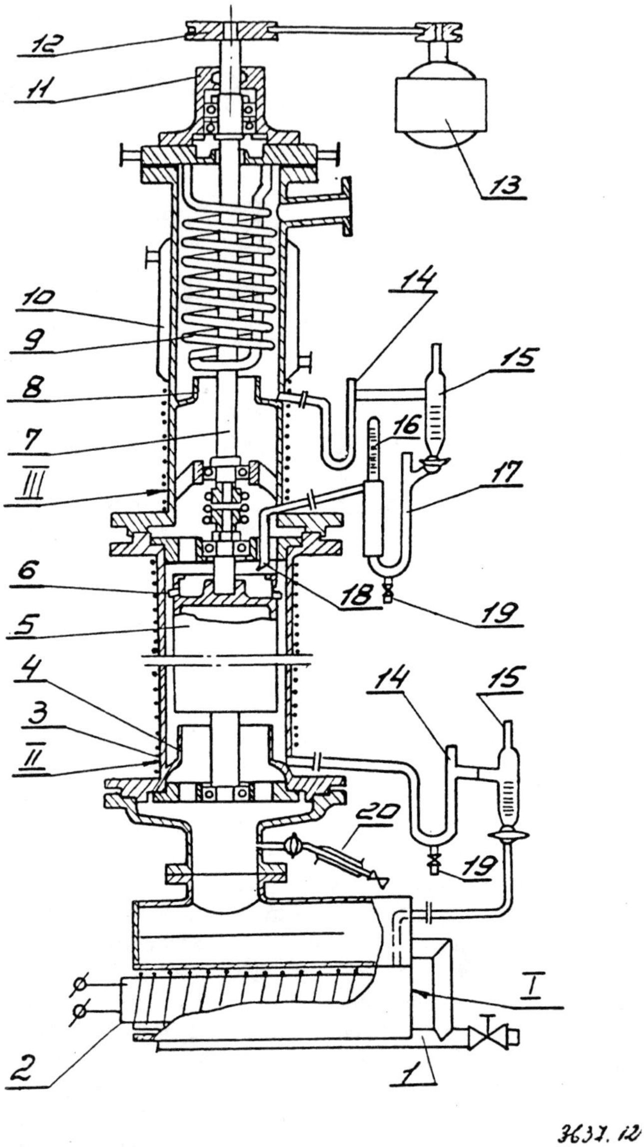

Figure 16. Schematic drawing of a unit with a column 0.1205 mm in diameter: I - still: 1 - recirculation line; 2 - electric heater; II - column: 3 - compensation heater; 4 - annular pocket; 5 - rotor; 6 - distribution pipes; III - condenser: 7 - drive shaft; 8 - annular pocket; 9 - cooling coil; 10 - cooling jacket; 11 - gland seal; 12 - pulley; Other components: 13-AO 314 electric motor, 0.6 kW, n=1410 rpm; 14 - glass hydraulic seal; 15 - calibrated vessel; 16 - mercury thermometer; 17 - metal hydraulic sea; 18 - tube; 19 - liquid sampler; 20 - vapor sampler.

surface of column II and flowed down the surface entering pocket 4. The liquid left the column through glass hydraulic seal 14, calibrated gas flow gauge 15, and entered still I.

The integrity of the film in column II was ensured even at low feed rates through forced distribution by the rotor at the inlet. The presence of the rotor mechanically affecting the phases in contact made it possible to alter the ratio of the phase mass transfer coefficients by changing the rotation speed. Of particular interest is a comparison of the rectification and absorption processes in the columns of this type enabling a more detailed analysis of mass transfer separately in the gas and liquid phases. For this reason, an absorption unit with a tubular film column was installed in addition to the rectification unit [23]. Hydrodynamic studies were also carried out on the absorption unit: the resulting directions of motion of the liquid film were measured. In this case the experimenters used a method based on the measurement of the deviation of a thread from the vertical position on its immersion into the film running down the wall of a stationary cylinder and featuring the development of a rotary motion [24].

The results of some representative sets of tests are shown in Figure 17 in the coordinates linear velocity of the surface of the cylinder vs. total height of a transfer unit (in mass transfer).

Curve 2a refers to vacuum rectification of a binary mixture n-heptane-methyl cyclohexane in a column with the diameter 0.12 m, height 0.44 m, and gap 0.003 m at an infinite reflux ratio and a residual pressure of 26.6 kPa (Re_g=600, $\nu_g=7.52\times10^{-6}$ m^2/s) has two sharp bends. The first bend (at U=2.7 m/s) is due to the formation of the Taylor eddies [22] in a laminar flow of the vapor phase enhancing its mixing which strongly intensifies vapor phase mass transfer. The critical Taylor number corresponding to gas (vapor) flow destabilization can be calculated from the formula [25]:

$$Ta_{cr} = \frac{71.6}{(1-0.865\Delta / r_m)(\Delta / r_m)^{0.38}}. \tag{10}$$

The second bend on curve 2a (at U=11.4 m/s) refers to the beginning of a very strong interaction between the rotating annular layer of the vapor phase and the liquid film in which the Taylor eddies arise. The critical velocity can be calculated from the empirical formula [24]

$$U_{cr} = 0.0181^{-0.177/\mu_x}. \tag{11}$$

The U_{cr} value is virtually unaffected by the gap width and the physical properties (density, viscosity) of the gas phase.

In ammonia absorption by water under atmospheric pressure in a column with the diameter 0.15 m, height 0.60 m, and gap 0.003 m at Re_l=200 and Re_g=1035 (curve 1b), the appearance of the Taylor eddies in the gas phase at U=5.7 m/s (see Equation (10)) intensifies mass transfer, i.e. h_{tg} decreases upon a further increase of the peripheral velocity of the rotor. However, this is not the case in CO_2 absorption by water (curve 1a) regardless of the identity of the hydrodynamic conditions in the gas phase in these cases. From this it follows that in ammonia absorption by water the resistance to mass transfer is due to the gas phase ($K_{cy} \approx \beta_y$) whereas in CO_2 absorption by water the gas phase does not control mass transfer ($K_{ox} = \beta_x$).

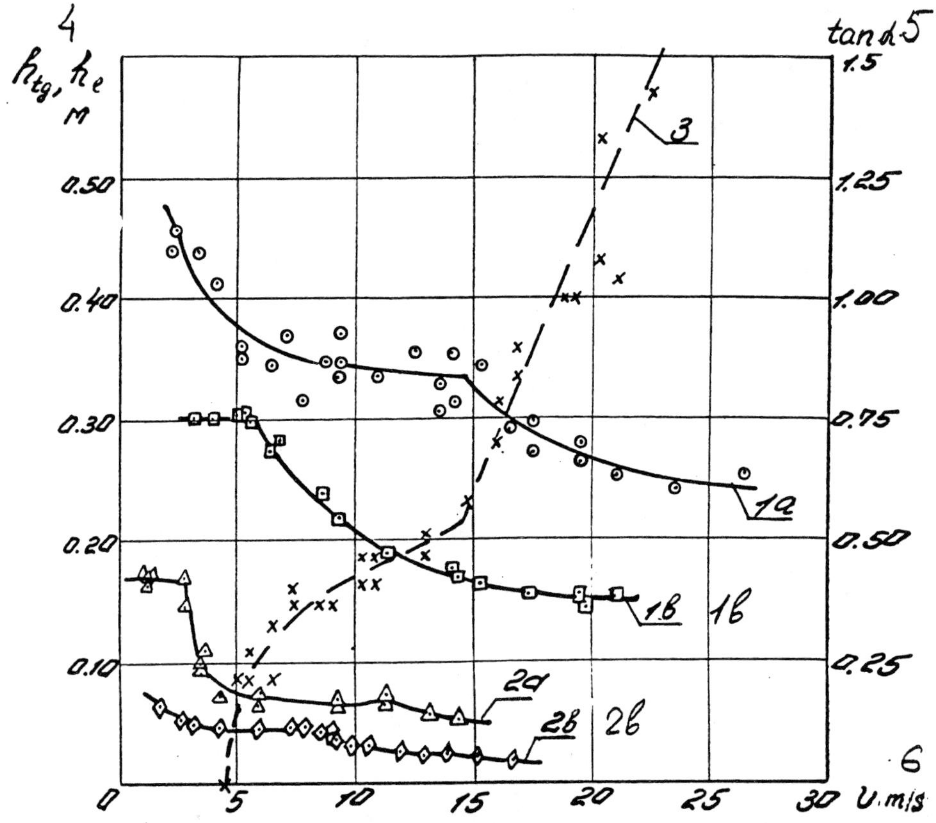

Figure 17. Effect of the peripheral velocity of the rotor, U, in columns comprising two vertical coaxial cylinders on h_x, h_{cy} in CO_2 (curve 1a) and ammonia (curve 1b) absorption by water under atmospheric pressure in a column with diameter d_c=0.15 m, height l=0.60 m, and gap Δ=0.003 m for Re_x=200 and Re_y=1035, on h_{cy} in rectification of a binary mixture n-heptane-methyl cyclohexane in a column with diameter d_c=0.120 m, height l=0.44 m, and gap Δ=0.003 m at an infinite reflux ratio under a residual pressure of 26.6 kPa for Re_y=600 and $v_g = 7.52 \times 10^{-6}$ m^2/s (curve 2a) and under atmospheric pressure for Re_y=500. $v_g = 2.43 \times 10^{-6}$ m^2/s (curve 2b), on the tangent of the swirl angle (tanα) of a film in a column with diameter d_c=0.15 and gap Δ=0.0058 m at distance l=0.27 m from the beginning of the film for Re_x=200 (curve 3). 4 - h_{tg}, h_l; 5 - tanα; 6 - U, m/s.

Curve 3 in Figure 17 depicts the dependence of the tangent of the swirl angle ($\tan\alpha$) of the film in a column with the diameter 0.15 m and gap 0.0058 m at a distance of 0.27 m from the beginning of the film at Re_l=200 on the peripheral velocity of the rotor.

The presence of characteristic bends in curves 1a and 3 (at U_{cr}=14.6 m/s, Equation (11)) and its absence in curve 1b, regardless of the identity of the hydrodynamic conditions in all three cases under discussion, indicates that a strong enhancement of mixing in the film caused by the appearance of the Taylor eddies intensifies mass transfer if it is controlled by the liquid phase, as in CO_2 absorption by water ($K_{ox}=\beta_x$) and has no effect on mass transfer efficiency when the major contribution to the resistance to mass transfer is made by the gas phase ($h_{tg}=h_g$).

The presence of both bends in curve 2a attests to the fact that the resistance to mass transfer in the annular gap with the inner rotating cylinder in rectification is controlled, in the general case, by both phases.

Curve 2b, refers to rectification of a binary mixture n-heptane-methyl cyclohexane in a column with a diameter of 0.12 m, height 0.44 m, and gap 0.003 m as does curve 2a. In this case rectification was run under atmospheric pressure (Re_g=500, $\nu_g=2.43\text{x}10^{-6}$ m^2/s.

A comparison between curves 2a and 2b shows that in film rectification in an annular gap with a rotating inner cylinder, the intensity of mass transfer decreases at lower residual pressures. This is explained by changes in the physico-chemical properties of the phases in contact caused by pressure and temperature changes; this affects Ta_{cr} and U_{cr} and hence, the intensity of phase mixing. The absence of a bend in curve 2b due to the formation of the Taylor eddies in the vapor phase is explained, according to the results of calculations from Equation (10), by its shift towards the zone of lower U values not covered by the experimental conditions.

The following equation was proposed for the description of mass transfer in the gas (vapor) phase in rectification in an annular gap with a rotating inner cylinder after the emergence of the Taylor eddies [26]

$$Sh_g = \frac{\beta_g d_e}{D_g} = 0.64\,\mathrm{Re}_e^{1/3}(1+18\frac{d_e}{l})(1+1.95\frac{Ta^{1/3}-Ta_{cr}^{1/3}}{Ta^{1/3}})Sc_g^{1/3}. \quad (12)$$

Equation (12) was derived as a result of experimental data processing; the major parameters of the atmospheric rectification process and the dimensions of the model were changed in a broad range: 0.032 m<d_c<0.1205 m; $2.57\text{x}10^{-3}$<d_e/l<$7.7\text{x}10^{-3}$; $1\text{x}10^{-3}$<Δ<$5\text{x}10^{-3}$ m; 0.125 m<l<1.165 m; 2 m/s<U<9 m/s; 150<Ta<7110; 11<Re_l<130; 300<Re_g<3000; 1060<Re_e<9250.

The mixture used: CCl_4-C_6H_6, acetone-C_6H_6, n-hexane-C_6H_6, CCl_4-cyclohexane, n-heptane-methyl cyclohexane.

Vacuum rectification experiments run on the same unit (Figure 16) and the experiments on ammonia absorption by water and hydrochloric acid carried out, as a continuation of these studies, in a tubular film column with the diameter 0.15 m containing a coaxial cylindrical rotor [27] also corroborated the validity of Equations (10) and (12) in vacuum rectification and absorption processes. Moreover, it was shown that Equation (12) adequately describes mass transfer in the gas (vapor) phase even at higher peripheral velocities of the rotor surface, (up to 20 m/s), featuring a very strong hydrodynamic effect of the rotating annular layer of the gas phase on the liquid film undergoing strong swirling. A further increase in the rotation speed of the rotor is not advisable because of a strong heat release in the gap due to the friction of the rotor against the gas phase leading to a substantial evaporation of the liquid film in the course of rectification and absorption and disrupting the adiabatic character of the process.

However, Equation (12) cannot be used for a quantitative description of mass transfer in the gas phase in the absence of the Taylor eddies, i.e. at $Ta<Ta_{cr}$. It is an experimentally established fact (Figure 17, curves 1b and 2a) that, in this case, rotation has no effect on mass transfer in the gas (vapor) phase, its role is actually being reduced to a uniform distribution of the liquid in the upper part of the wetted tube.

What was proposed for this case was the equation [28]

$$\frac{Sh_g}{Sc_g^{1/3}} = 0.6\,\mathrm{Re}_g^{1/3}\left(1+35\frac{d_e}{l}\right) \qquad (13)$$

obtained as a result of the processing of the experimental data for ammonia absorption by a weak (1N) hydrochloric acid solution in a column with the diameter 0.15 m and gap $\Delta=0.003$ m at the linear velocity of the surface of a cylindrical rotor U=5.1 m/s; the height was 0.11, 0.22 and 0.60 m; $\mathrm{Re}_l=100$.

These data can be processed from a different standpoint, considering the model as a slot channel with one flat wetted wall since, in the absence of the Taylor eddies, the curvature of the wetted wall does not exert any noticeable influence on the efficiency of mass transfer. Therefore, it is advisable to subject the experimental data obtained in ammonia absorption by a hydrochloric acid solution in the absence of the Taylor eddies to the same type of mathematical processing as in the experiments on mass transfer in rectification of an acetone-butanol mixture in a slot channel with one wetted wall, especially when the resistance to mass transfer is due to the gas phase in both cases.

In Figure 18 the experimental data on ammonia absorption by a hydrochloric acid solution in a laminar regime in the absence of the Taylor eddies [23] are plotted in the $\mathrm{Re}_g \frac{d_e}{l}$ vs. $\frac{Sh_g - 4.86}{Sc_g^{1/3}}$ coordinates. The graph also contains the data for mass transfer in a slot channel with one wetted wall obtained in rectification of an acetone-butanol mixture under cross-current, countercurrent, and cocurrent conditions already used in plotting the graph in Figure 14. The coordinate system was chosen based on the assumption that in

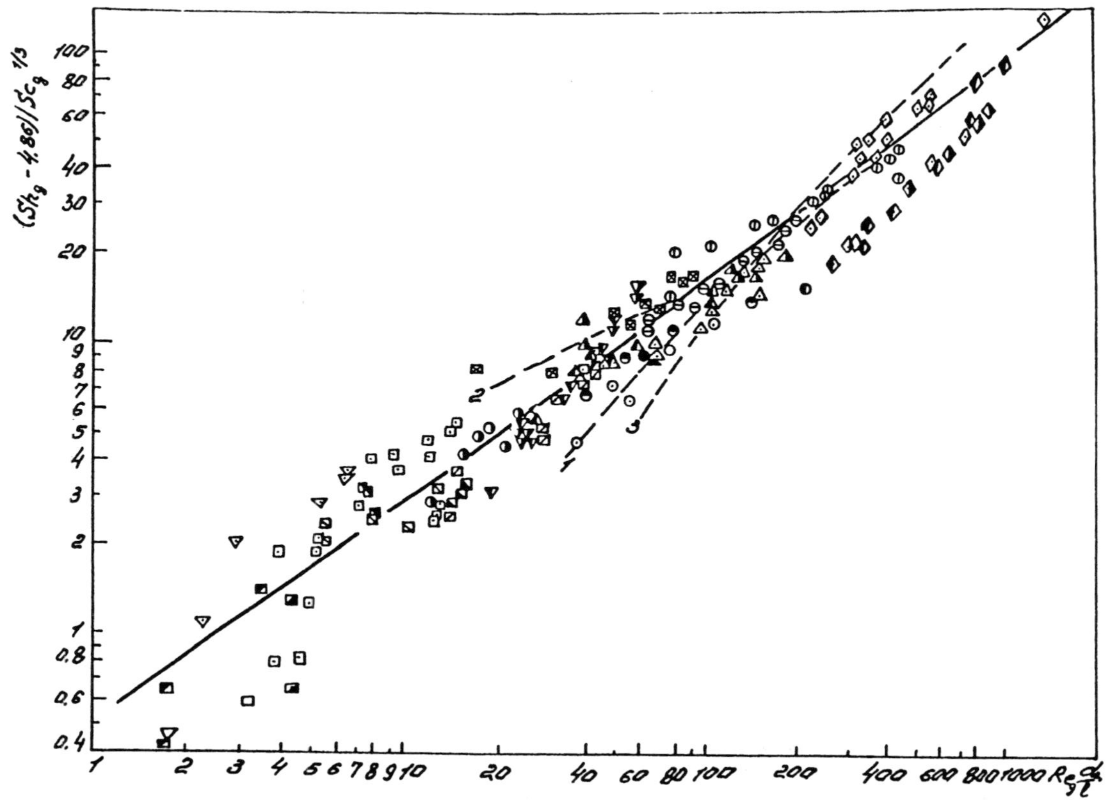

Figure 18. Effect of gas phase velocity and geometric dimensions on mass transfer efficiency in the gas phase in rectification and absorption in slot channels with one wetted wall. A duct with a 0.565 m length and a 0.122 m maximum length; atmospheric rectification at an infinite reflux ratio (Figure 12):

▽ - acetone-butanol (v_g=3.90x10^{-6} m^2/s, D_g=5.11x10^{-6} m^2/s, Sc_g=0.763), cocurrent flow, Δ=0.001 m, l=0.565 m, d_e/l=3.54x10^{-3}, angle 3°;

▼ - acetone-butanol, countercurrent flow, Δ=0.0055 m, l=0.565 m, d_e/l=1.95x10^{-3}, angle 3°; Δ - acetone-butanol, cross-current flow, Δ=0.012 m, l=0.122 m, d_e/l=1.97x10_{-3}, angle 3°;

▲ - n-heptane-ethyl benzene (v_g=2.46x10^{-6} m^2/s, Δ_g=3.475x10^{-6} m^2/s, Sc_g=0.708), cross-current flow, Δ=0.012 m, l=0.122 m, d_e/l=1.97x10^{-1}, angle 3°;

▲ - methanol-butanol (v_g=8.62x10^{-6} m^2/s, D_g=7.9x10^{-6} m^2/s, Sc_g=1.09), cross-current flow, Δ=0.012 m, l=0.122 m, d_e/l=1.97x10^{-1}, angle 3°;

○ - acetone-butanol, cross-current flow, Δ=0.0062 m, l=0.122 m, d_e/l=1.02x10^{-1}, angle 3°;

◑ - acetone-butanol, cross-current flow, Δ=0.003 m, l=0.122 m, d_e/l=4.92x10^{-2}, angle 9°;

◐ - acetone-butanol, cross-current flow, Δ=0.011 m, l=0.122 m, d_e/l=1.82x10^{-1}, angle 9°; -

◒ -acetone-butanol, cross-current flow, Δ=0.011 m, l=0.122 m, d_e/l=1.82x10^{-1}, angle 23°;

⊖ -acetone-butanol, cross-current flow, Δ=0.008 m, l=0.060 m, d_e/l=0.27, angle 3°;

⦶ -acetone-butanol, cross-current flow, Δ=0.007 m, l=0.030 m, d_e/l=0.46, angle 3°;

Packing made of horizontal annular flanged plates, Δ=0.009 m, l=0.028 m, d_e/l=0.643(Figure 15):

◊ - n-octane-ethyl benzene, pressure 760 Torr (v_g=2.51x10^{-6} m^2/s, D_g=3.44x10^{-6} m^2/s, Sc_g=0.73);

◆ - n-octane-ethyl benzene, residual pressure 50 Torr (v_g=2.56x10^{-5} m^2/s, D_g=3.34x10^{-5} m^2/s, Sc_g=0.766);

◊ - m-cresol-diethylene glycol, residual pressure 3-5 Torr (v_g=2.87x10^{-4} m^2/s, D_g=4.32x10^{-4} m^2/s, Sc_g=0.665).

Annular gap; rotating inner cylinder (Figure 16):

⊠ - absorption of ammonia from air by hydrochloric acid solution, 760 Torr (v_g=1.49x10^{-5} m^2/s, D_g=1.89x10^{-5} m^2/s, Sc_g=0.79), Δ=0.003 m, l=0.11 m, d_c=0.15 m, U=5.1 m/s;

◩ - same, l=0.22 m;

⧅ - same, l=0.60 m;

◪ - rectification, P=200 Torr, n-heptane-methyl cyclohexane (v_g=7.52x10^{-6} m^2/s, D_g=10x10^{-6} m^2/s, Sc_g=0.752), Δ=0.00125 m, l=0.44 m, d_c=0.1205 m, U=6.17 m/s;

◩ - same, Δ=0.00295 m, U=1.68 m/s;

□ - rectification, P=100 Torr, chlorobenzene-ethyl benzene, (v_g=17.8x10^{-6} m^2/s, D_g=22.3x10^{-6} m^2/s, Sc_g=0.8), Δ=0.00295 m, l=0.44 m, d_c=0.1205 m, U=2.58 m/s;

◪ - rectification, CCl_4-benzene, 760 Torr (v_g=2.67x10^{-6} m^2/s, D_g=4.63x10^{-6} m^2/s, Sc_g=0.564), Δ=0.001 m, l=0.105 m, d_c=0.032 m, U=0.4-2.4 m/s;

◩ - rectification, CCl_4-benzene, 760 Torr, Δ=0.00125 m, l=0.44 m, d_c=0.1205 m, U=2.18 m/s. 4 - $(Sh_g-4.86)/Sc_g^{1/3}$.

the limiting case, at $Re_g \frac{d_e}{l} \rightarrow 0$, i.e. in mass transfer in an infinitely long channel, the equality $Sh_g=4.86$ will be valid, by an analogy with heat transfer. Also, pushing the analogy further, an account was taken of the fact that the value of the Nusselt number, Nu, on heat transfer in an infinitely long channel is independent of the Prandtl number, Pr, while in a heat transfer in an infinitely short channel (Nu>>4.86) the proportionality $Nu \sim Pr^{1/3}$ is justified. Therefore, all intermediate cases have one common feature: as the value of the complex $Re \frac{d_e}{l}$ decreases, the effect of Pr on Nu becomes gradually less pronounced: from $Nu \sim Pr^{1/3}$ to $Nu \sim Pr^0$ [20].

Figure 18 also presents the experimental data for the rectification of various binary mixtures in an annular slot with a rotating inner cylinder under laminar flow conditions (no Taylor eddies) for the vapor phase [28]. The calculation of β_y made in an earlier study [23] was based on the experimentally obtained K_{cy} values taking account of the liquid phase diffusion resistance determined in such models from

$$\frac{\beta_l^v}{D_l} = 0.00345\, Re_l\, Sc_l^{0.5} \tag{14}$$

which holds true for the peripheral velocities of the cylindrical rotor being $2\ m/s \le U \le U_{cr}$ [22].

A calculation [23] shows that the diffusion resistance to mass transfer in this regime is mostly due to the vapor phase.

The plotted experimental points for mass transfer in the gas phase in absorption and rectification in an annular gap with a rotating inner cylinder do not differ from the set of points for mass transfer in rectification in an inclined slot channel with one wetted wall. Thus, there are grounds to believe that the kinetics of mass transfer in the gas phase in different cases discussed here obeys identical laws.

Curve 1 in Figure 18 represents Equation (9) proposed earlier [21] for the description of vapor phase mass transfer in an inclined slot channel with a wetted wall for a cross-current flow of the phases. Curve 1 was calculated for an acetone-butanol mixture ($Sc_g=0.763$).

Curve 2 in Figure 18 represents Equation (13) for the absorption of ammonia by a hydrochloric acid solution in a tubular film column with a rotating cylinder with the height l=0.11 m; the equivalent diameter of the annular gap was $d_e=0.006$ m.

Generally speaking both curves (1 and 2), if not coinciding, do not contradict the suggestion about the identity of the laws governing the mass transfer processes in the vapor (gas) phase in an annular gap with a rotating cylinder and in a flat slot channel which can be generalized as follows

$$Sh_g = 0.49\,(\mathrm{Re}_g \frac{d_e}{l})^{0.77} Sc_g^{0.33} + 4.86. \tag{15}$$

The conditions of motion of the gas and liquid phases in the gap between the neighboring wraps of the helical band in a rotor film column shown in Figure 7 are similar to those for a flat channel with one wetted wall shown in Figure 11. In both cases the cross-current flow of the phases occurred. In the rotor column the flow of the phases in contact is affected by the rotary motion of the flanged coiled bands. However, the rotation does not give rise to the Taylor eddies in the vapor phase since there is no relative motion between the wetted and nonwetted walls of the slot channel formed by the neighboring wraps of the coiled band. On the other hand, the liquid on the rotating coiled bands moves not by gravity, but under the action of centrifugal force; thus, the hydrodynamic conditions of motion of the liquid film are quite different. Nevertheless, this should not affect the hydrodynamic features of motion of the gas phase in the same gap. Therefore, it would be of interest to compare the data in Figure 18 with the equation for mass transfer in the gas phase

$$Sh_g = 7.4x10^{-2}\,\mathrm{Re}_g^{0.92}\,Sc_g^{0.33} \tag{16}$$

derived [2, p. 192] on the basis of the results of studies of NH_3 and CO_2 absorption by water and rectification of an acetone-CCl_4 mixture in a rotor column 250 mm in diameter (l=0.04 m, d_e=0.01 m).

K_{cy} values measured on NH_3 absorption by water and rectification and β_x values measured on CO_2 absorption by water under identical conditions of the motion of the liquid phase were used, by substituting them in the additivity equations for phase resistance (4) or (5), in calculating the β_y values necessary for the determination of Sh_g in Equation (16).

In so doing, an account was taken of the liquid phase diffusion coefficients in the CO_2-water and acetone-CCl_4 systems by employing the equations [2, p. 191]

$$\beta_x(NH_3 - H_2O) = \beta_x(CO_2 - H_2O)\left(\frac{D_l(CO_2 - H_2O)}{D_l(NH_3 - H_2O)}\right)0.5 \tag{17}$$

$$\beta_x(\text{acetone} - CCl_4) = \beta_x(CO_2 - H_2O)\left(\frac{D_l(CO_2 - H_2O)}{D_l(\text{acetone} - CCl_4)}\right)0.5 \tag{18}$$

Applying Equation (16) to the acetone-CCl_4 mixture for channel dimensions l=0.04 m and d_e=0.01 m and for a Reynolds number range of $250<\mathrm{Re}_g<2500$, curve 3 was plotted; the curve fitted the whole set of the experimental data already presented in Figure 18. Bearing in mind the fact that Equation (16) also describes the data obtained by the authors cited [2, p. 192] with a certain scatter, it can be said that Equation (15) is quite suitable for calculating mass transfer characteristics for the vapor (gas) phase in rectification and absorption in a rotor film type column of the GIAP design (see Figures 7-9).

Since in most instances the diffusion resistance to mass transfer in absorption and rectification in slot channels with one wetted wall under the conditions of laminar motion of the gas phase is essentially due to the gas phase, Equation (15) can also be used in calculating the overall mass transfer coefficient in such processes employing the approximate equality $K_{oy}=\beta_y$.

6. Comparative Evaluation of Different Methods of Effecting Mass Transfer in a Slot Channel with One Wetted Wall

At first, the best results, in terms of mass exchange efficiency, in rectification in a slot channel with one wetted wall are achieved in the case of additional mixing of the phases in contact by mechanical means. In this situation, the kinetics of mass transfer in the vapor phase is described by Equation (12) However, high values of mass transfer coefficients can be obtained for the same channel without mixing if the length of the channel is sufficiently small. In this case, the process is described by Equation (15). Let us suppose that it is required to design a film type mass exchange column whose packing contains slot channels with one wetted wall, the gap width in the channel being chosen to be 9 mm. One can use, for instance, the design of the packed column shown in Figure 15. Interaction between the film and vapor takes place in packing slots with width $\Delta=0.009$ m and length $l=0.028$ m (along the vapor flow). Another version may involve the use of a tubular film type column with an inner rotating cylinder forming a gap with $\Delta=0.009$ m. In this case, it is quite common that the length of the annular channel is large; the column houses moving parts, and in order to have the highest possible surface in one column the length of the rotor is chosen should also be large, e.g. 1 m. Other conditions chosen: column diameter $d_c=0.120$ m, rotor diameter $d_r=0.102$ m, rotation speed $n=3000$ rpm.

A comparative calculation of mass transfer coefficients in the vapor phase in both cases will be carried out on an example of rectification of a binary mixture CCl_4-C_6H_6 using Equations (12) and (15). For this mixture we have: $v_g=2.67\times10^{-6}$ m^2/s, $D_g=4.63\times10^{-6}$ m^2/s, $Sc_g=0.564$. Laminar flow in the gap at $Re_g=1555$ is assumed.

Perform the necessary sequence of calculations for a film type column with a rotating cylinder. Given: $\Delta=0.009$ m, $l=1$ m, $d_c=0.120$ m, $d_r=0.102$ m, $n=3000$ rpm. Calculated: $r_c=d_c/2=0.060$ m, $r_r=r_c-\Delta=0.051$ m, $r_m=\frac{r_c+r_r}{2}=0.0555$ m, $\frac{\Delta}{r_m}=0.162$, $d_e=2\Delta=0.018$ m, $d_e/l=0.018/1=0.018$.

The critical value of the Taylor number is found from Equation (10):

$$Ta_{cr} = \frac{71.6}{(1-0.865x0.162)x0.162^{0.38}} = 166.$$

$$U = \frac{\pi d_r n}{60} = 16\,m/s;\ \mathrm{Re}_c = \frac{Ud_e}{\nu_g} = 107865;$$

$$\omega = \frac{U}{r_r} = \frac{16}{0.051} = 313.7s^{-1};\ \mathrm{Re}_e = \sqrt{\mathrm{Re}_g^2 + 0.1915\,\mathrm{Re}_c^2} = 47228;$$

$$Ta = \frac{\omega r_m^{1/2}\Delta^{3/2}}{\nu_g} = 23633.$$

Equation (12) gives the diffusion Sherwood number for a film type column with a rotating cylinder

$$Sh_g = 0.64x47228^{1/3}(1+18x0.018)[1+1.95(1-\frac{166}{23633})]x0.564^{1/3} = 50.4$$

The diffusion Sherwood number for packing, consisting of flanged annular plates, (Δ=0.009 m; l=-0.028 m) is found from Equation (15)

$$Sh_g = 0.49(1555x2x0.009/0.028)^{0.77}x0.564^{0.33} + 4.86 = 87.5$$

Thus, the intensity of mass transfer in the vapor phase in the cases compared is virtually identical although in one case it is achieved by energetic mixing of the vapor phase, (film type column with a rotating cylinder), and in the other by fabricating the working surface of packing of small-size elements (a stack of flanged plates).

At a sufficiently high vacuum, the influence of mechanical mixing on vapor phase mass transfer efficiency becomes much weaker or vanishes altogether. Thus, in rectification of a binary mixture diethylene glycol-m-cresol under a residual pressure of 3-5 Torr in a film type column with a rotating cylinder with the dimensions indicated above, the critical value of the Taylor number (Ta_{cr}=166) is attained at the peripheral velocity

$$U = r_r\omega_{cr} = \frac{Ta_{cr}\nu_g}{r_m^{1/2}\Delta^{3/2}} = \frac{0.051x166x2.87x10^{-4}}{0.00555x0.009^{3/2}} =$$

=12.1 m/s. This means that at feasible, (for vacuum rectification), cylinder rotation speeds, Taylor eddies do not arise in the vapor phase under a residual pressure of 3-5 Torr so that the vapor moves in a laminar regime, as is the case with the ordinary slot channel in the absence of motion of one of its walls with respect to the other.

In a rotor film type column designed by GIAP (Figure 7) employed in caprolactam rectification under a residual pressure of 1-10 Torr, vapor phase mass exchange is intensified by making the wetted surface of short, (in the direction of vapor phase flow), elements. The role of the rotor in this case is reduced, not to artificial turbulization of the vapor phase, the absence of the motion of one wall of the channel with respect to the other one), but to the creation and preservation of a uniform liquid in the cross-section of the column due to the development of a centrifugal force field.

7. GIAP Design Film Type Rotor Evaporator

Not only is the selected type of the rectification column of great importance in the rectification of thermally unstable products but so is the evaporator design. Large residence time of the product in the heating zone and high hydraulic resistance of the evaporator can nullify the advantages of any type of rectification column.

Falling film evaporators are preferable but, in this case, it is difficult to attain a uniform distribution of the liquid over the inner surface of the tubes, particularly if the amount of the bottom product is small as compared with the quantity of the starting product. The presence of unwetted zones on the heat transfer surface may lead to local overheating and therefore to the enhancement of the thermal effect on the product coming in contact with these zones.

In the case of small liquid flow rates, best results are obtained in using film type rotor evaporators ensuring forced distribution of the liquid to be evaporated over the heating surface.

For such operating conditions, GIAP has developed a film type rotor evaporator with a corrugated rotor [29] which is presently being successfully used in some production processes. The evaporator (Figure 19) consists of a cylindrical shell 1 divided into several sections throughout its height, (the sections are fitted with heating jackets), and rotor 3 with several coaxial hollow drums 2 attached to it. The number of the drums is equal to the number of the sections. The surface of each drum is corrugated, the corrugations have holes 6 making a descending helix.

The principle of operation of the evaporator is as follows: The rotating distributing ring 5 sheds the liquid into the inner surface of the drum's corrugations where it forms a series of vertical flows. Holes 6 throw off jets of the liquid onto the heated surface of the shell. Since holes 6 make a helical line on the surface of the drum, uniform wetting is attained over the whole surface of the shell at a height equal to the height of the drum. The perforations in the drum can form a multithread helix. The unevaporated portion of the liquid flows from the walls of the shell to an annular collector 7 with baffles 8. After that, the liquid enters, via ducts 9, the distributing ring of the next section below where the process is repeated. The cavities between

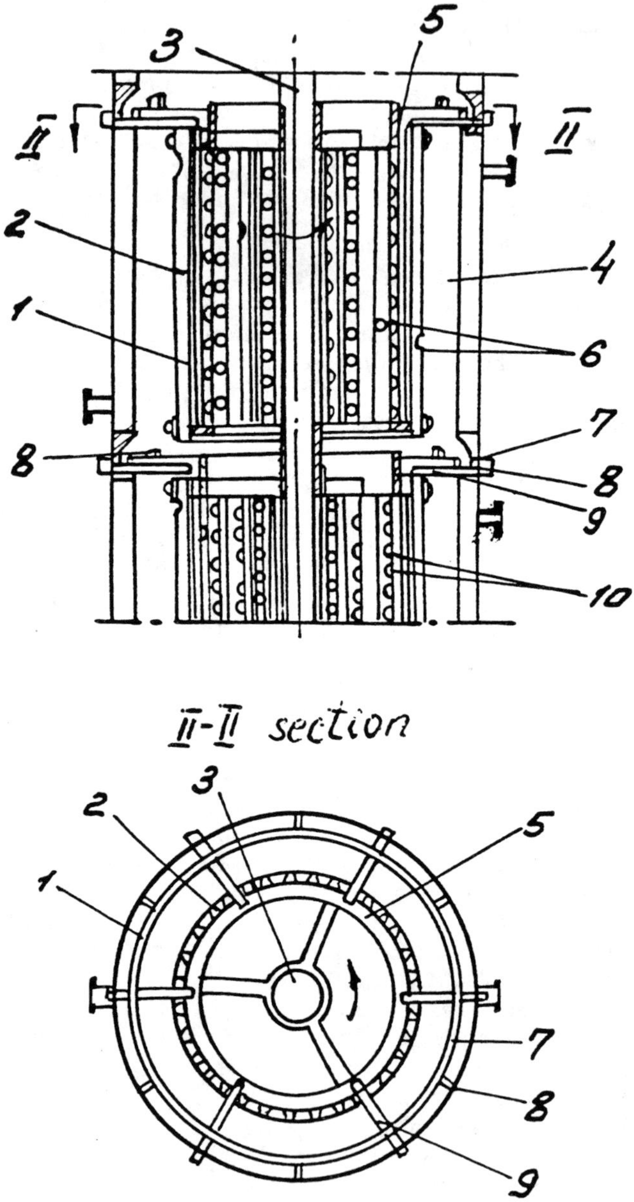

Figure 19. Film type rotor evaporator with a corrugated drum rotor: 1 - shell; 2 - drum; 3 - shaft; 4 - gap between the drum and shell; 5 - distributing ring; 6 - liquid ejection holes in the drum; 7 - annular collector; 8- baffle; 9 - duct; 10 - holes for the passage of vapor inside the drum; 11 - II-II section.

the corrugations have holes 10 serving for the passage of vapors into the space inside the drum which minimizes the pressure drop along the height of the evaporator. Due to the presence of holes 10, the vapors formed on evaporation inside the section rise not only in gap 4 but also throughout the whole cross-section of the evaporator.

The gap between the shell and drum can be several centimeters wide, virtually lifting the most stringent restrictions on the quality of treatment of the inner surface of the shell imposed for other designs of rotor-film evaporators. This point seems to be particularly important in the case of high-capacity equipment.

Another specific feature of the design is the use of the kinetic energy of the jets of the liquid discharged onto the heat transfer surface. The active turbulization of the liquid film occurring in this case makes it possible to attain high values of heat exchange coefficients.

Finally, the uniformity in height of liquid discharge onto the heat transfer surface should favor the efficient operation of the evaporator in a broad range of loads and thermal properties of the processed product.

Based on the results of extensive experimental studies [29], a formula for calculating heat transfer from the wall to the film in such an evaporator was proposed

$$Nu_l = (1+0.03P_{re}^{0.61}\xi)[1+0.22\frac{\sigma H_2O}{\sigma}(\frac{q}{qH_2O})^{0.5}] \tag{19}$$

where $\xi = V^* \sigma_l \rho_l / \mu_l$; V^* is a tangential velocity on the wall; σ, σ_{H_2O} is the surface tension of the evaporated substance and water, respectively; q, q_{H_2O} is the specific heat flux of the evaporated substance and water, respectively; δ_l is the liquid film thickness; ρ_l is density of the liquid; μ_l is the viscosity of the liquid.

REFERENCES

1. Inventor's Certificate of the USSR No. 230777. *Otkrytiya Izobreteniya.* 1968, No. 35 (in Russian).
2. V.M. Olevsky, V.R. Ruchinsky, A.M. Kashnikov and V.I. Chernyshev. *Plenochnaya Teplo-i Massoobmennaya Apparatura* (Film type heat and mass transfer equipment). Moscow: Khimiya, 1988, p. 240 (in Russian).
3. R. Billet *Verfahrenstechnik.* 1972, Bd. 6, No. 3, S. 71-79.
4. I.I. Beltser *Cand. Sci. (Techn. Thesis,* Moscow: GIAP, 1981 (in Russian).
5. B.I. Vinogradsky, V.M. Olevsky, V.R. Ruchinsky and G.I. Evstafieva. *Tr. GIAP Khimiya i Tekhnologiya Produktov Organicheskogo Sinteza,* 1969, Issue 2, Part 2, 22-30 (in Russian).
6. V.M. Olevsky and V.R. Ruchinsky, *Khim. Prom.* 1961, No. 1, 57-62 (in Russian).
7. GDR Patent No. 146689, 1981.

8. Inventor's Certificate of the USSR No. 729881. *Otkrytiya Izobreteniya,* 1978, No. 32 (in Russian).
9. Inventor's Certificate of the USSR No. 729881. *Otkrytiya. Izobreteniya.* 1983, No. 37 (in Russian).
10. US Patent No. 4256674, 1981.
11. Inventor's Certificate of the USSR No. 702564. *Otkrytiya. Izobreteniya.* 1982, No. 17 (in Russian).
12. Inventor's Certificate of the USSR No. 1084035. *Otkrytia. Izobreteniya.* 1982, No. 39(in Russian).
13. A.V. Trubachev and A.M. Lushin *Stroitel'stvo i Osvoenie Proizvodstvennykh Moshchnostei.* 1987, No. 6 (in Russian).
14. Inventor's Certificate of the USSR No. 699700, *Otkrytiya. Izobreteniya,* 1983, No. 14 (in Russian).
15. US Patent No. 4038353, 1977.
16. F. Neimann *Chem. Ing. Techn.* 1961, Bd. 33, No. 7, S. 485-491.
17. V.B. Kogan and M.A. Kharisov, *Oborudovanie dlya Razdeleniya Smesei pod Vakuumon* (Equipment for vacuum separation of mixtures). Leningrad: Mashinostroenie, 1976, 146-147 (in Russian).
18. V.V. Kurkovskaya and A.V. Shafranovsky *Nauchno-Tekhnicheski Sbornik. Seriya: Azotnaya Promyshlennost'.* Moscow: NIITEKhim, 1979, Issue 1, 32-35 (in Russian).
19. V.M. Olevsky and V.R. Ruchinsy *Rektifikatsiya Termicheski Nestoikikh Veshchestv.* Rectification of thermally unstable substances). Moscow: Khimiya, 1972, p. 200. (in Russian).
20. B.S. Ptukhov *Teploobmen i Soprotivlenie pri Laminarnon Techenii Zhidkosti v Trubackh* (Heat exchange and resistance in laminar flow of liquids in pipes). Moscow: Energiya, 1967, 225 (in Russian).
21. V.M. Olevsky, A.V. Shafranovksy, L.N. Amelin and V.R. Ruchinsky, *Khim. Prom.* 1982, No. 8, 44-50 (in Russian).
22. A.V. Shafranovsky, *Cand. Sci. (Techn.) Thesis.* Moscow: GIAP, 1969 (in Russian).
23. V.N. Gromoglasova *Cand. Sci. (Techn.) Thesis.* Moscow: GIAP, 1973 (in Russian).
24. V.N. Gromoglasova, A.V. Shafranovsky and V.R. Ruchinsky *K Voprosu o Kriticheskoi Skorosti pri Massoobmene v Kol'tsevom Zazore s Vnutrennim Vrashchayushchimsya Tsilindrom* (on the critical velocity problem in mass transfer in an annular gap with an inner rotating cylinder). Manuscript deposited with NIITEKhim Division, Cherkassy, No. 192/74, 1974, 1-13 (in Russian).
25. A.V. Shafranovsky and V.R. Ruchinsky *Teor. Osn. Khim. Tekhn.* 1971, Vol. 5, No. 1, 34-37 (in Russian).
26. A.V. Shafranovsky and V.R. Ruchinsky *Teor. Osn. Khim. Tekhn.* 1971, Vol. 5, No. 2, 187-196 (in Russian).
27. V.N. Gromoglasova, V.R. Ruchinsky and A.V. Shafranovsky *Tr. GIAP. Khimiya i Tekhnologiya Azotnykh Udobrenii.* 1972, Issue 14, 127-138 (in Russian).

28. V.N. Gromoglasova, A.V. Shafranovsky and V.R. Ruchinsky *Tr. GIAP. Khimiya i Tedkhnologiya Produktov Organicheskogo Sinteza.* 1973, Issue 20, 60-65 (in Russian).
29. V.A. Petrov, V.R. Ruchinsky and I.F. Evkin *Teor. Osn. Khim. Tekhn.* 1970, Vol. 4, No. 4, 515-519 (in Russian).

Hydrodynamics and Mass Transfer of Phase Inversion Under Conditions

Z.N. Memedlyaev

Abstract—A new method of stabilization and control of gas-liquid flow under conditions of phase inversion is presented, the concept of which lies in the availability of main and bypass gas flows. The hydrodynamics of the process in an irrigated packing and wetted-wall column is investigated. A new method of calculating the pressure drop in an irrigated packing, fully or partially filled with a layer of gas-liquid emulsion is determined. Mass transfer in liquid and gas phases in a layer of irrigated Raschig's or Pall's rings has been studied and it is shown that a change from the film regime to the phase inversion regime is accompanied by a decrease in the height of one liquid phase transfer unit by 1.9 to 2.6 times whereas no such sharp relationship has been observed for a gas phase. For studied packings the correlations for heights of transfer units in liquid and gas phases are proposed. A series of packed apparatuses realizing a new method of stabilization and control of the phase inversion regime is determined. Information on practical applications of such apparatuses for absorption, desorption and absorption with chemical reaction is included.

Symbols

a	surface area per unit volume of packing, m^2/m^3
d_{eq}	equivalent diameter of packing, m
d	inner diameter of pipe, m
D_{AB}	molecular diffusion coefficient, m^2/s
G, L	gas and liquid mass velocity, $kg/(m^2.s)$
g	acceleration due to gravity, m/s^2
H, H_{GL}	height of layer of packing and gas-liquid emulsion, m
h	height of a transfer unit, m
m	equilibrium line slope
q	liquid rate per unit area of cross-section, $m^3/(m^2.h)$
t	time, s
V	gas flow rate, m^3/s
W_G, W_L	superficial velocities of gas and liquid, m/s; m/h
$W_{G, inv}$	flooding velocity, m/s
$W_{G, ax}$	gas velocity at axis of wetted-wall column, m/s
α_0, α	void fraction in non flow and flow conditions

β	mass transfer coefficient, m/s
δ	mean liquid film thickness, m
$\varepsilon_0, \varepsilon$	fractional gas holdup for dry and wet packing
θ	reduced thickness of film, m
λ_0, λ	friction factor for gas flow through dry and wet packings
ΔP	pressure drop, Pa
μ	viscosity, mPa.s
ν	kinematic viscosity, m^2/s
$\rho_G, \rho_L, \rho_{GL}$	density of gas, liquid and gas-liquid emulsion, kg/m^3
τ	shear stress, N/m^2

$$X = (L/G)^{1/4}(\rho_G/\rho_L)^{1/8}; Y = \frac{W_{G,inv}^2\, a}{g\varepsilon_0^3}\frac{\rho_G}{\rho_L}\mu_L^{0.16};$$

$$j_G^* = W_G\rho_G^{1/2}[gd(\rho_L-\rho_G)]^{-1/2}; j_L^* = W_L\rho_L^{1/2}[gd(\rho_L-\rho_G)]^{-1/2}$$

$Re_G=W_G d_{eq}/(\varepsilon_0 \nu_G)$	Reynolds number for gas flow
$Re_L=4q/(3600a\nu_L)$	Reynolds number for liquid flow
$Sh=\beta\theta/D_{AB}$	Sherwood number
$Sc=\nu/D_{AB}$	Schmidt number

Subscriptions:

inv	inversion (flooding point)
l.p.	loading point
f	film regime
G, L	gas, liquid

INTRODUCTION

The phenomenon of a sharp rise of heat and mass transfer intensity between gas and liquid phases under the conditions of phase inversion has been described in [1].

The investigation of hydrodynamic structure of flows in packed towers has shown [2, 3] that a transition to an inverted flow of liquid is preceded by the earlier unknown hydrodynamic regime in which the counter-current flow remains, but the nature of phase interactions sharply changes. An inversion of phases occurs with the formation of a movable emulsion system which is permeated with gas, liquid and gas-liquid eddies of many sizes rolling and mixing with each other.

A transition from a separated flow of phases to the new regime is illustrated in Figure 1 as far as some process parameters are concerned [4]. It has been proven that a sharp change of flow conditions also ensures a sudden rise in the heat rate [5,6] and mass transfer [3,5] which is attributed to the intensive interaction of phases accompanied by a sharp development and continuous renewal of interphase surface. The practical importance of this regime lies in the fact that it has shown a new method of intensifying heat and mass transfer processes. The use of the phenomenon, however, calls for the stabilization and control of the phase inversion conditions as it emerges at only one relative velocity of flows [5] whereas a layer of gas-liquid emulsion tends to continuous expansion.

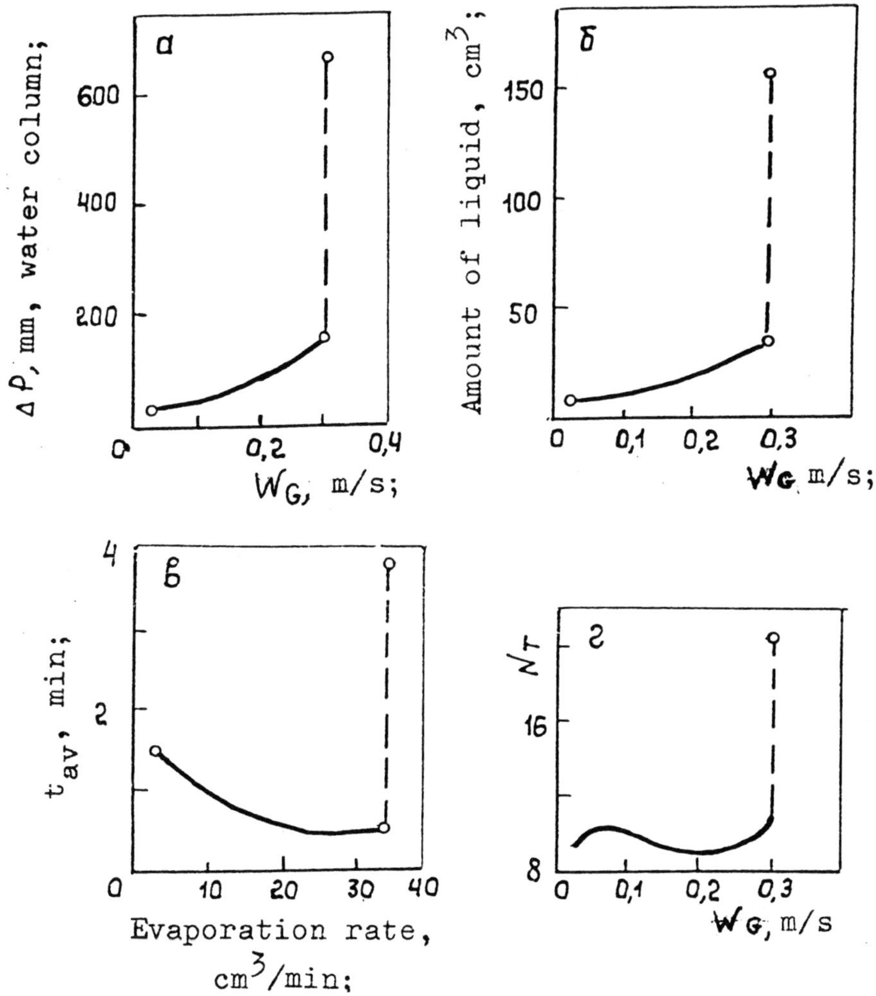

Figure 1. Spasmodic transition from separate phase flow regime to phase inversion regime in a rectification packed column [4]: a - pressure difference; b - amount of liquid held; c - mean time of liquid staying; d - separability.

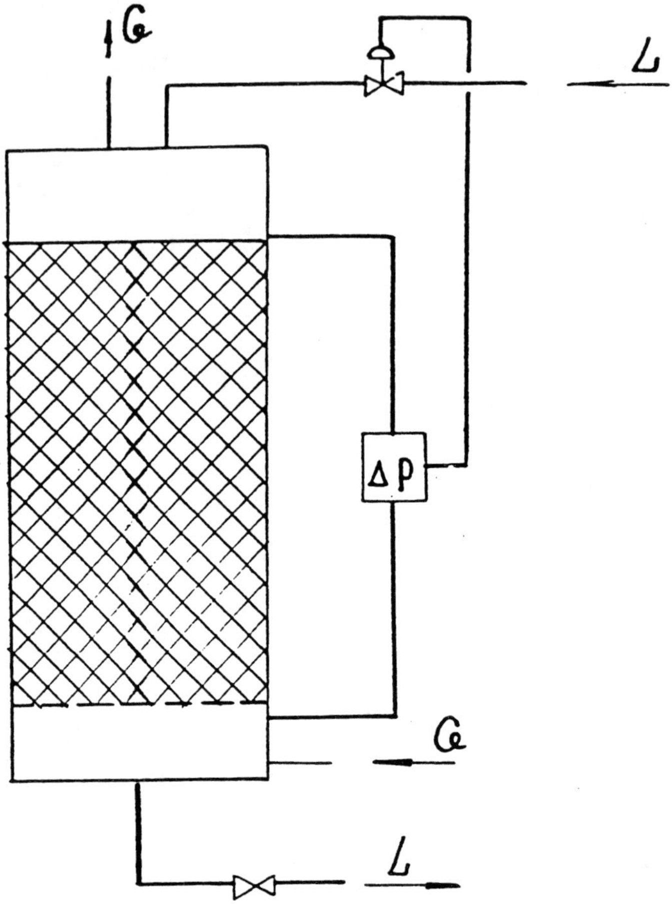

Figure 2. Scheme of continuous control of liquid influx [7].

One of the possible schemes to continuously regulate the column liquid in flow, regarding pressure difference, is presented in Figure 2 [7].

A scheme has also been proposed for the recirculation and continuous regulation of a part of the column liquid outflow [8] (Figure 3). A principle of compulsory holding of a layer of gas-liquid emulsion in the packing layer by creating a resistance to liquid run-off (Figure 4) with the help of a hydraulic lock has been proposed in [9].

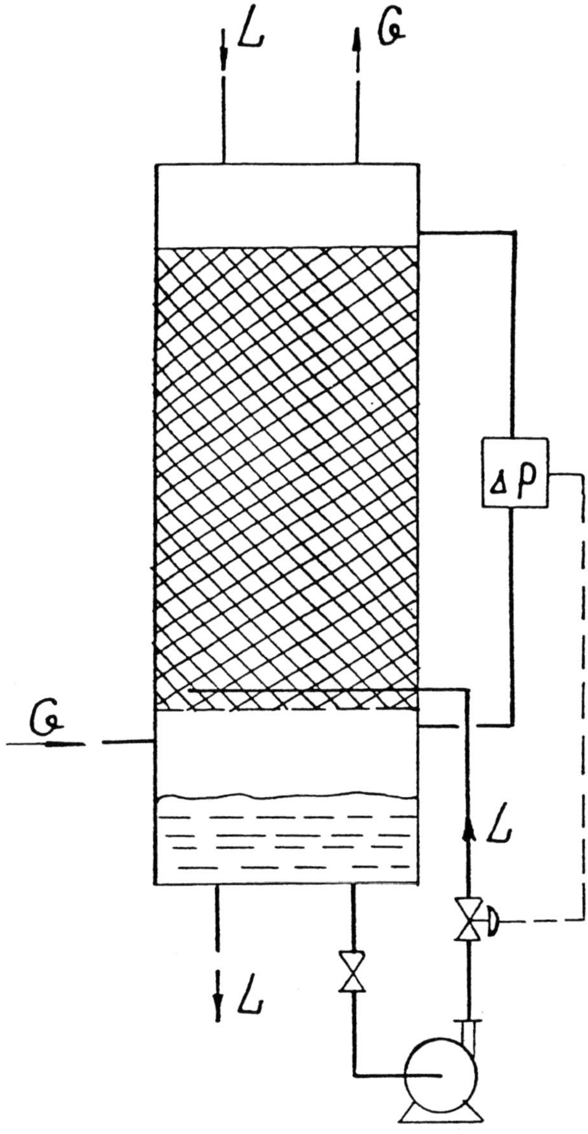

Figure 3. Scheme of continuous control of recirculating liquid flow [8].

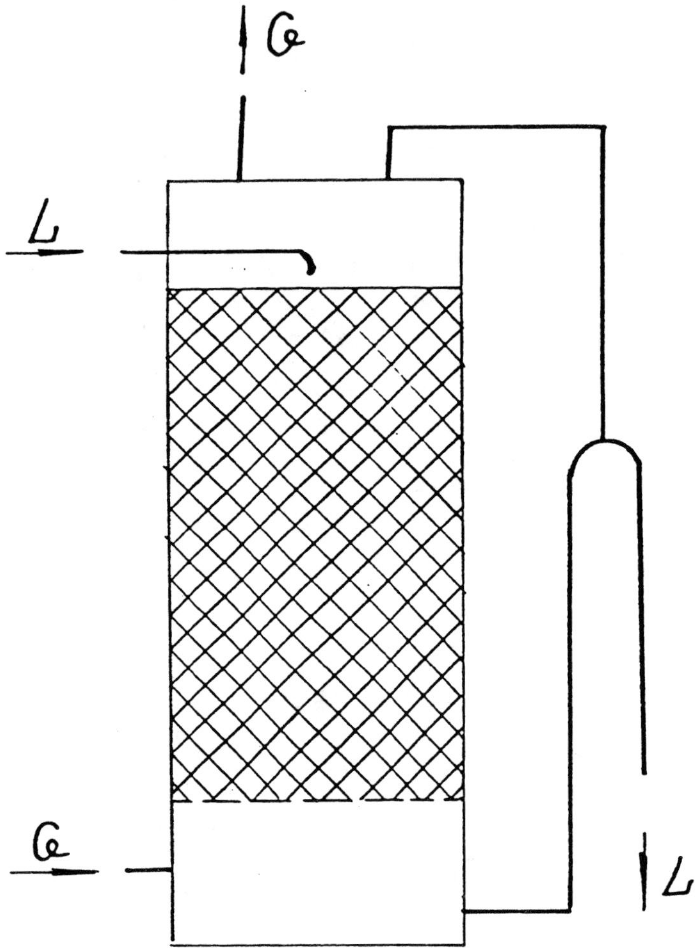

Figure 4. Scheme of a column with submerged packing [9].

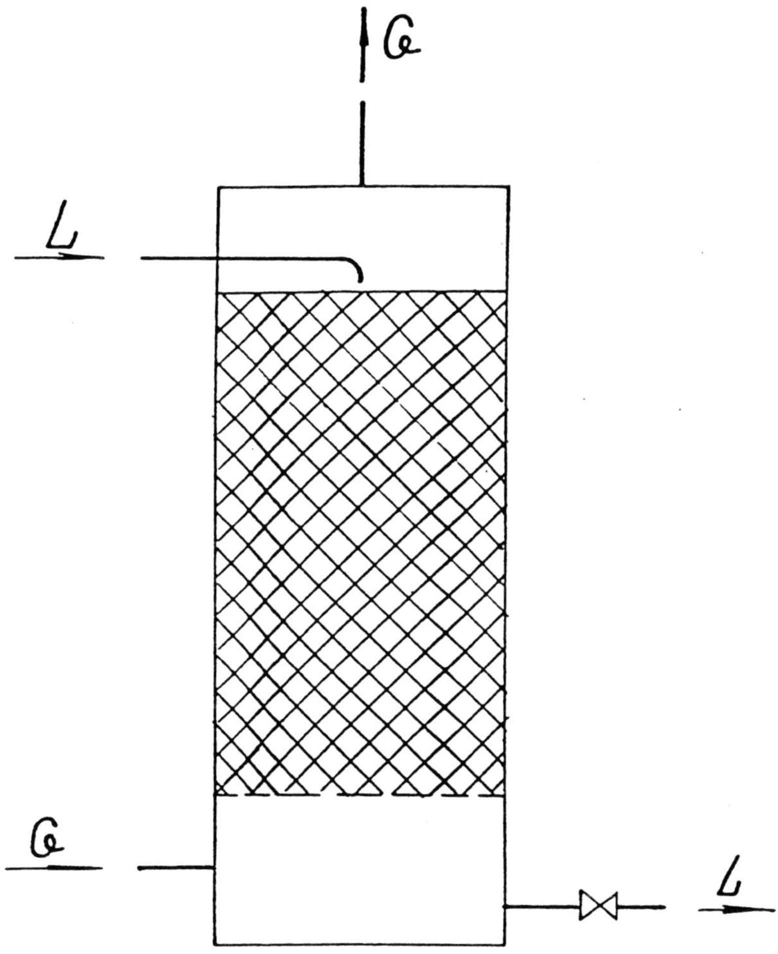

Figure 5. Scheme of a column with submerged packing [10].

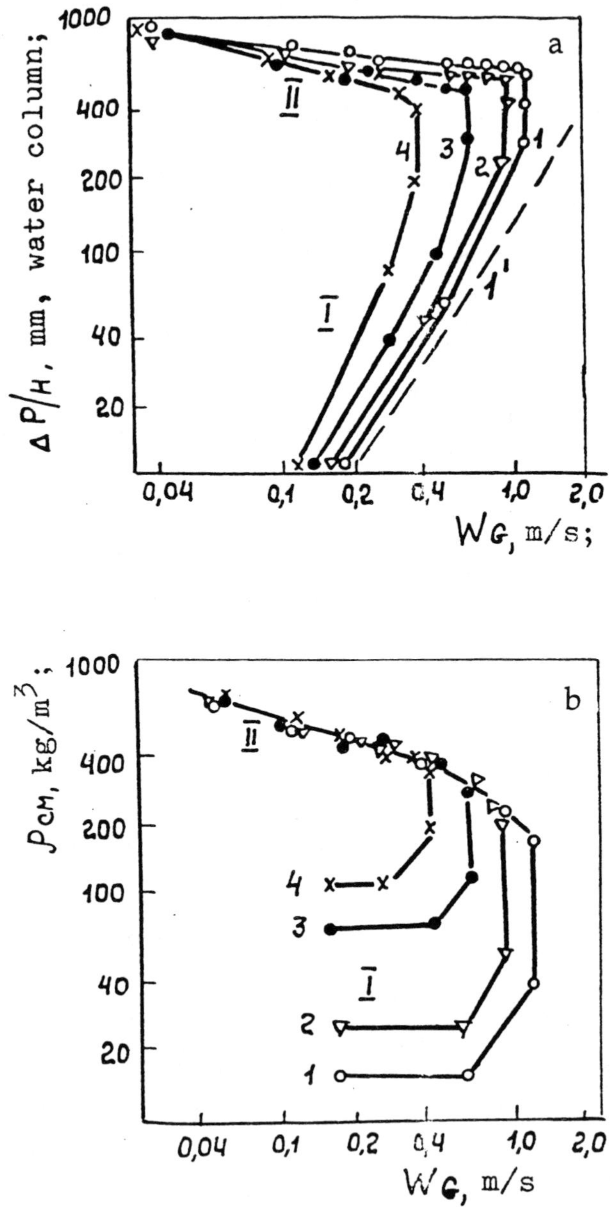

Figure 6. Resistance (a) and retentivity (b) of irrigated (I) and flooded (II) layers of Raschig porcelain rings measuring 10x10 mm 5, 11: I'-q=0; I-1.04 m^3/m^2; 2-1.875; 3-7.08; 4-16.25.

In the column presented in Figure 5 this principle is realized by installing a shutter or valve in the line of liquid cutlet which can be regulated manually or automatically by the readings of a level gauge or differential manometer [10].

The columns described in [9, 10] are known as submerged packing columns. The comparison of the characteristics of irrigated and submerged packings (Figure 6) has led to the conclusion that the characteristics of a submerged packing reach those of an irrigated packing only for $W_G=W_{G,\, inv}$ [5, 11], i.e. under conditions of phase inversion. But this regime is unstable [5, 12, 13]. Therefore, columns with both irrigated and submerged packings are operated at $W_G<W_{G,\, inv}$ to avoid flooding, though maximum output and effectiveness are reached and the scale-up coefficient [14] tends toward unity [5, 15] at $W_G=W_{G,\, inv}$.

So, a desire for optimal utilization of the phenomenon of sharp rise of heat and mass transfer intensity between gas and liquid phases under the conditions of phase inversion fosters a search for new solutions for the stabilization and control of this regime.

1. Stabilization and Control of Phase Inversion Conditions by Competitive Bypass Gas Flow

A new method has been proposed which ensures the reliable operation of columns at $W_{G,inv}$ in the conditions of both constant and variable loads due to a phenomenon of self-organization of the regime of phase inversion.

1.1. The Concept of the Technique

The essence of the method lies in the fact that there are main and bypass flows of a gas (Figure 7) with a common gas inlet and outlet. This allows one to write the following equations [16] for a loss of gas head

$$\Delta P = \xi_1 \frac{\rho_G W_{G,inv}^2}{2} + \rho_{GL} g H_{GL} = \xi_2 \frac{\rho_G w^2}{2} \qquad (1)$$

From this equation it follows that the height H_{GL} of a layer of gas-liquid emulsion takes a fixed value, i.e. the phase inversion regime gets stabilized. A variation of the gas and liquid flow rates gives rise to a change in the velocity w of the competitive gas flow by the value of $\pm\Delta w$ which leads to the automatic setting of new equilibrium values of ΔP and H_{GL}, the initial or required values of which can be set by changing the hydraulic resistance coefficients ξ_1 and/or ξ_2 with the help of, say, shuters or valves. This way the phase inversion process may be controlled.

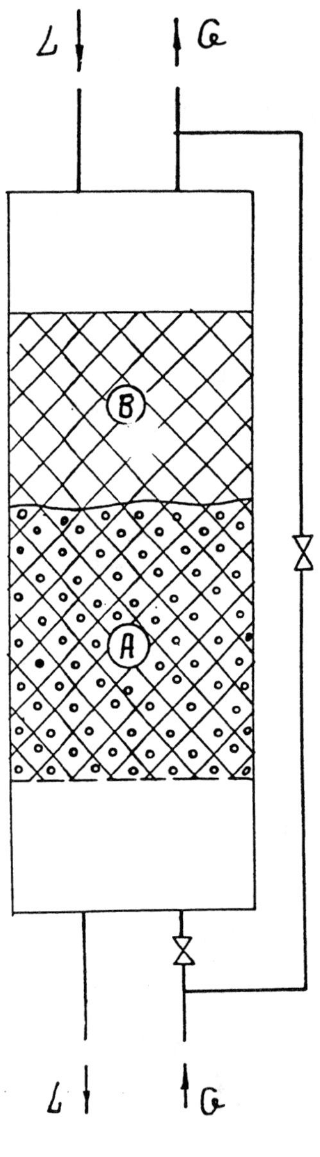

Figure 7. Scheme of an irrigated packing layer with concurrent gas flow.

1.2. Apparatus with Self-Organized Phase Inversion

The idea of stabilizing and controlling phase inversion regime in an irrigated packing with the help of a competitive gas flow has been implemented in a number of designs of heat-and-mass exchange columns which [17, 18] are schematically presented in Figure 8.

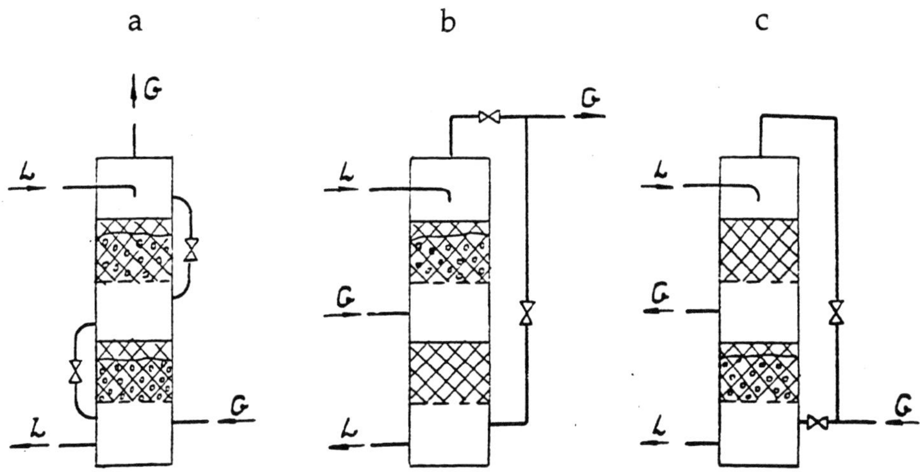

Figure 8. Schemes of columns with self-organization of phase inversion regime in an irrigated packing [17, 18].

The first of the presented columns works (see Figure 8a) as follows. Gas and liquid are supplied to the column, respectively, from the bottom and top in amounts ensuring the excess of phase inversion rate. A larger part of the gas, having contacted the liquid in the phase inversion regime in the bottom layer of the packing gets in the interlayer space. Due to a positive pressure difference before and after the top layer of the packing and availability of a by-pass, a part of the gas bypasses this layer while as a rule the remaining larger part of the gas interacting with falling jets and drops of the liquid continues moving upward and before it enters the top layer of the packing the part of the gas that has by-passed the bottom layer of the packing joins the upstream to interact, (together with the main gas flow), with the liquid in the phase inversion regime in the top layer of the packing.

The column presented in Figure 8b operates as follows; Liquid is supplied into the column through an inlet over the top layer of the packing and removed under the bottom layer of the packing. Gas is supplied into the column between the packing layers where two floors are automatically formed without external influences: the ascending gas flow in the top layer of the packing and descending gas flow in the bottom layer. The latter gas flow helps stabilize and control the phase inversion regime in the top layer of the packing.

For the column scheme presented in Figure 8c liquid is supplied and removed as described above. Gas is supplied into the column in two flows through the input under the bottom layer of the packing and partially through the input over the top layer of the packing. Owing to such organization of the flows the phase inversion regime and descending straight-flow occur in the bottom and top layers of the packing, respectively. It should be stressed that the stabilization and control of the phase inversion regime in this column as in earlier ones are affected by the competitive bypass gas flow.

2. HYDRODYNAMICS

2.1. IRRIGATED PACKING

Hydrodynamic investigations were carried out on a water-air system in an experimental unit consisting of a glass column with an inner diameter of 105 and 200 mm and packing rings dumped to a height of 600 and 800 mm. The packing characteristics are given in Table 1 (values of ε were measured experimentally)

Table 1.

Type	Dimensions, mm	Material	ε_0	ε	$a, m^2/m^3$
Raschig rings	15x15x3	Ceramic	0.645	0.609	294.0
Raschig rings	25x25x4	Ceramic	0.680	0.668	221.6
Pall rings	25x25x1	Steel	0.925	0.912	176.0

Figure 9 exemplifies the stable values of the heights of gas-liquid emulsion in the packing layer composed of Raschig rings measuring 15x15x3 mm. This, or another value of H_{GL} was achieved by changing the total gas flow rate or by-pass hydraulic resistance coefficient ξ_2 with the help of a valve. It turns out that the invariable liquid rate per unit area of different stable heights of the layer of gas-liquid emulsion exists almost at one and the same speed of the main gas flow, as it follows from Equation (1). Figure 10 presents the generalization of experimental data on ultimate loads for various packings in the form of the Bain-Hougen equation [13, 19].

$$l_g y = b - cX \tag{2}$$

The same figure provides the data on ultimate loads for Pall rings measuring 50 mm [20].

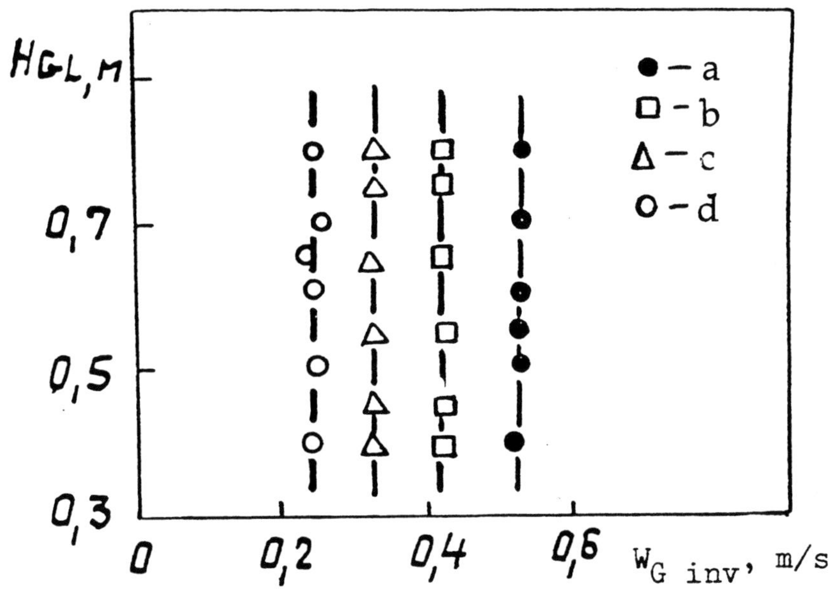

Figure 9. Stable heights of a layer of gas-liquid emulsion for different liquid rate per unit area of cross-section different by-pass resistance for Raschig rings measuring 15x15x3 mm: a - q=29.47 $m^3/m^2 \cdot h$; b -41.37; c -53.15; d - 65.05

It is clearly seen that the position of straight lines describing the phase inversion regime depends on a type of packing and the registered values of $W_{G,inv}$ agrees well with the known correlations [13, 19, 20]. At the same time, it should be admitted that the known correlations, though they correctly describe the interrelation of gas and liquid ultimate loads, bear little information on the mechanism of origination of such complicated phenomenon as phase inversion.

On the other hand, the analysis of the known hydrodynamic regimes at the counterflow of gas and liquid in the packing layer [13]-film, hold-up, overflooding (inversion)- and causes of their emergence makes it possible to conclude that the role of shear stress at the phase interface is decisive. In the authors' opinion, the knowledge of laws governing the interphase shear stress, lays the groundwork for a deeper understanding of the phase inversion mechanism.

The method of competitive bypass gas flow [16] ensuring the stabilization and control of the phase inversion regime in zone A (Figure 7) at the same time provides for the stabilization and control of the pre-inversion regime of phase interaction in zone B (Figure 7). This fact allows one to receive information on the interphase shear stress just near the point of phase inversion.

The pressure drop, in zone B is related to the relative speed of gas as follows [13]:

$$\frac{\Delta P}{H_B} = \lambda_B \frac{1}{d_{eq}} \frac{\rho_G W_{Grel}^2}{2\varepsilon^2 \alpha_B^2} = \lambda_B \frac{\rho_G W_{Grel}^2 a}{8\varepsilon^3 \alpha_8^3} \tag{3}$$

Combining Equation (3) and Equation [21]

$$\tau_B = \lambda_B \frac{\rho_G W_{Grel}^2}{8\varepsilon^2 \alpha_B^2} \tag{4}$$

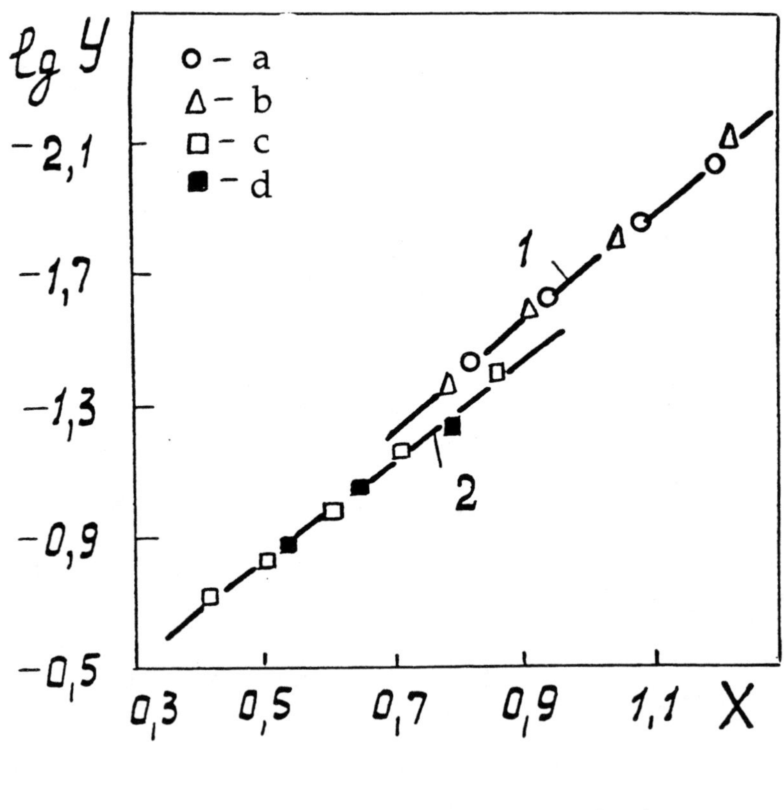

Figure 10. Colleration of data on phase inversion rate: a - Raschig rings measuring 15x15x3 mm; b - Raschig rings measuring 25x25x4 mm; c - Pall rings measuring 25x25x1 mm; d - Pall rings measuring 50 mm 20; 1 and 2 - correlations [13, 19] and [20].

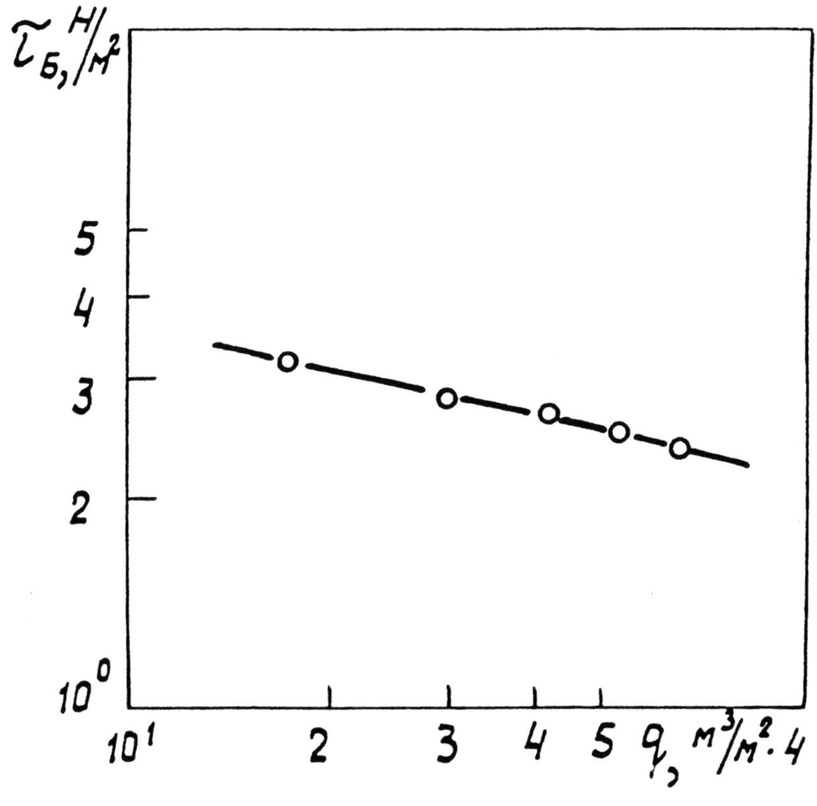

Figure 11. Dependence of preinversion shear stress from liquid rate per unit area of cross-section for Raschig rings measuring 15x15x3 mm.

we come to the relationship

$$\tau_B = \frac{\Delta P}{H_B} \frac{\varepsilon \alpha_B}{a} \tag{5}$$

Figure 11 illustrates a variation of the preinversion shear stress, calculated by formula (5), with an increase of liquid rate per unit area of cross-section for Raschig rings measuring 15x15x3 mm.

The revealed route of relationship τ_b—f(q) indicates that:

- the thicker the liquid film, the less its stability;
- one should expect a change in the intensity of such phenomenon as phase inversion at variation of regime parameters.

It is evident that the pressure drops of the designed columns (Figure 8) is determined by the pressure drop in a layer of an irrigated packing with steady phase inversion regime.

Generally, the layer of an irrigated packing with steady phase inversion regime consists of two zones (Figure 7). If a layer of gas-liquid emulsion exists in the bottom zone (zone A), then a separate flow of phases is realized in the top zone (zone B).

It is easy to express the pressure drop of such a system as a sum of pressure drops of both zones

$$\Delta P = \Delta P_A + \Delta P_B \tag{6}$$

In its turn ΔP_A can be expressed as

$$\Delta P_A = \Delta P_1 + \Delta P_2 \tag{7}$$

where: ΔP_1 is the pressure drop at wet packing, ΔP_2 is the pressure drop to maintain a gas-liquid emulsion column.

The pressure drop of a packing dry layer is usually expressed by the following relationship

$$\Delta P / H = \lambda_0 \rho_G W_G^2 a / (8\varepsilon_0^3) \tag{8}$$

For a wet layer, λ_0 should be replaced with λ in this relationship and $\varepsilon<\varepsilon_0$ should be used instead of ε_0 because of a decrease of the free volume of packing by the value of static delay. Since free volume in the layer of packing filled with gas-liquid emulsion equals $\varepsilon\alpha$, then the first summand of the right member of Equation (7) may be expressed by an analogy with Equation (8) as follows:

$$\Delta P_1 / H_A = \lambda \rho_G W_{G,inv}^2 a / (8\varepsilon^3\alpha^3)$$

while the second summand for $\rho_L >> \rho_G$ will have the form

$$\Delta P_2 / H_A = \rho_L g(1-\alpha)$$

Then

$$\frac{\Delta P_A}{H_A} = \lambda \frac{\rho_G W_{G,inv}^2 a}{8\varepsilon^3\alpha^3} + \rho_L g(1-\alpha) \tag{9}$$

To make calculations by Equation (9) one must know the values of λ and α. The specific pressure drop of dry and wet packings in all cases meets the relationship $\Delta P \sim W_G^2$, i.e. in the studied range of Reynolds numbers for gas phase the pressure drop coefficients λ_0 and λ of dry and wet packings, respectively, are constant and do not depend on Re_G.

The processing of experimental data made it possible to find that λ=6 for Raschig wet rings measuring 15x15x3 (Re_G=250-1000) and λ=5 for Raschig rings measuring 25x25x4 mm (Re_G=450-2850) and λ=3.5 for Pall rings measuring 25x25x1 mm (Re_G=1480-4800). It should be noted that the value of λ_0=4.5 found in this work for Raschig dry rings measuring 15 and 25 mm agrees with the published data [13] for ceramic rings measuring 25 and 50 mm, and the value of λ_0=3 for Pall dry rings measuring 25 mm coincides with the value obtained in work [22] for Pall plastic rings measuring 50 mm.

To describe void fraction in the phase inversion regime let us use the approach reported in work [23]. According to this approach, the average actual velocity of gas relative to the walls in a flow-type column with submerged packing equals that in a non-flow column less the velocity of a liquid which differs from the average one by factor C:

$$\frac{W_G}{\varepsilon\alpha} = \frac{W_G}{\varepsilon\alpha_0} - C\frac{W_L}{\varepsilon(1-\alpha)} \tag{10}$$

For conditions of phase inversion ($\alpha_0=\alpha^2$ [23]) Equation (10) may be rewritten as follows:

$$\frac{W_{G,inv}}{\varepsilon\alpha} = \frac{W_{G,inv}}{\varepsilon\alpha^2} - C\frac{W_L}{\varepsilon(1-\alpha)} \tag{11}$$

Solving Equation (11) with respect to α we will have

$$\alpha = \sqrt{W_{G,inv}} \Big/ \left(\sqrt{W_{G,inv}} + \sqrt{CW_L}\right) \tag{12}$$

Void fraction in the packing layer was determined by both a traditional method of cutoff of flows [13] and the new method [24] according to which, (after cutoff of irrigation, with invariable gas supply), the height of the decreasing level of gas-liquid emulsion was measured at successive time moments. Solving the equation that interrelates the known, measured and unknown parameters $W_L/\varepsilon(1-\alpha)=\Delta H_{GL}/\Delta t$, we will find

$$\alpha = 1 - \frac{W_L \Delta t}{\varepsilon \Delta H_{GL}} \tag{13}$$

The experimental data proves that the ultimate void fraction decreases with the increase of liquid rate per unit area of cross-section whereas for an invariable liquid rate per unit area it is larger for packings with larger free volume. It is apparently attributed to the fact the inversion velocity, other conditions being equal, decreases with the increase of liquid rate per unit area of cross-section and increases with the free volume of the packing.

The relation of factor C and regime parameters of phase inversion for the investigated packings is shown in Figure 12.

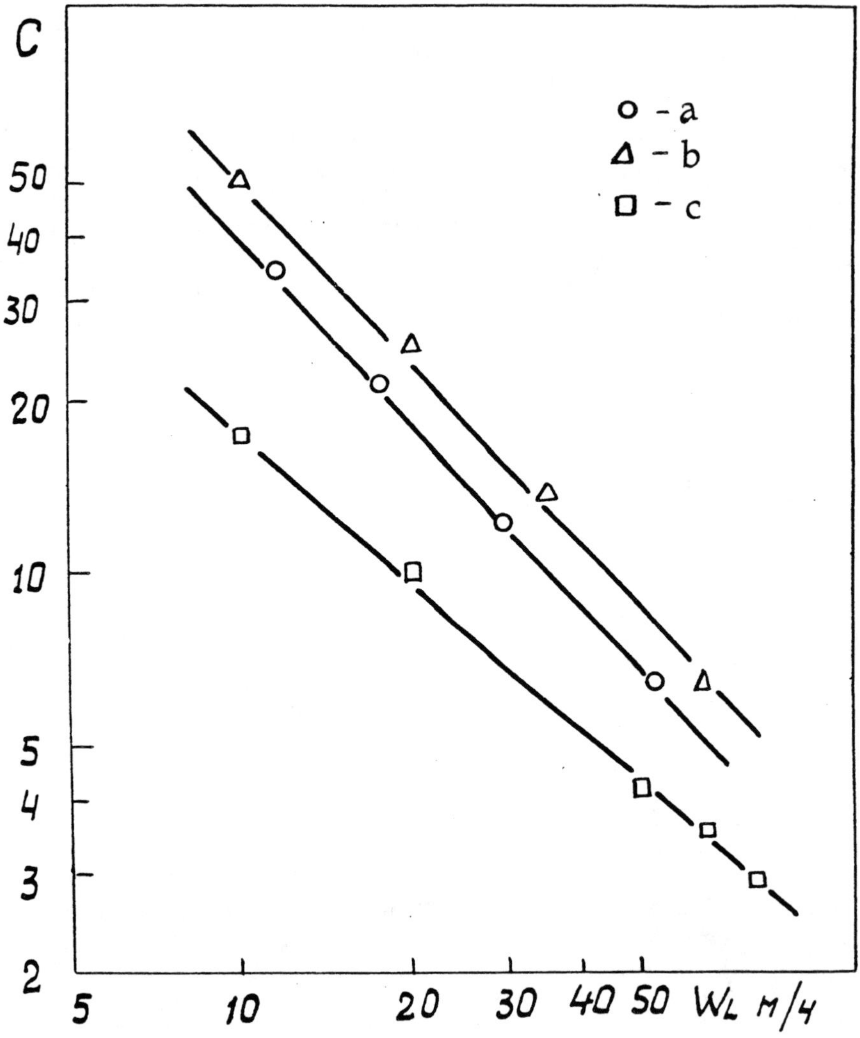

Figure 12. Dependence of C for studied packings from superficial liquid velocity: a - Raschig rings measuring 15x5x3 mm; b - Raschig rings measuring 25x25x4 mm; c - Pall rings measuring 25x25x1 mm.

Knowing the values of α one can use Equation (9) to calculate the pressure drop in the packing layer filled with gas-liquid emulsion (zone A in Figure 7). With values of ε and λ not available, one may use values of ε_0 and λ_0

which leads to an error of no more than 10 to 15%. The pressure drop of a packing layer with separate flow of phases (zone B in Figure 7), according to our experimental data, can be found with the help of Eduljee graphical dependence [13, 25]. By summing up the pressure drops of zones A and B according to Equation (6), we will find the total pressure drop of an irrigated packing with a layer of gas-liquid emulsion (Figures 7,8).

2.2. Wetted-Wall Column

The new method of stabilization and control of phase inversion regimes may be spread to other counterflow gas-liquid apparatuses, in particular to a vertical wetted wall column. It was experimentally confirmed in a glass column with an inner diameter of 27 mm and a length of 1500 mm. The column had a straight bottom cut and was provided at the top with a special nozzle reducing the pressure drop and ensuring the uniform irrigation in the conditions of a pipe bunch [26].

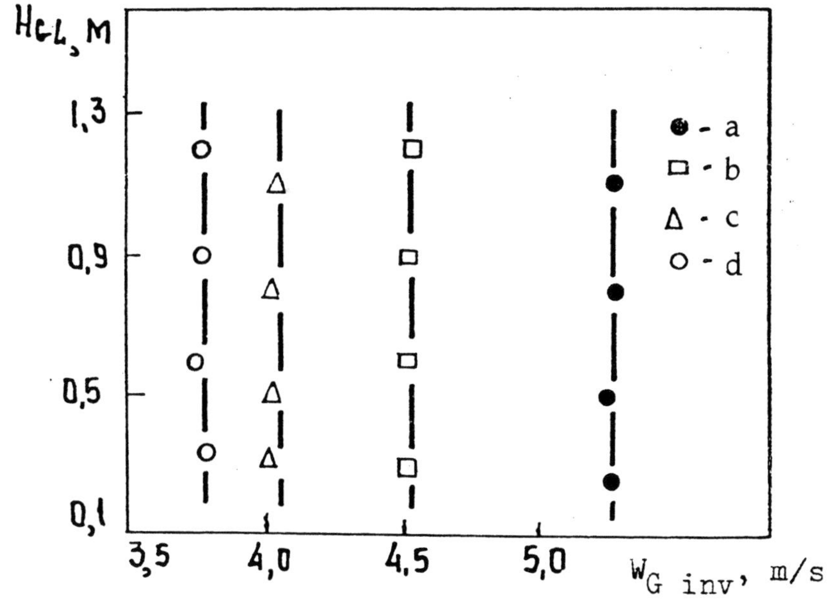

Figure 13. Coexistence of separate phase flows and phase inversion.

Various average stable values of the heights of a layer of gas-liquid emulsion in a vertical wetted-wall column are presented in Figure 13a. Figure 13b shows the photograph illustrating the existance of regimes of separate flows of phases (top column) and their inversion (bottom column). Here we speak about average values of H_{GL} since the value of the height of a layer of gas-liquid emulsion and related parameters continuously and sharply fluctuate around the average values which is almost uncharacteristic of an irrigated packing.

The registered ultimate loads with respect to water and air obey Walles's correlation [27] and in coordinates $\sqrt{J_G^*}$ and $\sqrt{J_L^*}$ group together around the straight line

$$\sqrt{J_G^*} + \sqrt{J_L^*} = 0,8 \tag{14}$$

It is evident that the original cause of flooding a wetted-wall column as an irrigated packing is an interphase shear stress. To determine the interphase shear stress in circular gas-liquid flows we propose a method which is based on the known fact [21, 28] that the relationship is fulfilled, with gas flowing in smooth, rough and wetted-wall columns:

$$W_{G,ax} - W_G = const \cdot W_{G*} \tag{15}$$

with the value of the constant being the same for different kinds of flow. Having taken the value of the constant equal to 4.07 [21], measured the axial velocity and determined the average thickness of liquid film, we may calculate $\tau = w_G^2 P_G$ with the help of the expression [29]

$$\tau = \rho_G \left[W_{G,ax} - \frac{4V}{\pi (d - 2\sigma)^2} \right]^2 / 16,56 \tag{16}$$

Axial velocity of gas was measured with the help of a thin L-shaped Pitot tube [30] and the average thickness of a liquid film was determined by the method of cutoff of flows [13].

It turned out that for a vertical column, like an irrigated packing, the preinversion shear stress decreases with the growth of a liquid rate per unit area of cross-section. So, the thicker the liquid film, the less the gas velocities at which phase inversion starts.

3. Mass Transfer

An investigation of mass transfer kinetics in created gas-liquid system is of theoretical and practical interest [31] (Figures 7,8).

3.1. Mass Transfer in Liquid Phase

Desorption of carbon dioxide from water to air was investigated when the process rate was almost fully controlled by the liquid phase resistance [13,

32, 33]. Water at 12±0.5°C was saturated with carbon dioxide in an additional column with a submerged packing, went to a pressure tank communicating with an atmosphere and through an irrigator was fed to the nozzle. To reduce the end effects related to the mass transfer in the space above and under the packing layer, the irrigator and sample taking funnel for draining the liquid were installed near the top and bottom ends of the packing layer, respectively.

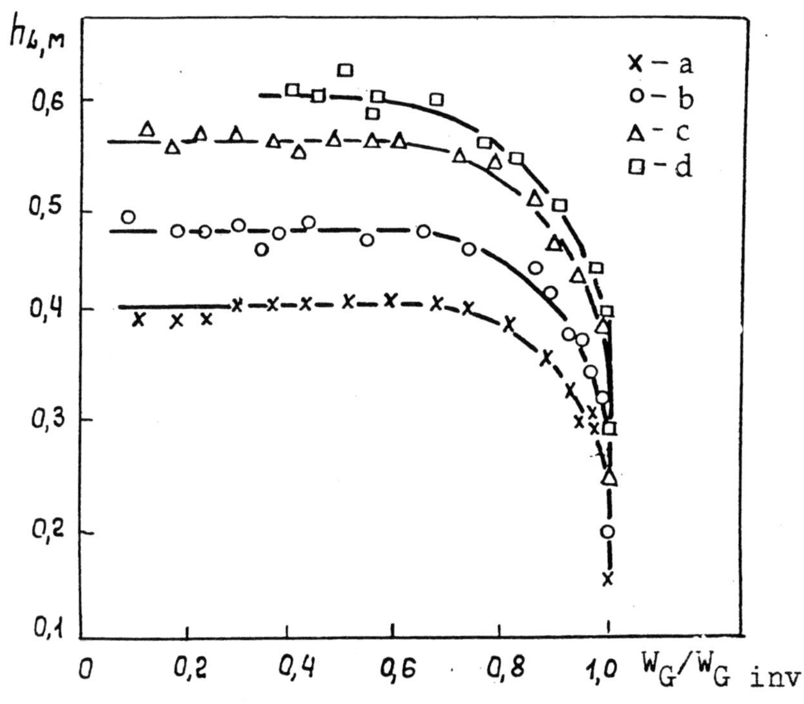

Figure 14. Dependence of height of a transfer unit from $W_G/W_{G,inv}$: a - q=17.33 $m^3/m^2 \cdot h$; b - 29.47; c - 41.37; d - 53.15.

Figure 14 illustrates a change in the height h_D of a transfer unit in a liquid phase with the growth of superficial air velocity W_G for a different liquid rate unit area of cross-section q=17.33 to 53.15 $m^3/(m^2.h)$ for a packing layer composed of Raschig rings measuring 15x15x3 mm.

The curves clearly reveal the known [13] hydrodynamic modes of interaction of gas and liquid at their countercurrent: film, horizontal sections of curves, loading point, drop region h_L, flooding (inversion), vertical sections of the curves. The latter mode occurs when $W_{G,inv}$ is reached, the value of which may be found by the known empirical correlations [13, 19].

As expected [13, 34-37], within the film mode the gas velocity does not affect the intensity of mass transfer which may be characterized by the value of h_L.

For processing, experimental data, use was made of the value of $\theta = (\nu_L^2 / g)^{1/3} = 5.38 \cdot 10^{-5}\ m$ and $D_{AB}=1.51 \cdot 10^{-9}\ m^2/s$. Schmidt's number did not vary in the experiments. According to the results of a large number of works discussed in [13] and conclusions made in works [38-41] the dependence $Sh_L \sim Sc_L^{0.5}$ was accepted which for a height of a transfer unit was transformed to $h_{L,f}/\theta \sim Sc_L^{0.5}$.

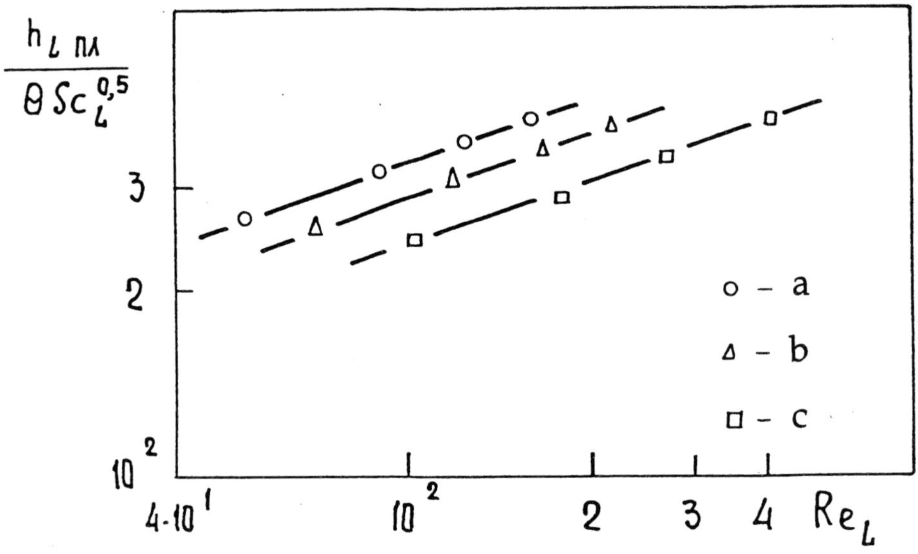

Figure 15. Correlation of experimental data on mass transfer in a liquid phase for the film regime: a - Raschig rings measuring 15x15x3 mm; b - Raschig rings measuring 25x25x4 mm; c - Pall rings measuring 25x25x1 mm.

Figure 15 shows the dependence of $H_{L,f}/(\theta Sc_L^{0.5})$ from Re_L which may be presented by equation $h_{L,f}= A\theta Re_L^n Sc_L^{0.5}$ [13]. The same value of coefficient n=0.35 was found in the experiments [13, 35] with Raschig rings measuring 12.5 mm. As to the values of coefficient A, they equal 65.8, 57.6 and 47.7 for Raschig small and large rings and Pall rings, respectively.

Upon attaining a certain value of $W_G/W_{G,inv}$ (Figure 14) the gas velocity starts influencing the hydrodynamic situation in the layer of the irrigated plating causing a liquid hold-up accompanied by the intensification of mass transfer process [13, 36, 42-44]. Further growth of W_G intensifies this process and, when the value of $(W_G/W_{G,inv})=1$, leads to an emergence of phase inversion regime accompanied with sharp rise of the intensity of heat and mass transfer [1,4,5].

It is evident that the leading regime exists within the velocity range from $W_{G,l.p.}$ to $W_{G,inv}$. Both these characteristic velocities depend upon the type and size of packing, the liquid rate per unit area of cross-section and physical-chemical properties of interacting phases [13, 22, 25]. The values of $W_{G,l.p.}$ and $W_{G,inv}$ for Raschig rings are recommended to be estimated by the empirical equations [13, 19, 45]

$$\lg Y = -0{,}073 - 1{,}750X \tag{17}$$

$$\lg Y = 0{,}022 - 1{,}750X \tag{18}$$

Thus, the velocity range related to the hold-up regime may be described by a set of equations (17), (18) that generalize the experimental data obtained with the help of hydrodynamic dependences like $\lg\Delta P \sim \lg W_G$. It is, however, known that the point of loading beginning cannot be clearly determined [46, 47] by such dependencies. Therefore, the estimation of the loading point range by the data on mass transfer is of interest. It should be noted that the values of $W_{G,l.p.}$ are to be refined since the values of $W_{G,inv}$, as was noted above, obey the equation (18).

Values of $W_{G,l.p.}$ were found by the point where the graphic dependence $h_L \sim W_G/W_{G,inv}$ starts deviating from the horizontal (Figure 14). The experimental and calculated (with reference to Equations (17) and (18) ranges of velocities ($W_{G,inv}$-$W_{G,l.p.}$) covering the loading point for Raschig rings measuring 15x15x3 mm, in per cent of maximum possible range (from 0 to $W_{G,inv}$) are given in Table 2.

Table 2.

q, $m^3/(m^2 \cdot h)$	17.33	29.47	41.37	53.15
$[(W_{G,inv}-W_{G,l.p.})/W_{G,inv}]_e \cdot 100\%$	32.00	35.00	40.00	41.00
$[(W_{G,inv}-W_{G,l.p.})/W_{G,l.p.})W_{g,inv}]_c \cdot 100\%$	15.10	17.00	22.50	24.10

It is clearly seen that the mass transfer data indicate a considerably large range of velocities covering the loading-point regime than it is predicted by Equations (17) and (18).

The intensity of mass transfer in the loading-point and phase inversion regimes may be characterized by values of $h_{L,\ell p}$ and $h_{L,inv}$ respectively. But then the relations of $h_{L,f}/h_{L,\ell.p.}$ and $h_{L,f}/h_{L,inv}$ show the increase of mass transfer intensity in these regimes as compared with an ordinary film regime.

Figure 16 shows the dependencies of $h_{L,f}/h_{L,l.p.}$ from Re_L for Raschig rings for various values of $W_G/W_{G,inv}$. The presented data testifies to the fact that the mass exchange intensity in the liquid phase increases as W_G

approaches $W_{G,inv}$, while the W_G and $W_{G,inv}$ being equal the increase of mass exchange intensity does not depend upon Re_L and is larger for large rings. Thus, for $(W_G/W_{G,inv})=0.99$ the value of $h_{L,f}/h_{L,l.p.}$ for Raschig rings measuring 15x15x3 mm is equal to 1.65 and for Raschig rings measuring 25x25x4 mm it equals 1.90. It should be noted that the extrapolation made in work [44] for the hypothetical hold-up regime $(W_G/W_{G,inv}=1)$ on Raschig rings measuring 25x25x3 mm has predicted an increase of mass exchange intensity by 2.15 times as compared with the film regime, which is rather close to the experimental value. The experimental data given in Figure 16 are described by the equation

$$h_{L,f}/h_{L,l.p.}=K(1-W_G/W_{G,inv})^{-0.15} \quad (19)$$

with coefficient K being equal to 0.83 and 0.91 for Raschig small and large rings, respectively.

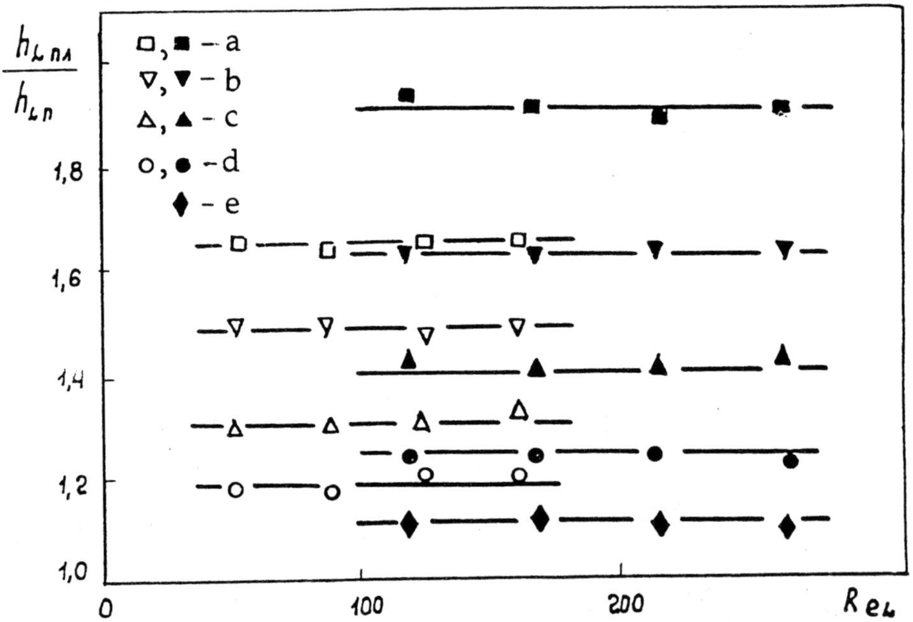

Figure 16. Dependence of $h_{L,f}/h_{L,l.p.}$ from Re_L: a - $W_G/W_{G,inv}$=0.99; b - 0.98; c - 0.95; d - 0.90; e -0.80; light points - data obtained for Raschig rings measuring 15x15x3 mm and dark points - data obtained for Raschig rings measuring 25x25x4 mm.

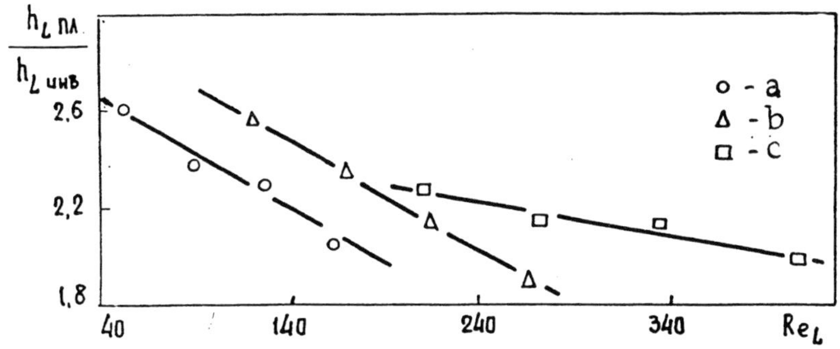

Figure 17. Dependence of $h_{L,f}/h_{L,inv}$ from Re_L for different packings: a -Raschig rings measuring 15x15x3 mm; b - Raschig rings measuring 25x25x4 mm; c - Pall rings measuring 25x25x1 mm.

Figure 17 shows the dependencies of $h_{L,f}/h_{L,inv}$ from Re_L.

For all tested packings, the increase of mass exchange intensity, as compared with the film regime, depends on the regime parameters of phase inversion. The degree of intensification is higher for lesser liquid rate per unit area of cross-section and, accordingly, larger values of $W_{G,inv}$, i.e. for larger values of preinversion shear stress (Figure 11). The mass exchange intensity drops with an increase of liquid loads, (when preinversion shear stress drops), the rate of drop being higher for Raschig rings than for Pall rings.

Thus, a conclusion on the different activity of the phase inversion regime, that was formulated on the basis of investigation of hydrodynamics of a layer of irrigated packing under ultimate load conditions, has been confirmed. The experimental data presented in Figure 17 are described by the equation

$$h_{L,f} / h_{L,inv} = B - k\,Re_L \tag{20}$$

with coefficient B values for small and large Raschig and Pall rings being equal to 2.88, 3.12 and 2.57, respectively, and coefficient $K \cdot 10^3$ being equal to 4.77 and 1.46 for Raschig and Pall rings.

It should be noted that Equation (19) was obtained in the experiments with Raschig rings, with the main parameters varying within the following ranges: $50 \le Re_L \le 270$. $W_{G,l.p.} < W_G < W_{G,inv}$. Equation (20) was derived in the experiments with Raschig and Pall rings for Re_L=50-400. The difference between the experimental and calculated data does not exceed ±1% and ±4% for the loading-point regime and phase inversion regime, respectively.

3.2. Mass Transfer in Gas Phase

Desorption of ammonia from water to air in a layer of irrigated Raschig rings, the characteristic of which is given above, was investigated. The height of a packing layer amounted to 900 mm. The experiments were conducted at water temperatures of 21 to 23°C. A portion of resistance to mass transfer on the part of the liquid phase was calculated by the equation [48]

$$h_G = h_{0G} - \frac{mG}{L} h_L \left(\frac{D_{NH_3}}{D_{CO_2}} \right)^{0,5}$$

where h_L=height of a transfer unit obtained at desorption of carbon dioxide under similar hydrodynamic conditions. The resistance on the side of the liquid phase in the experiments amounted to 3-17% of the total resistance to mass transfer.

As in the case of mass exchange in the liquid phase one might expect that a change in the hydrodynamic regime of phase interaction would have a considerable effect on mass exchange intensity in the gas phase. However, there is not a common opinion on the part of a nature of influence of a change of hydrodynamic regime of gas and liquid interaction upon the mass exchange intensity in the gas phase. Some researchers [49] have found such an influence, others have not [50].

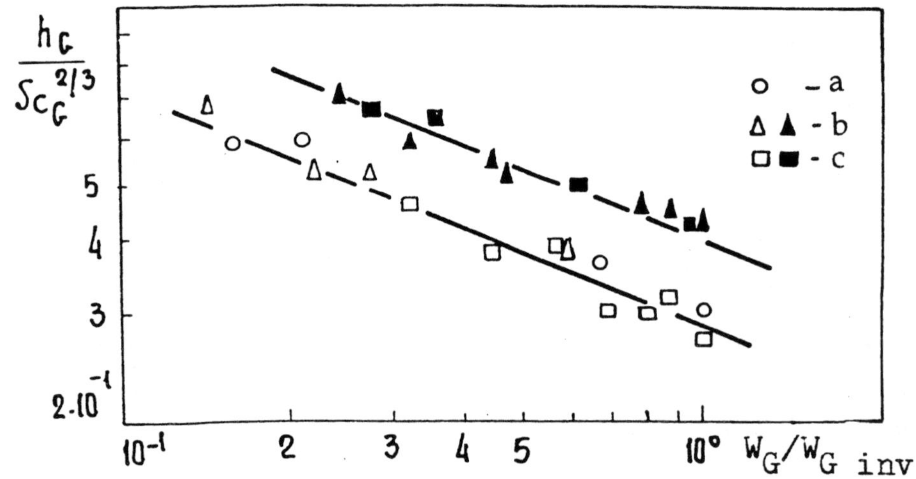

Figure 18. Correlation of experimental data on mass transfer in gas phase for Raschig rings measuring 15x15x3 mm (light points) and 25x25x4 mm (dark points): a - q=17.33 $m^3/m^2 \cdot h$; b - 29.47; c - 41.37.

In processing our experimental data we used the value of a diffusion coefficient of ammonia in air $D_{AB}=19.29\cdot10^{-6}$ m^2/s. Schmidt's number was invariable. On the basis of a large number of works discussed in [13] the dependence $Sh_G \sim Sc_G^{\frac{1}{3}}$ was accepted which for the height of a transfer unit is reduced to the following form: $h_G \sim Sc_G^{\frac{2}{3}}$.

Figure 18 presents the dependence of mass transfer effectiveness in a gas phase on Raschig rings from the gas relative velocity. The processing of the experimental data gives the equation

$$h_G = 33d_{eq}(W_G / W_{G,inv})^{-2/5} Sc_G^{2/3} \tag{21}$$

From the data in Figure 18 and Equation (21) it follows that the height of a transfer unit in the conditions of a film regime, hold-up regime and phase inversion regime can be derived from one and the same equation. It turned out that under all conditions h_G decreases with a rise in gas velocity and increases with the increment of sizes of packings.

Equation (21) is obtained within the following range of variation of the main parameters: $50 \leq Re_L \leq 170$, $0.1<(W_G/W_{G,inv})\leq 1$. The difference between the experimental and calculated data is not more than ±10%. So, Equations (19), (20) and (21) can be used for the kinetic calculations of a layer of irrigated packing with steady phase inversion regime (Figures 7,8). Finally, the new method of stabilization and control of phase inversion process [16] and related apparatus [17,18] have shown a good advantage in the course of experimental checkups and experimental-industrial operation in a number of production process at the stages of absorption, desorption and chemisorption. Presently, the sphere of practical application of such apparatus is being expanded.

The author deeply appreciates the continuous interest and active participation of Professor N.N. Kulov in the works of this trend.

REFERENCES

1. *Discovery*, No. 141, "Discoveries, Inventions, Designs, Trade Marks," 1974, No. 30, p. 3.
2. V.V. Kafarov, L.I. Blyakhman, *Zhur. prikladn. Khimii*, V. 23, No. 3, 1950, pp. 244-255.
3. V.V. Kafarov, L.I. Blyakhman, *Zhur. prikladn. khimii.*, 1951, V. 24, No. 12, pp. 1274-1290.
4. V.V. Kafarov, L.I. Blyakhman, A.N. Planovsky, *Teor. osnovy khimich.* Tekhnologii, 1976, V. 10, No. 3, pp. 331-339.
5. V.V. Kafarov, Osnovy massoperedachi, *M. yssh.* shkola publishers, 1979, p. 439.
6. H. Schrader, *Kaltetechnik*, 1958, No. 9, pp. 290-295.

7. P.C. Brooks, B.E.T. Hutton, L.S. Leung, D.S. Ficklin, *Chem. Eng.*, 1976, V. 4, p. 152.
8. L.S. Leung, D.J. Nicklin, P.C. Brooks, B.E.T. Hutton, *Chem. Eng.*, 1976, V. 83, No. 27, p. 88.
9. Author's certificate No. 104499, *Inventions Bulletin*, 1956, No. 11, p. 9.
10. Author's certificate No. 319548, *Discoveries, Inventions, Designs, Trade Marks*, 1971, No. 33, p. 66.
11. L.I. Blyakhman, A.M. Yakubson, In col.: *Protsessy khimicheskoi tekhnologii.* Gidrodinamika, teplo i massaperedacha, L., Nauka publishers, 1965, pp. 67-71.
12. *Raschety osnovykh protsessov i apparatov neftepererabotki,* edited by A.N. Sudakov, Khimiya publishers, 1979, p. 568.
13. V.M. Ramm, *Absorbtsiya gazov,* M., Khimiya publishers, 1976, p. 656.
14. *Masshtabnyi perekhod v khimicheskoi tekhnologii,* edited by A.M. Rozen M., Khimiya publishers, 1980, p. 320.
15. L.I. Blyakham, S.G. Sorokin, A.I. Lumpov. *Khim. prom.*, 1983, No. 4, pp. 231-235.
16. A.A. Ilinykh, Z.N. Memedlyaev, N.N. Kulov, V.A. Mayusov, *Teor. osnovy khim. tekhnol.*, 1987, V. 21, No. 2, pp. 184-190.
17. Author's certificate No. 1095917, *Discoveries, Inventions*, 1984, No. 21, p. 7.
18. W.A. Bain, O.A. Hougen, *Trans. Am. Inst. Chem. Engrs.*, 1944, V. 40, No. 1, pp. 29-49.
18. Author's certificate No. 1140821, *Discoveries, Inventions*, 1985, No. 7, p. 28.
19. W.A. Bain, O.A. Hougen, *Trans. Am. Inst. Chem. Engrs.*, 1944, V. 40, No. 1, pp. 29-49.
20. I.I. Gelperin, A.M. Kogan, V.V. Dilman, et al. *All-Union conference on aerohydrodynamics of chemical apparatus, part 3,* Severodonetsk, 1981, pp. 29-34.
21. G. Shlikhting, *Teoriya pogranichnogo sloya,* M., Nauka publishers, 1974, p. 712.
22. L.V. Alekperova, Yu.V. Akselrod, V.V. Dilman, et al. *Khim. prom.*, 1974, No. 5, p. 60-64.
23. L.I. Titelman, Yu.V. Akselrod, V.V. Dilman, et al., *Teor. osnovy khim. tekhnol.* 1977, V. 11, No. 3, pp. 389-392.
24. Author's certificate No. 1265542, *Discoveries, Inventions*, 1986, No. 39, p. 144.
25. H.E. Eduljee, *Brit. Chem. Engng.*, 1960, V. 5, No. 5, pp. 330-334.
26. Author's certificate No. 971388, *Discoveries, Inventions, Designs, Trade Marks*, 1982, No. 41, p. 42.
27. G. Wallis, *Odnomernye dvykhfaznye techeniya, M.* Mir publishers, 1972, p. 440.
28. S.K. Myasnikov, N.N. Kulov, *Teor. osnovy khim. tekhnol.*, 1982, V. 16, No. 5, pp. 663-669.
29. Author's certificate No. 1341545, *Discoveries, Inventions*, 1987, No. 36, p. 200.

30. Z.N. Memedlyaev, *Teor. osnovy khim. tekhnol.*, 1980, V. 14, No. 4, pp. 620-622.
31. A.A. Ilinykh, Z.N. Memedlyaev, N.N. Kulov, *Teor. osnovy khim. tekhnol.*, 1989, V. 23, No. 5, pp. 569-574.
32. V.M. Olevsky, V.R. Ruchinsky, *Rektifikatsiya termicheski nestoikikh produktov*, M., Khimiya publishers, 1972, p. 200.
33. S. Kamei, J. Oishi, *Met. Fac. Eng.*, Kyoto Univ., 1955, V. 17, No. 4, p. 227.
34. T. Sherwood, R. Pigford, Ch. Wilky, *Massoperedacha*, m. Khimiya publishers, 1982, p. 696.
35. T.K. Scherwood, F. Holloway, *Trans. Am. Inst. Chem. Engrs.*, 1940, V. 36, No. 1, pp. 21-70.
36. A.G. Kasatkin, I.N. Tsiparis, *Khim. prom.*, 1952, No. 7, pp. 203-210.
37. I.A. Gildenblat, N.M. Gurova, V.M. Ramm, *Khim. prom.*, 1968, No. 3, pp. 219-222.
38. V.V. Dilman, *Teor. osnovy khim. tekhnol.*, V. 1, No. 4, pp. 438-445.
39. I.A. Gildenblat, A.I. Rodionov, A.L. Loshakov, *Teor. osnovy khim. tekhnol.*, 1969, V. 3, No. 3, pp. 334-354.
40. I.A. Gildenblat, A.I. Rodionov, B.I. Demchenko, *Dokl. Akad. Nauk SSSR*, 1971, V. 198, No. 5, pp. 1149-1151.
41. N.N. Kulov, V.V. Maksimov, V.A. Malyusov, N.M. Zhavoronkov, *Teor. osnovy khim. tekhnol.*, 1983, V. 17, NO. 3, PP. 291-306.
42. T. Teutsch, *Chem. Eng. Techn.*, 1965, V. 37, No. 12, pp. 1224-1229..
43. J.E. Vivian, P.L.T. Brian, V.J. Krukonis, *A. IChE Journal*, 1967, V. 13, No. 1, pp. 174-175.
44. T. Yilmar, *Chem. Ing. Techn.*, 1973, V. 45, No. 5, pp. 253-259.
45. A.N. Planovsky, V.V. Kafarov, *Khim. prom.*, 1946, No. 4, pp. 13-35.
46. D. Perry, *Spravochnik inzhenera-khimika*, L. Khimiya publishers, 1969, V. 2, pp. 504.
47. N.M. Zhavoronkov, *Khim. prom.*, 1949, No. 3, pp. 68-75.
48. Yu.A. Avdonin, V.M. Olevsky, D.M. Popov, *Khim. prom.*, 1949, No. 6, pp. 171-177.
50. V.N. Tychinin, V.P. Tolstykh, E.D. Vertuzaev, *Teor. osnovy khim. tekhnol.*, 1973, V. 7, No. 6, pp. 916-918.

HYDRODYNAMICS AND MASS TRANSFER IN AGITATED GAS-LIQUID SYSTEMS

V.M. Barabash

Abstract—Design relationships for a quantitative assessment of sizes of bubbles inside an agitated vessel, the axial distribution of the gas holdup and the flow rate of the gas entrained from the surface are obtained on the basis of the analysis of the physical laws governing the processes of formation and transport of elements of the dispersed phase. The method for analysis of the mass transfer process in agitated gas-liquid systems is developed with the laws of the turbulent flow near deformable and rigid surfaces taken into account along with the conclusions of the local isotropic turbulence theory. The operation of the conventional ring-shaped sparger was analyzed and its modified construction is proposed, which contributes to a higher intensity of mass transfer. The results of the theoretical analysis are compared with experimental data obtained in research work and in the published work as well.

Symbols

A	constant
B	constant
C	concentration of absorbed component in solution, kg/m^3
C_{eq}	equilibrium concentration of absorbed component in solution, kg/m^3
d_s	diameter os stirrer, m;
d_b	diameter of bubble, m
d_{cr}	minimum diameter of oscillating bubble, m
D	diameter of vessel, m
D_E	eddy-diffusion coefficient, m^2/s
D_M	molecular-diffusion coefficient, m^2/s
f_n	natural frequency of bubble oscillation, s^{-1}
f_w	pulsation frequency in wake, s^{-1}
F_{sp}	specific phase contact surface, m^{-1}
g	acceleration due to gravity, m/s^2
G_g	gas flow rate, m^3/s
h	current coordinate of height, m

h_b	height of stirrer blade, m
h_s	height of stirrer above vessel bottom, m
H	height of vessel filling, m
$H_{g\text{-}l}$	height of gas-liquid layer, m
K_N	power factor
j	flux of substance from bubble surface, $kg/m^2 \cdot s$
n	rotary speed of stirrer, s^{-1}
N	power consumed for mixing bulk liquid, W
N_G	power consumed for mixing gas-liquid medium, W
P	fraction of pulsations resulting in catching bubble
q	amount of absorbed gaseous component, kg/m^3
r	current value of radius, m
R	radius of vessel, m
R_D	diffusion resistance on bubble surface, s/m
v'_o	pulsation speed in scale δ_0, m/s
$v'_{o\lambda}$	pulsation speed in scale λ, m/s
v_r	bubble rise velocity within operating zone of vessel, m/s
v_r^s	bubble rise velocity within near-surface zone, m/s
V	operating volume of vessel, m^3
W'	reduced gas velocity, m/s
W'_s	reduced velocity of gas entrained from surface, m/s
y	current distance from bubble surface, m
α	fraction of front bubble surface where mass transfer is based on averaged motion
α_b	fraction of back bubble surface where mass transfer is based on pulsation motion
α_p	coefficient of proportionality
β	linear mass-transfer coefficient, m/s
β_v	volumetric mass-transfer coefficient, s^{-1}
δ_0	thickness of layer with main diffusion resistance, m
ε_0	mean energy dissipation rate in stirred vessel, W/kg
ε_w	energy dissipation rate in wake zone, W/kg
ε_0^1	local value of energy dissipation rate, W/kg
λ	pulsations scale, m
μ	dynamic viscosity of bulk liquid, Ns/m^2
ν	kinematic viscosity of bulk liquid, m^2/s
ρ	density of bulk liquid, kg/m^3
ρ_g	density of dispersed phase, kg/m^3
σ	surface tension, N/m
τ	current time, s
ψ	local gas holdup
φ_m	mean gas holdup
φ_s	gas holdup in near-surface zone

$$h = \frac{h}{H_{g-1}}$$

$$\Gamma_D = \frac{D}{d_s}$$

$$r = \frac{2r}{d_s}$$

$$We = \frac{\rho \cdot n^2 \cdot d_s}{\sigma} \text{ – Weber number}$$

$$Re = \frac{v_r d_b}{v} \text{ – Reynolds number}$$

$$Sc = \frac{v}{D_s} \text{ – Schmidt number}$$

INTRODUCTION

In cases where use is made of the gas phase in agitated vessels the most characteristic phenomenon for all the processes occuring in the gas-liquid system, is the mass transfer which can be intensified by considerable power input per unit volume, (up to 100 kW and upwards per cubic meter of liquid), applied to the two-phase layer.

The theory and practical experience accumulated over the years shows that the use for this purpose of the volumetric-type apparatuses such as agitated vessels, rather than mass-exchange columns, is expedient in cases where considerable quantities of sparingly-soluble gases are to be absorbed, usually with the subsequent chemical reaction between the absorbed gas and the substances dissolved in the liquid. These processes include oxidation, chlorination, etc.

Now the mass transfer processes are also being used extensively in three-phase gas-liquid-solid systems where the gaseous reagent absorbed by the liquid reacts with the solid particles suspended in the liquid.

The processes agitating gas-liquid systems is used for the physical absortion or chemisorption in chemical reactors [1], where the reaction proceeds between the gas and the liquid and where part of the components is delivered with the gas phase.

The demand for the use of the processes occurring in the gas-liquid systems has grown considerably lately with the advent of the microbiological industry [2] and the fact that these processes are very promising for the purification of utility and industrial waters [3]. In the chemical industry the volumes of the apparatuses used for mixing gas-liquid systems seldom exceeds 50 to 100 m^3, and those used in the microbiological industries, a few hundreds of cubic meters, the volume of the aerated reservoirs in the latter case reaches a million and more cubic meters.

The processes of mass transfer in the gas-liquid system, just as in the cases of process mixing in the solid-liquid and liquid-liquid systems, are governed not only by the properties of the phases being mixed, but also by the hydrodynamic situation in the vessel [4]. The most important characteristics affecting the mass transfer between the gas and liquid phases are the power spent for agitation, the average gas holdup, the size of the bubbles and the character of their distribution within the volume being mixed.

In addition to this, a considerable effect on the intensity of the process of mass transfer is exerted by not only the rate of the gas flow through the sparger, which is usually mounted under the stirrer, but also by the phenomenon of entrainment of gas from the surface of the liquid.

It should be noted that where the analysis of the experimental data concerning the power input and mean gas holdup is dealth with in numerous publications [5-11], little information is available on the analysis of the data concerning the distribution of gas holdup, [12] bubble sizes [7] throughout the volume of the vessel, and the flow of the gas entrained from the surface of the liquid. As a rule, most of the empirical relationships have been obtained within a narrow range of variations in the operating conditions and vessel design characteristics which considerably limit their practical use in designing industrial facilities.

Nonetheless, the available data on studying the process of mixing in the gas-liquid systems and results obtained in the physical-and-chemical hydrodynamics [14] and in the theory of turbulent flows [15] make it possible to conduct a more detailed analysis of some physical phenomena underlying this process and to approach the development of the generalized methods of its engineering analysis more closely [16, 17].

The structure of the gas-liquid mixture in agitated vessels depends, to a great extent, on the intensity of agitation, gas introduction conditions and gas flow rate.

When the gas is introduced through a sparger, located under a stirrer, practically no gas dispersion occurs at a low agitation intensity, and in this case the gas rises in the form of large-size bubbles in the central section of the vessel. The beginning of the gas dispersion and an increase in the gas holdup (Figure 1) corresponds to a certain minimum level of mixing intensity which depends on the gas flow rate, the size of the vessel and the properties of the phases. As shown in publication [18] in the case of agitation with a turbine-type stirrer this minimum level of mixing intensity in the aqueous solutions - air systems is determined by the equations:

$$\frac{n_o d_s}{D}(gD)^{1/2} = 0.07 \text{ at } D < l\, m \tag{1}$$

$$\frac{\mu n_0 d_s^2}{D\sigma}\left(\frac{\rho\sigma^3}{g\mu^4}\right)^{1/4}\left(\frac{H - h_s}{D}\right)^{1/2} \quad \text{at D} < \text{l m} \tag{2}$$

where σ is the surface tension at the interface between the phases, N/m;
μ is the viscosity of the continuous phase, Pa·s;
g is the acceleration due to gravity, m/s^2;
n_0 is the rotary speed of the stirrer (s^{-1}) corresponding to the beginning of dispersion;
$h_s = \frac{H}{3}$ is the height at which the stirrer is installed above the vessel bottom, m;
D is the vessel diameter, m;
H is the height to which the vessel is filled with the liquid, m;
d_s is the stirrer diameter, m.

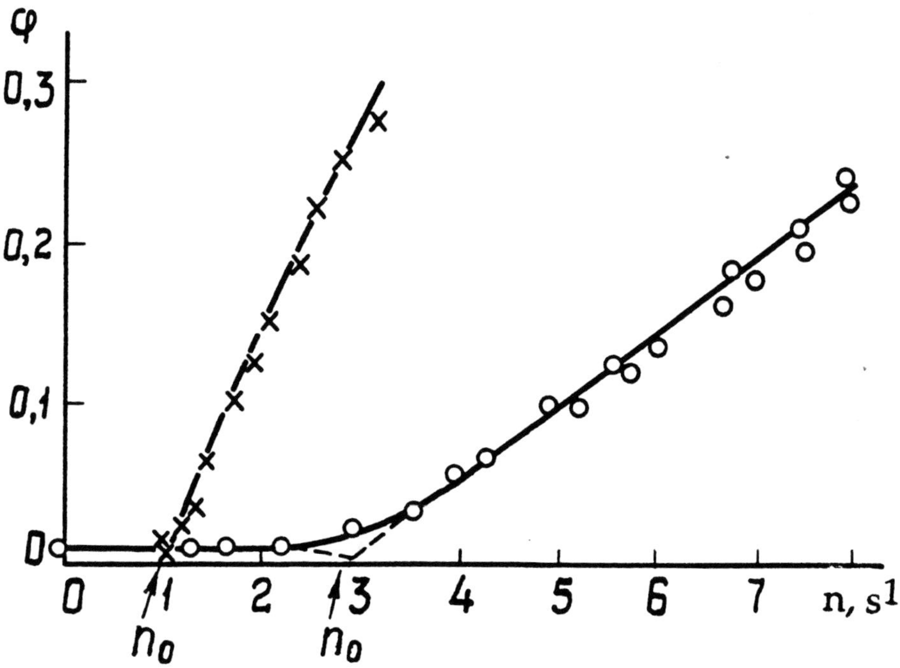

Figure 1. Gas holdup as a function of stirrer rotational speed. System: aqueous solution of aminocaprolactam-air [18, 20]: W'=0.005 m/s; H=D; x-d_s=0.3; D=0.45 m; 0-d_s=0.14 m; D=0.29 m.

The applicability of equations (1) and (2) is limited by the relationships:

$$W' \leq 0.03,\ 0.1 < \frac{H - h_s}{D} < 0.6 \tag{3}$$

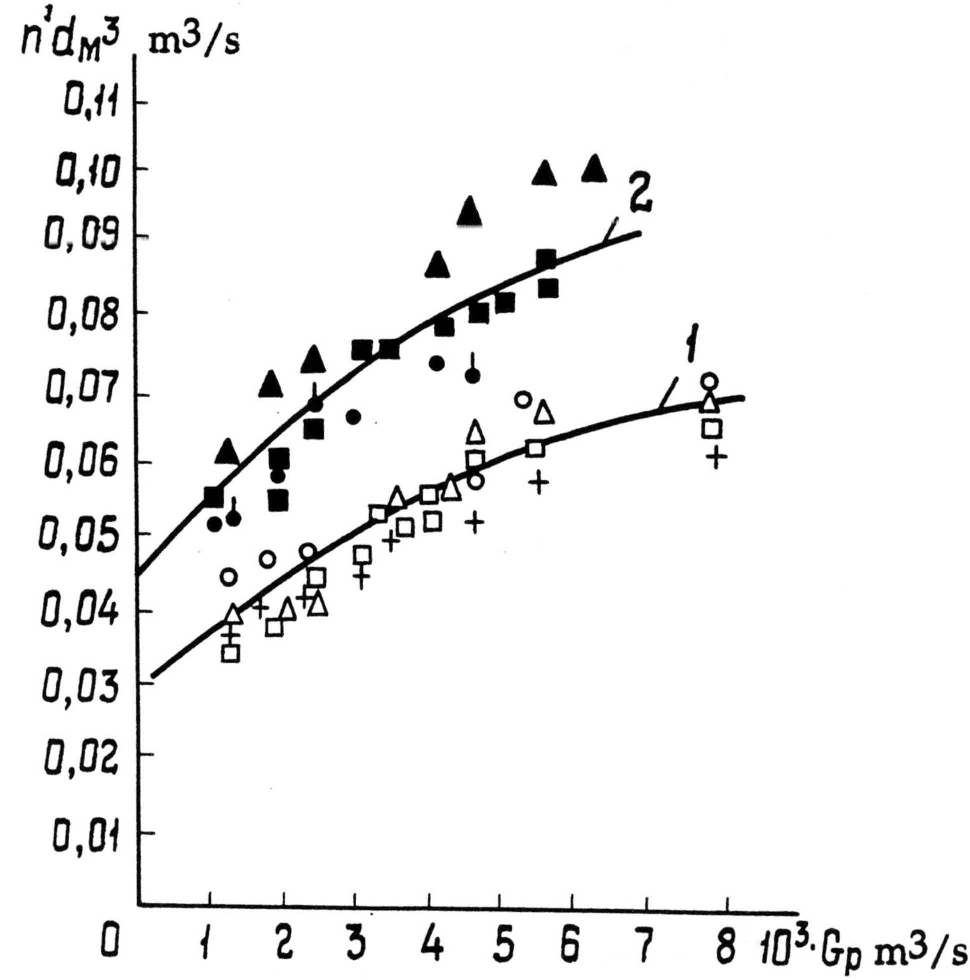

Figure 2. Value of $n'\cdot d_s^3$ as a function of rate of gas flow through vessel; 1-$\frac{h_s}{D}$=0.5; 2-$\frac{h_s}{D}$=0.25; D=0.8 m; =1.0 m; -d_s=0.25 m; -d_s=0.2 m; -d_s=0.182 m -d_s=0.16 m (water-air); Solid lines-calculations from Equation (5).

where W' is the reduced velocity of the gas delivered through the sparger, m/s.

The equation for determining n_0 obtained in publication [19] is of the form:

$$\frac{n_0 d_s}{(\sigma g/\rho)^{1/4}} = A + B(D/d_s) \tag{4}$$

where A=1.22 and B=1.25 for turbine-type stirrers; A=2.25 and B=0.6 for paddle stirrers.

In situations where the intensity of agitation just barely exceeds the dispersion threshold, the bulk mass of the gas delivered into the vessel becomes dispersed and distributed as small bubbles throughout the vessel volume located above the plane of stirrer rotation, with the volume lying below the stirrer being free from aeration. As the stirrer rotational speed increases the discrimination between the aerated and unaerated zones becomes less stable, and as soon as a certain critical value of the rotational speed is exceeded the entire volume of the vessel becomes filled with gas-liquid mixture.

This critical value of rotational speed of the stirrer increases with the increasing gas flow rate and height at which the stirrer is located above the vessel bottom (Figure 2).

In publication [21] the following equation has been obtained for open turbine-type stirrers

$$n' d_s^3 = \frac{0.0075 g D^2 (1 + 4h_s / D)}{0.35 + 0.65 \exp[-0.138\left(\frac{G_g}{n}\frac{10^2}{d_s^3}\right)^{1,1}} \tag{5}$$

where Gg is the volumetric flow rate of the gas delivered through the sparger, m3/s.

At values of rotational speed of stirrer n>n' the medium agitated in the vessel is essentially a gas-liquid mixture consisting of gas bubbles and the continuous phase. The principal characteristics of this sytem which affect the mass transfer in gas-liquid systems directly or indirectly are: the size of the bubbles, the gas volume fraction, the intensity of spontaneous gas entrainment from the surface of the liquid, energy distribution throughout the working volume of the vessel and the method of gas delivery through the sparger.

This paper is devoted to studying these characteristics and their effect on mass transfer.

1. Size of Bubbles in Agitated Gas-Liquid Systems

The principal element of any gas-liquid flow are the bubbles whose size and number are determined by the local value of the gas holdup and the specific interphase contact surface.

A number of relationships are suggested in the literature for determining the size of the bubbles as a function of the properties of the working media and agitation conditions. Nearly all these relationships are based on the mechanism of breakup of bubbles in a turbulent flow suggested by Calderbank

[22] by an analogy with the Kolmogorov's model for breakup of drops [23]. In this model, the breakup effect is caused by the fact that in a turbulent stream the liquid velocity varies from one point to another and, hence, two different points are subjected to different dynamic heads. Once the difference between these heads is in excess of the capillary pressure, the bubble deformation and breakup occur.

Having analyzed the bubble breakup in uniform and isotropic turbulence field on the basis of the suggested mechanism, the authors of publication [22] have obtained the relationship relating the stable size of the bubble to the characteristics of the working medium and to the local rate of energy dissipation:

$$d_8 \sim \left(\frac{\sigma}{\delta}\right)^{0,6} \left(\varepsilon_0^l\right)^{-0,4} \tag{1.1}$$

The experimental data obtained in the agitated vessels with turbine-type stirrers for the water-air system were processed on basis of this relationship and, as a result, the following relationship has been obtained for the value of the "sauter" diameter of bubbles averaged throughout the volume of the vessel:

$$d_8 = 1,15\left(\frac{\sigma}{\rho}\right)^{0,6} \varepsilon_0^{-0,4} \varphi_m^{0,5} + 0,0009 \tag{1.2}$$

For the gas-aqueous solutions systems of electrolytes the following relationship was suggested in publication [22].

$$d_8 = 2,25\left(\frac{\sigma}{\rho}\right)^{0,6} \varepsilon_0^{-0,4} \varphi_m^{0,4} \frac{\mu_8}{\mu} \tag{1.3}$$

and for aqueous alcohol solutions:

$$d_8 = 1,9\left(\frac{\sigma}{\rho}\right)^{0,6} \varepsilon_0^{-0,4} \varphi_m^{-0,4} \varphi_m^{0,65} \left(\frac{\mu_g}{\mu}\right)^{0,25} \tag{1.4}$$

where: σ is the surface tension, N/m;

ρ is the density of continuous phase, kg/m^3;

ε_0 is the average energy dissipation rate, W/kg;

φ_m is the mean volume fraction of the gas phase (holdup) in the vessel.

The calculations conducted for equations (1.2) to (1.4) show that the mean bubble diameter within the working volume of the vessel is equal to 5 to 6 mm, i.e. 2 to 4 times the bubble diameter within the stirrer zone for which, in taking of the local rate of energy dissipation in this zone into account with the normal law of distribution of pulsation velocities [15], the bubble diameter can be determined [4] from the formula

$$d_8 = 0{,}155\left(\frac{\sigma}{\rho}\right)^{0{,}6}\left(\frac{\rho}{\rho_{\bar{y}}}\right)^{0{,}2}\varepsilon_0^{-0{,}4} \tag{1.5}$$

where ρ_g is the density of the gas phase, kg/m^3.

Apart from physical and chemical characteristics of working media equations (1.2) to (1.5) contain the average-over-volume characteristics of the gas-liquid flow - the energy dissipation rate ε_0 and the gas holdup φ_0.

Since, despite the local uniformity and isotropy of the flow [24, 25] in agitated vessels, the rate of energy dissipation is found to be essentially non-uniform within the entire working volume [26, 27], then a wide spectrum of bubble sizes should take place in conformity with (1.1), which is not the case in reality [22].

To resolve these contradictions an attempt to conduct the theoretical analysis of the process of bubble breakup was undertaken for the cases where the gas-liquid systems are mixed under conditions of a well-developed turbulent flow of the working medium.

In analyzing problems relating to the determination of bubble sizes one should consider the two cases encountered in practice which differ from each other in the character of formation and distribution of the gas in the vessel:

a) the working medium does not contain the additives preventing coalescence;
b) the working medium contains the surface-active substances (SAS) or the electrolytes preventing coalescence.

In the theoretical analysis of the bubble breakup in the presence of coalescence (version "a") the following assumptions have been made:

1) The working volume of the vessel consists of two zones, (the bulk volume of the vessel and the radial stream formed by the stirrer blades), each of which is characterized by its own, specific only for this zone, bubble size resulting from the breakup and coalescence.
2) Gas bubbles take part in oscillatory motion.

First, consider the zone of the bulk volume of the vessel which is characterized by a uniform distribution of the gas phase [17].

Within the framework of the mechanism assumed in this paper the breakup of the bubbles within this zone is caused by the behavior of the continuous phase in the presence of the resonance between the natural bubble oscillation frequency and the frequency of the pulsative motion of the working medium resulting from the breakaway of eddies from the surface of an oscillating body washed by fluid flow.

Figure 3. Flow structures in bubble rise in stagnant liquid.

In the conditions normally encountered in practice the gas holdup in the vessel ranges from 5 to 20%, which means that the average distance between the bubbles and the bubble size are values of the same order of magnitude.

Taking this fact into account, the zigzag pattern followed by the rising bubble [22] and the presence of the wake zone formed behind every bubble [29] (Figure 3) it can be assumed that the bubble size is determined mainly by the characteristcis of the flow in the wake zone.

To assess the value of the rate of energy dissipation within this zone let us write down the expression for the power consumed when the spherical-shape bubble is washed by fluid flow

$$N_8 = \xi_6 \frac{\rho v_r^2}{2} v_r \frac{d_8^2}{4} \tag{1.6}$$

Since the Reynolds number for the rising bubbles of the above-mentioned sizes is not less than 10^3, the value ξ_b may be assumed to be constant and equal to 0.8 [28].

Since for all practical purposes power N_b is dissipated completely in the nearest wake [30], i.e. in the zone whose volume is approximately equal to the volume of the bubble, then, taking the expression for the bubble rise velocity into account [28]:

$$v_r = \sqrt{\frac{2\sigma}{d_8(\rho - \rho_g)} + \frac{g d_8}{2}(1 - \rho_g/\rho)} \tag{1.7}$$

the rate of energy dissipation within the breakaway zone of the flow can be determined as follows:

$$\varepsilon_w = \frac{N_8}{\rho V_w} = \frac{0,6}{d_8}\left[\frac{2\sigma}{d_8(\rho - \rho_g)} + \frac{g d_8}{2}\left(1 - \frac{\rho_g}{\rho}\right)\right]^{3/2} \tag{1.8}$$

The common characteristic for sufficiently large-size bubbles in both moving and still liquid is their oscillatory motion. The amplitude of these oscillations turns out to be the greater as the bubble size increases. For instance, the amplitude of oscillations of very large bubbles may be so great that at the very first oscillation they burst, (as if they puncture themselves), becoming transformed into torus [31]. The bubbles of a smaller diameter oscillate with smaller amplitudes and, when they're not subjected to external effects, retain their integrity. Small bubbles retain their integrity in the liquid in both the presence or absence of oscillations even when subjected to intensive external effects (for instance, pulsations in a turbulent liquid flow).

Let us assess the critical (minimum) size of the bubble that is susceptible to deformation.

In the most general case two real causes of the bubble deformation exist:

a) the dynamic head, acting on the bubble due to turbulent pulsations of velocity;
b) the dynamic head, acting on the bubble due to the bubble rise.

Taking into account the fact that v_r·0.23 to 0.3 m/s and the size of the liquid pulsation velocity v' for the bubble encountered in the currently-used agitated vessels determined from Kolmogorov's - Obukhov's relationship [32, 33]

$$v' = (\varepsilon_0 d_8)^{1/3} \tag{1.9}$$

normally does not exceed 0.2 m/s, the minimum size of the oscillating bubble may be determined from the relationship

$$\frac{2\sigma}{d_{cr}} = \frac{\rho v_r^2}{2} \tag{1.10}$$

Substituting the well-known relationship for determining the rise velocity of single small-size bubbles [28]

$$v_r = \frac{1}{18}\frac{g d_8}{v} \tag{1.11}$$

into expression (1.10) shall yield:

$$d_{cr} = 2,55\left(\frac{\sigma v^2}{\rho}\right)^{0,2} \tag{1.12}$$

For the working water-air medium the value of d_{cr} is equal to 1.5 mm, and for SAS solutions (δ=3·10^{-2} N/m) d_{cr}ı1.25 mm. Thus, in the bulk volume of the agitated vessels employing working media without SAS or electrolytes the bubbles are in a constant oscillatory motion.

According to [31] the minimum possible natural oscillation frequency of the bubble can be determined from the formula:

$$f_n = \frac{1}{2?}\sqrt{\frac{g?\sigma}{\rho d_8^3}} \tag{1.13}$$

Assume that, while rising, the bubble is washed by the wake formed by the bubble that is rising immediately ahead. Within this zone, the eddy breakaway frequency turns out to be equal to the natural oscillation frequency of the body being washed (bubble), instead of being determined from the known expression Sh=0.2 [34], i.e. the so-called frequency synchronization effect takes place [35]. In this case the energy dissipation rate in the wake is determined by the pulsating motion caused by the breakaway of the eddies from the surface of the oscillating bubble being washed by fluid flow. The frequency of pulsations in the wake may be written as follows with the use of the "two thirds" law [32, 33]

$$f_w = \frac{v'_w}{d_8} = \left(\frac{\varepsilon_w}{d_8^2}\right)^{1/3} \tag{1.14}$$

From the frequency synchronization conditions ($f_b=f_w$) we shall obtain:

$$d_8 \cong 1{,}7\left(\frac{\sigma}{\rho}\right)^{0{,}6} \varepsilon_w^{-0{,}4} \tag{1.15}$$

The analysis of the equation (1.15), with an account taken of the expression for the energy dissipation rate in the wake (1.8), shows that under conditions of intensive turbulent mixing, without additives preventing coalescence, the size of the bubbles in the main volume of the vessel turns out to be a constant value dependent only on properties of the medium.

In the presence of SAS or electrolytes in the working medium the size of the bubbles in the main volume of the vessel turns out to be equal to the size of the bubble within the stirrer zone. This effect is accounted for by the peculiarities of formation of the dispersed phase in the vessel. Actually, the delivery of the gas through a conventional sparger located under a standard-design stirrer is organized in such a way that in the absence of "choking" the entire amount of the gas delivered into the vessel passes through the surface formed by the external edge of the blade of the rotating stirrer. Since the zone in the vicinity of this surface is characterized with the highest rate of energy dissipation [27] the size of the bubble in the working volume of the vessel without coalescence is determined by the conditions of the bubble breakup within this zone.

In this case the bubble breakup process [14] is regarded as the result of the action of the difference between the dynamic heads deforming the bubble which should be in excess of the capillary pressure resisting deformation.

Taking this condition into account along with the information on the maximum value of the rate of dissipation near the stirrer blade tip [27]

$$\varepsilon_0^{\ell} \approx \frac{0{,}16(\pi n d_s)^3}{h_p} \tag{1.16}$$

and taking the maximum value of the pulsation velocity into consideration in the case of normal distribution of pulsation velocities [15]

$$(v')_{\max} \approx 3v' \tag{1.17}$$

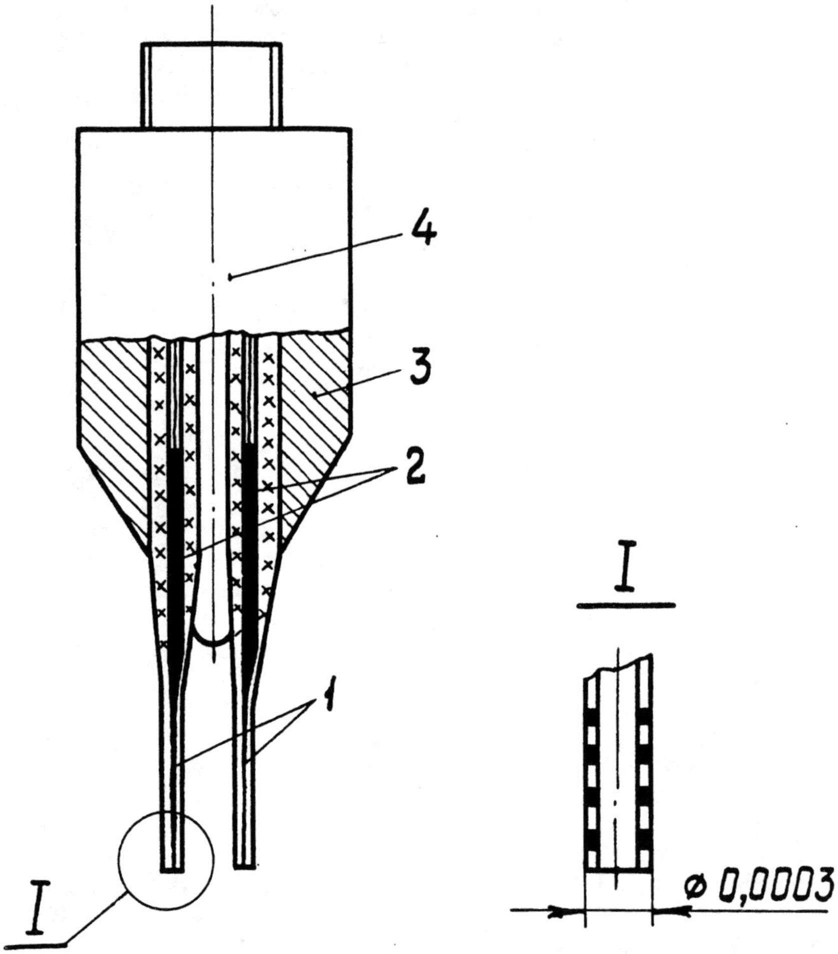

Figure 4. Transducer for measuring gas-liquid flow parameters; 1 - electrodes from Wood's alley; 2 - glass tubes; 3 - transducer holder; 4 - axis of symetry of transducer.

we obtain the relationship for the calculation of the bubble diameter in the working volume of the vessel

$$d_g \approx 0{,}706\left(\frac{\sigma}{\rho}\right)^{0,6}\left(\frac{\rho}{\rho_g}\right)^{0,2}\left(\varepsilon_0^{\ell}\right)^{-0,4} \tag{1.18}$$

For checking the results of the theoretical analysis the experimental studies have been conducted for determining the sizes of the bubbles in agitated vessels by means of the instrument employing the needle-type conductive transducer as its principal sensing element (Figure 4).

The measurements were conducted on the vessels that were 0.2, 0.25 and 0.5 m in diameter with the filling height of (0.7 to 0.3) D. The volume of the vessels ranged from 15 to 100 l. Agitation was performed by standard six-blade turbine stirrers in the conditions of the developed turbulent flow of the working medium ($Re>10^4$). The stirrer was located at a height of (0.5 to l)d_s above the vessel bottom, the ratio of the vessel diameter to the stirrer diameter ranged from 2 to 4.

The determination of the size bubbles in the bulk volume of the gas-liquid phase makes it possible to draw the conclusion that within the range of the mean value variations and the energy dissipation rate of 0.4 W/kg to 0.3 W/kg the diameter of bubbles in the water-air system remains practically unchanged and may be assumed to be equal to 4.5 mm. (Figure 5).

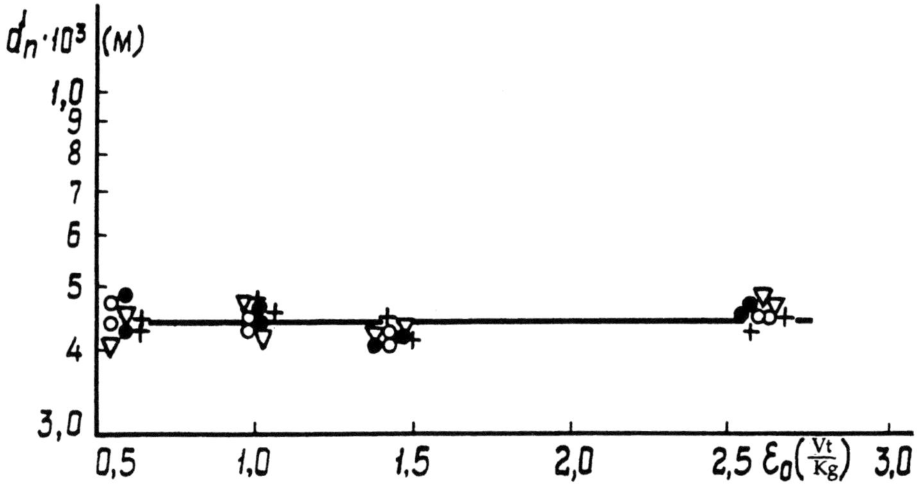

Figure 5. Bubble diameter in bulk volume of apparatus as a function of agitation intensity. D=0.5 m; H=0.5 m; d_s=0.15 m; open turbine stirrer o-G_g=4.22 m^3/h; o-G_g=12.0 m^3/h; p-G_g=18.15 m^3/h; +-G_g=16.0 m^3/h. Solid line - calculations

The formerly-conducted measurements of sizes of bubbles of water-air systems in the bulk volume of the baffle-free vessel with a propeller stirrer showed that, just as with vessels with standard-type stirrers, the bubble diameter may also be assumed to be constant and equal to 4.5 mm. The analysis of characteristics of two-phase flows conducted for bubble columns [36] also points to the constancy of the bubble size and to its independence from equipment design and intensity of agitation. The data for the water-air system presented in this publication indicates that at relatively high gas flow rates ensuring the gas holdup exceeding 10% the mean size of bubbles may be assumed to be constant and equal to 4.5 mm.

2. Gas Holdup in Agitated Vessel

At high rotational speeds of the stirrer the agitated medium in the vessel is essentially the gas-liquid mixture with the mean specific gas holdup:

$$\varphi = \frac{H_{g-l} - H}{H_{g-l}} \tag{2.1}$$

where H, $H_{g\text{-}l}$ is the height of layers of the liquid and gas liquid mixture in the vessel, m.

In real conditions the gas holdup ranges from 2 to 20%. The value of φ increases with an increase in the rotational speed of the stirrer or in power input per unit volume N/V and, in the operating conditions prior to "choking," with an increase in the gas flow rate. In the industrial agitated vessels no effect of the size of the sparger orifices on the gas holdup is usually observed.

For estimating the mean gas holdup the authors of Soviet and foreign publications [6, 7, 37-39] have suggested a number of diverse empirical formulae based on laboratory studies.

More recently, the gas holdup is presented as a function of the energy dissipation rate averaged over the volume of the vessel at a given rotational speed and diameter of the stirrer

$$\varepsilon_0 = \frac{N_g}{\rho V} \tag{2.2}$$

(or power input per unit volume N_g/V) and reduced gas velocity

$$W' = 4\,G_g\,/(\pi D^2) \tag{2.3}$$

where N_g is the power spent for agitating the gas-liquid mixture.

According to [37-39] this relationship can be expressed as

$$\varphi \sim< N_g / V)^{0,47} W^{0.5} \tag{2.4}$$

Along with the power input per unit volume and the gas flow rate factors affecting the value of the gas holdup also include the physical properties of phases. The equation which allows for this effect is obtained in publication [7]

$$\varphi = \left(\varphi \frac{W'}{v_r}\right)^{0,5} + 0.000216\left(\frac{\sigma}{\rho}\right)^{0,6} \varepsilon_0^{0,4}\left(\frac{W'}{v_r}\right)^{0,5} \tag{2.5}$$

where v_r is the velocity of bubble rise which is assumed in the present paper to be equal to 0.265 m/s; σ is the surface tnesion, N/m.

For the cases of high gas loads (W'>0.018 m/s) the following relationship has been suggested [5, 8] on the basis of the equation (2.5):

(2.6)

where ρ_{go}, ρ_g are the densities of the gas at atmospheric and working pressures, kg/m^3

N_t is the total power spent for agitation and determined with account taken of three components - power Ng spent for agitation of the gas-liquid mixture and the powers introduced with the gas due to its kinetic (Nk) and potential (Np) energies

$$N_K = G_g \rho_g W_{sp}^2 / 2 \tag{2.7}$$

$$N_p = G_g \rho_g H \tag{2.8}$$

where G_g is the gas flow rate, m^3/s; ω_{sp} is the gas velocity at sparger outlet, m/s; ρ is the density of the liquid, kg/m^3; ρ_g is the density of the gas, kg/m^3.

The power spent for agitation of the gas-liquid system was calculated by the authors [5, 8] from the equation:

$$N_g = 0,706(N^2 n d_s^3 / G_g^{0,55})^{0,45} \tag{2.9}$$

The effect of agitation conditions on the gas holdup was also investigated in publication [40] in the vessels 0.25 and 0.5 m in diameter with open turbine-type stirrers at liquid viscosities up to 22.4 mPa·s.

The reduced speed in the experiments ranged from 0.015 to 0.03 m/s and the power input per unit volume - from 2.5 to 12.0 kW/m^3. As a result of processing the experimental results equations have been obtained, which describe the effects of the gas flow rate, power input per unit volume and viscosity of the medium on the gas holdup:

1) at μ=0.0010 to 0.0025 Pa·s

$$\varphi_m = 0,33\left(\frac{W'}{v_r}\right)^{0,55}\left(\frac{N_g}{V}\right)^{0,19}\left(\frac{W'}{v_r}\right)^{-0,3} \tag{2.10}$$

2) at μ=0.0025 to 0.0224 Pa·s

$$\varphi_m = 0,43\left(\frac{W'}{v_r}\right)^{0,55}\left(\frac{N_g}{V}\right)^{0,17}\left(\frac{W'}{v_r}\right)^{-0,3}\left(\frac{\mu}{\mu_w}\right)^{-0,2} \tag{2.11}$$

In addition to the equation $\varphi=f(\varepsilon_0, W')$ a number of other relationships have been suggested in the literature which relate the gas holdup directly to the stirrer design parameters, its diameter and rotational speed.

Out of these correlations the most general one is the relationship obtained in [6] and presented in Figure 6.

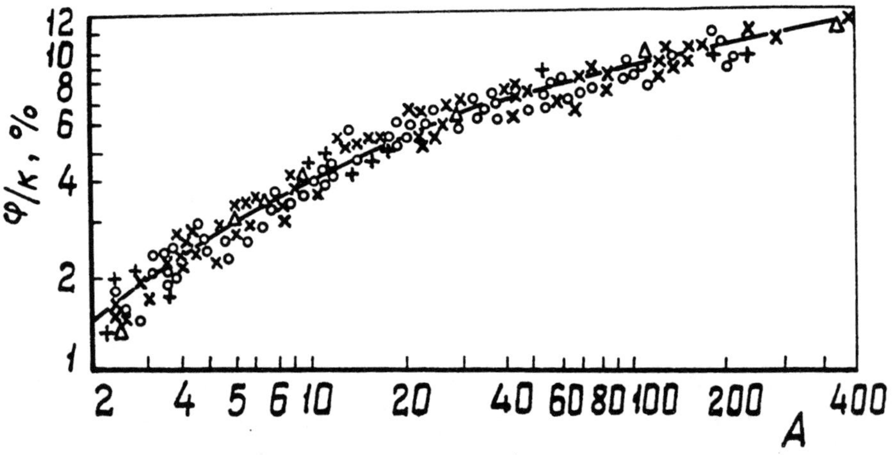

Figure 6. Function Y/K=f(A). Experimental and calculated values are obtained in ref. [6].

$$\varphi_m = K\,f(A) \tag{2.12}$$

where

$$K = 10^{0,0063}(1-\frac{\mu}{\mu_w}) \tag{2.13}$$

$$A = \left(\frac{H}{D}\right)^{0,4}\left(\frac{G}{nd_s^3}We\right)^{0,67\sqrt{Fr}} \tag{2.14}$$

$We=\frac{\rho \cdot n^2 \cdot ds}{\sigma}$ is the Weber number;

μ_W is the viscosity of water at 293K, Pa·s.

The comparison of the values of the gas holdup estimated with the use of equations (2.6), (2.10) and (2.12) applied to the 10-m^3-capacity vessel and presented in Figure 7 shows not only a considerable quantitative difference but also the great qualitative discrepancy between the obtained data.

The foregoing equations have been obtained from the experiments with pure liquids. In the cases of gas dispersion in electrolyte solutions the gas holdup normally proves to be much higher (sometimes reaching as high as 40% [41]).

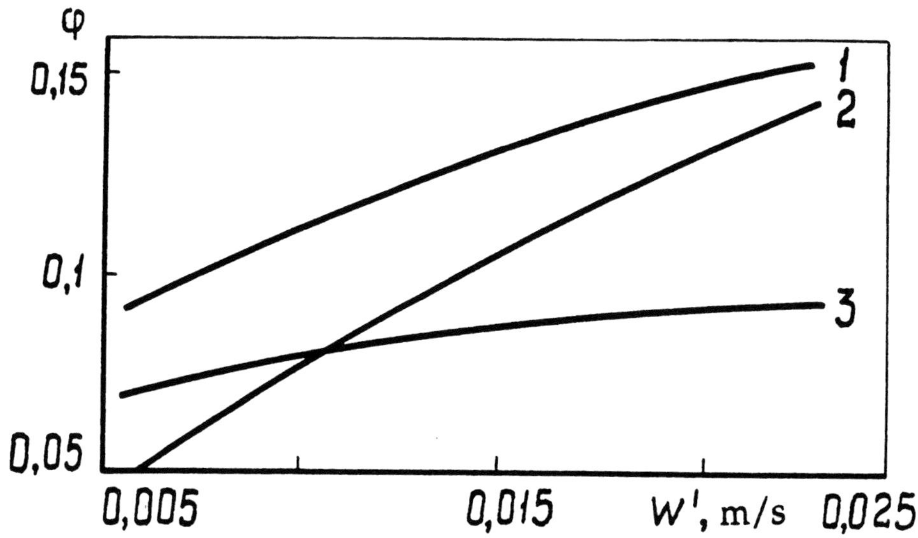

Figure 7. Gas holdup as a function of reduced gas velocity in 10-m^3-capacity vessel. D=2.2 m; d_s=1 m; Ng/V=5 kW/m^3. Curves comply with Equstions (2.6), (2.10), (2.12).

It should be noted that relationships (2.6), (2.10) and (2.12) make it possible to only determine the value of the gas holdup averaged throughout the volume of the vessel, whereas the actual distribution of this characteristic of the gas-liquid phase is, as a rule, far from uniform.

For determining the local values of the gas holdup the analysis was conducted, allowing for the peculiarities of the gas-phase transport in the working volume of the vessel.

Mixing of gas-liquid systems in agitated vessels is performed only if these vessels are provided with baffles and under intensive turbulence conditions. Taking the results of the investigations into the processes of mixing of suspensions in such vessels [42] into consideration, along with mixing of gas-liquid system in bubble-column-type vessels [43] featuring the flows similar in structure to those in the baffled agitated vessels, for describing the processes of the gas-phase transport we shall use the one-parametrical diffusion model where no gas-phase concentration gradient is assumed to exist along the radius of the vessel and non-uniformity of gas holdup is assumed to take place only in the axial direction.

$$v_r \frac{d\varphi}{dh} - D_E \frac{d^2\varphi}{dh^2} = 0 \tag{2.15}$$

In the cases of low gas holdup in the agitated vessel the velocity of the bubble rise is calculated without allowance for the restrained conditions from the formula (1.7) [28].

As has been shown above (Section 1), for the water-air systems the diameter of bubbles d_b in the working volume of the vessel may be constant and equal to 4.5 mm. The average value of the eddy-diffusion coefficient D_E in baffled agitated vessels [4] is determined from the formula:

$$D_E = 0,435 n d_s D \left(\frac{\xi_s Z_s}{\Gamma_D X} \right)^{1/3} \tag{2.16}$$

where $\gamma = 2\frac{H}{R} + 1; \Gamma_D = \frac{D}{d_s};$

H is the height to which the vessel is filled with liquid, m;
n is the rotational speed of the stirrer, s^{-1};
D is the diameter of the vessel;
d_s is the diameter of the stirrer;
ξ_s is the stirrer drag coefficient;
Z_s is the number of stirrers on the shaft.
In the conditions of uniqueness:

$$v_r \varphi - D \frac{d\varphi}{dh} = 0 \tag{2.17}$$

and

$$\varphi = \varphi_s \text{ at } h = H_{g-1} \tag{2.18}$$

reflecting the gas balance in every cross-section perpendicular to the axis of the vessel and the gas holdup in the near-surface zone.

We shall write down the solution of the equation (2.15) in the following form:

$$\varphi(h) = (\varphi_s - \frac{W'}{v_r}) e \frac{v_r (h - H_{g-\ell})}{D_E} + \frac{W'}{v_r} \tag{2.19}$$

where H_{g-1} is the height of the gas-liquid layer, m;
h is the current height coordinate, m.

By integrating the equation (2.19) we can determine the value of the mean gas holdup in the agitated vessel.

$$\varphi_m = (\varphi_s - \frac{W'}{v_r})\left(1 - e^{\frac{-v_r H_{g-\ell}}{D_E}}\right)\frac{D_E}{v_r H_{g-\ell}} + \frac{W'}{v_r} \tag{2.20}$$

From equations (2.19) and (2.20) it follows that the distribution of the dispersed phase within the working volume of the vessel and the mean value of gas holdup are affected not only by the design of the vessel and stirrer, the properties of the working medium and the gas flow delivered through the sparger, but also by the surface gas holdup φ_s which is determined to a considerable extent by the laws governing the process of the gas entrainment from the liquid surface.

For determining the value of φ_s we shall consider the process of the gas phase motion in the near-surface zone.

The gas transport in this zone shall satisfy the balance conditons

$$v_r^s \varphi_s - \frac{D_E^\ell}{H_{g-\ell}} \frac{d\varphi}{d\bar{h}}\Big|_{\bar{h}\to 1} = W' + W_s' \tag{2.21}$$

where $\bar{h} = h / H_{g-l}$.

In accordance with [14], the local eddy-diffusion coefficient D_E^l can be determined through the local value of energy dissipation in the near-surface zone ε_0^1:

$$D_E^l = (\varepsilon_0^l)^{1/3}(d_D R)^{4/3} \tag{2.22}$$

The value of ε_0^1 shall be determined as the sum of the values of the rate of dissipation of the energy introduced into the working volume of the agitated vessel (ε_0^1) and the energy introduced into the liquid with th egas delivered through the sparger (ε_0^1).

Taking into account the fact that with the stirrer located in the lower section of the vessel the value of (ε_0^1) in the near-surface zone accounts for about 10% of the mean value of the rate of dissipation of the energy ε_0 [26, 27] introduced into the working zone of the vessel by the stirrer, and that the energy introduced by the gas is practically dissipated uniformly throughout the vessel volume, the value can be written as follows:

$$\varepsilon_0^1 \approx 0,1\varepsilon_0 + W'g \tag{2.23}$$

The intensity of the gas entrainment into the vessel from the gas cushion above the fluid surface is characterized by the reduced gas velocity W'_S. It should be noted that the phenomenon of the gas backflow from the liquid surface in agitated vessels was investigated in a number of publications [7, 18, 44-46].

For instance, in publication [18] it is shown that the gas entrainment takes place only at the rotational speeds of the stirrer higher than a certain critical value n*. For calculation of this value the equation

$$\frac{\mu n^* d_s^2}{D\sigma}\left(\frac{\rho\sigma}{g\mu}\right)^{1/4} = 2\left(\frac{H-h_s}{D}\right)^{1/2} \tag{2.24}$$

was suggested which was obtained for open turbine-type stirrers at $0.1 \le \frac{H-h_s}{D} \le 0.25 + 1.75\frac{ds}{D}$.

The variation of the value of n* with the distance between the stirrer and the liquid surface h_S is also shown in the publication [44].

From the experiments conducted on the water-air system in agitated vessels ranging from 4 to 100 m^3 in capacity with single- and multi-row turbine stirrers the authors of [44] have obtained the following relationship

$$Fr\frac{ds}{h_s} = 0,36 \tag{2.25}$$

where $Fr = \frac{(n^*)^2 \cdot ds}{g}$ isthe Froude criterion.

However, the experimental relationships obtained in the analysis of the gas entrainment on the surface, including those obtained in publications [45, 46], are qualitative in character, and several quantitative estimates, which have so far been obtained yield contradictory results.

Let us first consider the mechanics of the formation of bubbles on the surface of pools during rainfall [31]. At first this phenomenon seems to bear little resemblance to the process of the surface seration of interest for us. However, on closer investigation of these phenomena one can find that they have much in common. The filming of the bubble formation process during rainfall [31] makes it possible to observe the actual mechanism of this phenomenon without the necessity for modelling the process.

Falling onto the surface of the pool the raindrop forms beneath itself a dent which on straightening out turns into a bulge. The formation of the dent occurs in all cases irrespective of the size and velocity of the raindrop. If the falling raindrop possesses a high kinetic energy, a deep recess (channel), may form which, instead of straightening out, closes, thereby forming a closed bubble under the surface (the so-called negative bubble). On rising, this bubble emerges on the water surface and becomes visible during rainfall. The filmed processes of the fall of a solid ball represented in Figure 8 enables us to understand and simulate the process of the fall of a raindrop onto the water's surface, (the initial stage of this process). The difference between the behavior of the ball and that of the drop lies in the fact that the fallen drop finally merges into the bulk liquid, whereas the ball sinks. However, the initial stage of the fall, (the formation of the dent and channel), proves identical for both of the bodies. If, in the case of the fall of a drop, the length of the formed channel is considerably larger than its diameter, the channel closes instead of straightening out, thus forming the negative, (undersurface), bubble in which the volume of the gas phase is equal to the volume of the gas phase in the channel.

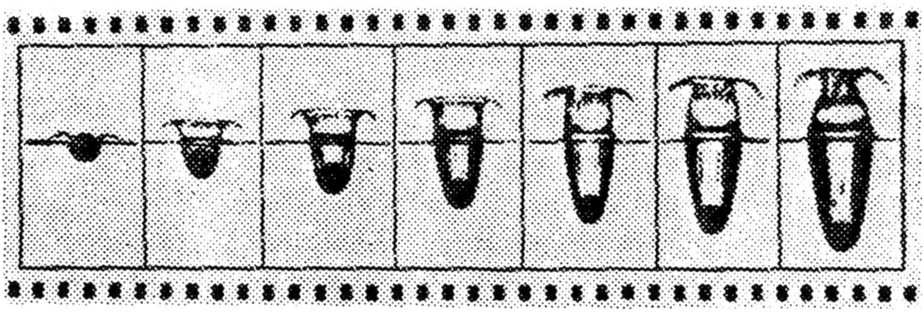

Figure 8. Progressive initial stages of free fall of heavy-weight ball in water.

Returning to the phenomenon of the gas entrainment from the surface in agitated vessels it should be noted that the observations reveal the periodic occurrence on the surface and a shutting of the low-intensity turbulent eddy-like formations, "caverns," accompanied by the entrainment of the formed bubbles into the bulk liquid. In most publications devoted to this problem the formation of these bubbles is regarded as the result of surface deformations caused by the presence of turbulence in the near-surface zone [18, 45, 46].

Drawing on the analogy between the bubble entrainment in agitated vessels and bubble formation during rainfall a conclusion can be made that the turbulent pulsations generated by the stirrer rotating in the working medium play the part of the drops deforming the liquid surface.

In the case of small sizes of the zones, (caverns), being deformed, i.e. in the conditions where the effect of the gravitational forces is inconsiderable, the scale of minimum surface deformations λ is determined by the equation of the equilibrium between the capillary and pulsative pressures

$$\frac{\rho v_{0,\lambda}'^2}{2} = \frac{2\sigma}{\lambda} \tag{2.26}$$

Hence,

$$\lambda = \frac{4\sigma}{\rho v'_{0,\lambda}} \tag{2.27}$$

Assuming that within the near-surface zone, just as throughout the entire volume of the vessel, turbulence can be regarded as locally-isotropic and determining $v_{0,\lambda}^2$ in conformity with the Kolmogorov's-Obukhov's "law of 2/3" [32, 33] from the equation

$$v'^2_{0,\lambda} = (\varepsilon_0^l \lambda)^{2/3} \tag{2.28}$$

we shall obtain

$$\lambda = 2,3\left(\frac{\sigma}{\rho}\right)^{0,6} (\varepsilon_0^l)^{-0,4} \tag{2.29}$$

Assuming further that the frequency of surface deformations in the given point is equal to the frequency of velocity pulsations of the scale l:

$$v_{0,\lambda} = \frac{v_{0,\lambda}}{\lambda} = \frac{(\varepsilon_0^l)^{1/3}}{\lambda^{2/3}} \tag{2.30}$$

we shall have the following equation for unit surface allowing for the local gas holdup

$$v_{0,s} = \frac{v_{0,\lambda}(1-\varphi_s)}{\tau\lambda^2} = 0,32(\varepsilon_0^l)^{1/3}\lambda^{2/3}(1-\varphi_s) \tag{2.31}$$

Not every pulsation results in the entrainment of the bubble formed due to the davern shutting, but only those whose velocity is higher than the velocity of bubble rise v_r^s. The fraction of such pulsations within the spectrum of pulsation velocities of the scale (taking their normal distribution law into account [15]), it can be determined as

$$P = 1 - 2\phi\left(\frac{v_r^s}{v_{0,\lambda}^1}\right) \tag{2.32}$$

where $2\Phi\left(\frac{v_r^s}{v'_{0,\lambda}}\right)$ is the probability of the event where the pulsation velocity $v'_{0,\lambda}$ exceeds the velocity of the rise of the bubble formed on the surface in the case of the validity of the normal distribution law [47].

From expressions (2.31) and (2.32) we shall obtain the following equation for the frequency of bubble entrainment from the surface

$$v_s = v_{0,s}P = \frac{v'_{0,\lambda}}{?\lambda^3}(1-\varphi_s)\left[1-2\phi\left(\frac{v_r^s}{v'_{0,\lambda}}\right)\right] \tag{2.33}$$

Hence, assuming the volume of the entrained bubble to be approximately equal to the volume of the cavern we can write

$$W'_s \approx 0,88\left(\frac{\sigma}{\rho}\right)^{0,2}\left(\varepsilon_0^l\right)^{0,2}(1-\varphi_s)\left[1-2\varphi\left(\frac{v_r^s}{v'_{0,\lambda}}\right)\right] \tag{2.34}$$

From expression (2.21), taking (2.34) into account, we can obtain the following relationship for determining the gas holdup in the near-surface zone:

$$\varphi_s = \frac{W' + 0,88\left(\frac{\sigma}{\rho}\right)^{0,2}(\varepsilon_o^l)^{0,2}[1-2\phi\left(\frac{v_r^s}{v'_{0,\lambda}}\right)] + \frac{D_E^l}{H_{g-l}}\frac{d\varphi}{d\bar{h}}\Big|_{\bar{h}\to 1}}{v_r^s + 0,88\left(\frac{\sigma}{\rho}\right)^{0,2}(\varepsilon_0^l)^{0,2}[1-2\phi\left(\frac{v_r^s}{v'_{0,\lambda}}\right)} \tag{2.35}$$

The obtained equations form the system which makes it possible to determine the gas holdup and its distribution in height as functions of the basic design characteristics of the vessel and stirrer, the properties of the phases and the flow rate of the gas delivered through the sparger. However, it should be noted that the analysis with the use of these equations cannot be conducted without the information available on the coefficient of proportionality α_f in expression (2.22) and on the concentration gradient in the near-surface zone $\left|\frac{d\varphi}{d\bar{h}}\right|_{\bar{h}\to 1}$ that can be obtained experimentally.

It should also be noted that in cases where the surface gas holdup, the gas holdup distribution in vessel height and the mean gas holdup are determined by calculations, the height of the gas-liquid-phase layer is not known *a priory* and, hence the final analysis is essentially an iterative procedure with respect to the variable $H_{g\text{-}l}$ the initial value of which is assumed to be equal to the height of the liquid layer prior to the delivery of the gas into the vessel. In this procedure each subsequent approximation for the height of the gas-liquid-phase layer is determined from the expression

$$H_{g-l} = \frac{H}{1-\varphi_m} \tag{2.36}$$

where the value of φ_m is found from in the equation (2.20) using the characteristics corresponding to the preceding iteration.

The calculation procedure is schematically shown in Figure 9.

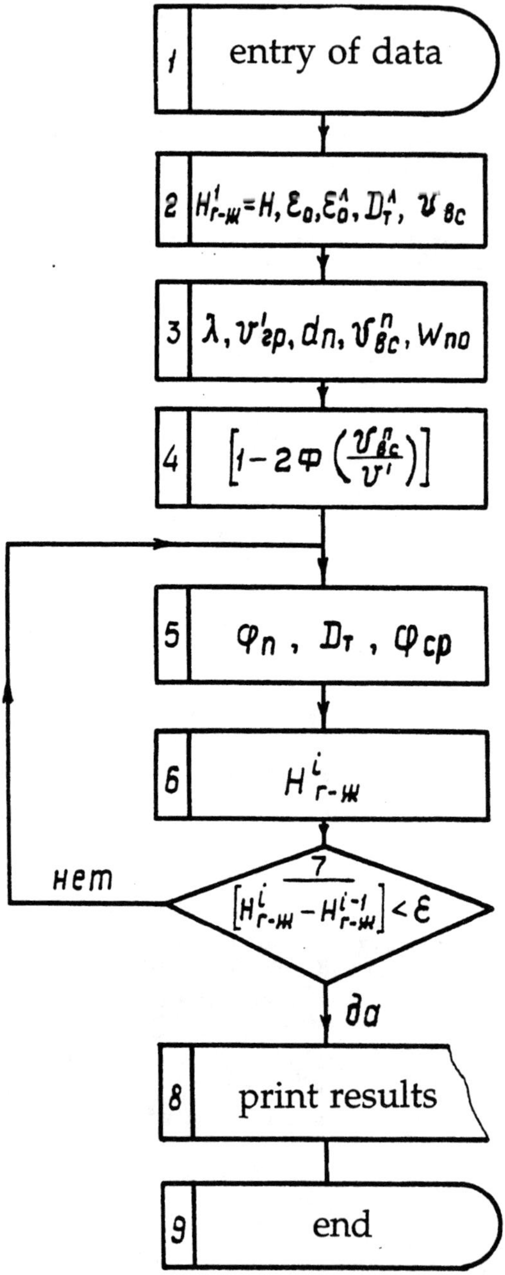

Figure 9. Schematic diagram for calculation of gas holdup in stirred vessel.

a

b

Figure 10. Visualization of surface aeration: D=0.24 m; d_s=0.1 m; n=250 rpm. Open turbine - type stirrer.

a

b

Figure 11. Visualization of surface aeration: D=0.24 m; d_S=0.1 m; n=496 rpm. Open turbine - type stirrer.

For determining the unknoqn parameters of the model and verifying the validity of the obtained relationships the experimental studies have been conducted.

The measurements of the gas holdup were conducted with the use of the needle-type conductive transducer (Figure 4) in the agitated vessels ranging in volume from 15 to 100 l and using six-blade turbine-type stirrers. The ratio between the vessel and stirrer diameters was varied within 2 and 4, and the height at which the stirrer was located above the vessel bottom ranged from (0.5 to 1)$\cdot d_S$.

For determining the mechanics of the gas entrainment from the surface the process was observed visually and photographed without forced gas delivery through the sparger.

The obtained photographs are shown in Figures 10 and 11. Each pair of the photographs corresponds to a certain rotational speed of the stirrer and presents the side view (a) allowing the viewer to judge about the extent to which the vessel is filled with gas and the top view (b) showing the turbulence intensity on the surface of the gas-liquid phase.

At low rotational speeds of the stirrer corresponding to the energy dissipation ε_0<l W/kg (Figure 10) the liquid surface is relatively calm.

Light ripple and surf are visible in the vicinity of the baffles. No mass entrainment of the gas is observed. Beginning from the speeds at which the energy dissipation ε_0ª0.5 W/kg individual wandering vortices periodically appear here and there through which individual large-size bubbles are drawn into the bulk liquid and subsequently break up within the stirrer zone into many smaller-size bubbles. However, since the frequency with which these vortices appear on the surface is small enough it allows the gas entering the bulk liquid through these vortices to leave the vessel prior to the formation of new vortices, and so in these cases it is too early to speak about whatever stable gas holdup.

As the stirrer speed increases (ε_0>1 W/kg), heavy turbulence is observed on the surface (Figure 11), there are no longer wandering vortices and the gas penetrates into the working volume of the vessel from the entire surface of the liquid in the form of sufficienctly large bubbles, which within the stirrer zone break up into small bubbles. Since at these stirrer speeds the operation of the vessel is characteritzed by a continuous flow of the gas from the surface and the time of the residence of the gas bubbles within the mixing volume is a certain finite value, the gas holdup grows progressively with increasing stirrer speed.

An increase in the intensity of surface turbulence and the presence of the gas entrainment from the entire area of the surface is a qualitative confirmation of the assumed mechanism of the surface aeration phenomenon based on the surface deformation under the action of the turbulent pulsations.

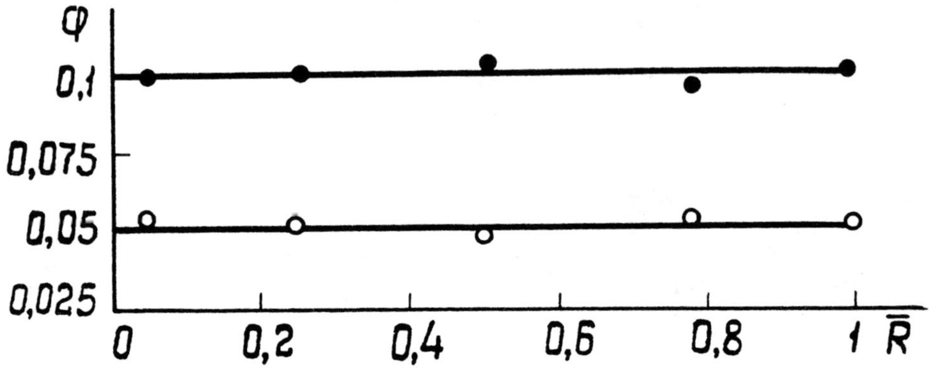

Figure 12. Radial gas holdup distribution in vessel: D=0.5 m; d_s=0.15 m; ξ_s=8.4; H=0.5m; G_g=5.7 m^3/h; n=10 s^{-1}; •- $\bar{h}$ =1.0; o- $\bar{h}$ =0.5.

The results of experimental studies of the gas holdup distribution over the volume of the agitated vessels have made it possible to confirm the validity of the assumption of the zero gas phase concentration gradient in the radial direction (Figure 12). It should also be noted that from the results of measurements of the gas holdup distribution in the height of the vessel it follows that in the immediate vicinity of the surface the gas holdup gradient, expressed through the relative height coordinate $\bar{h}$ within the accuracy

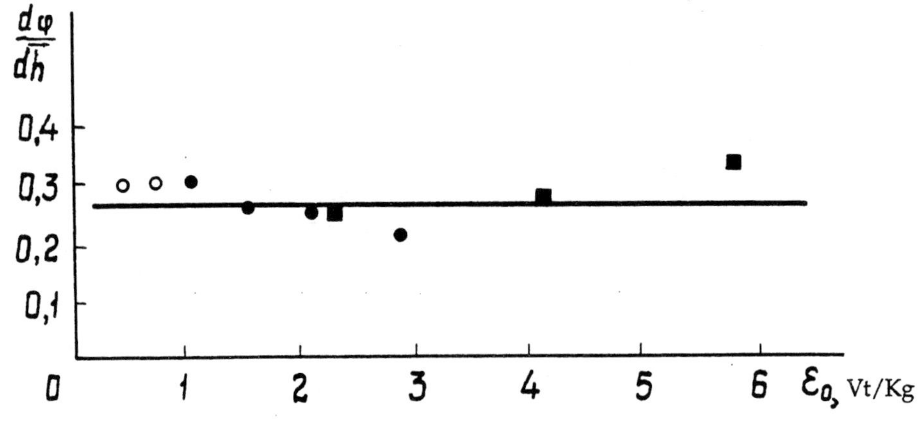

Figure 13. Gas holdup in near-surface zone as a function of mean energ y dissipation: •- D=0.5 open turbine-type stirrer; o-D=0.24 open turbine-type stirrer; -D=0.24 3-paddle stirrer.

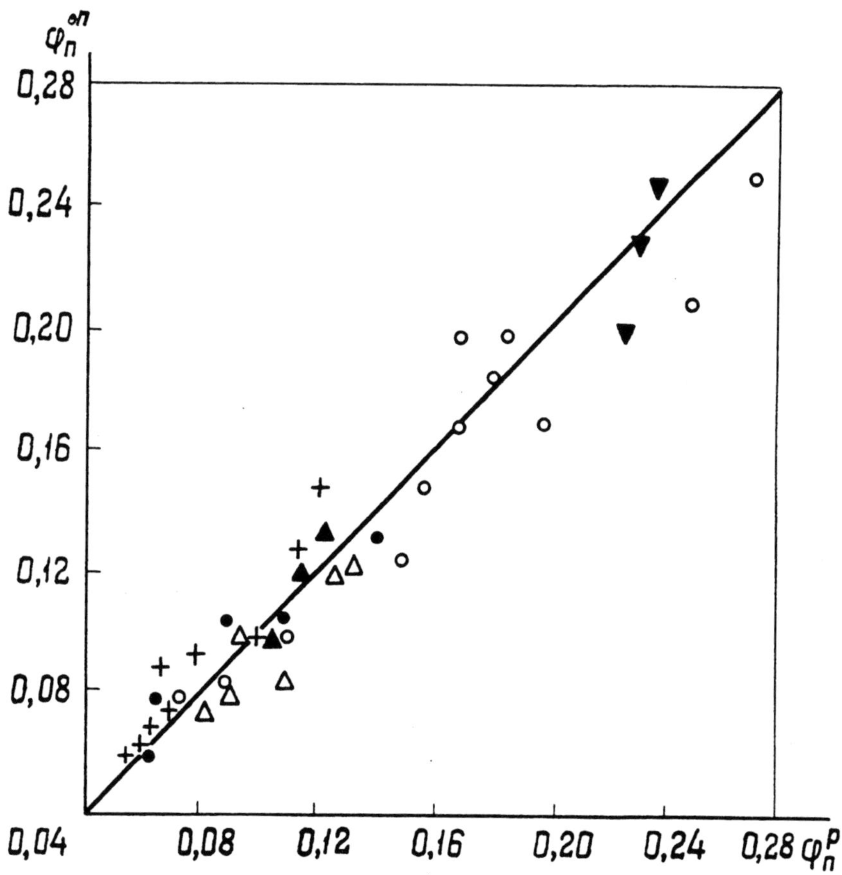

Figure 14. Comparison of analytical and experimental values of surface gas holdup:

$\bullet - D = 0.5\ m;\ d_s = 0.15\ m$
$\text{o} - D = 0.24\ m;\ d_s = 0.1\ m$
$v - D = 0.24\ m;\ d_s = 0.125\ m$
$+ - D = 0.24\ m;\ d_s = 0.08\ m$ mill stirrer;
$\Delta - D = 0.24\ m;\ d_s = 0.125\ m$ two – paddle stirrer;
$\blacktriangle - D = 0.24\ m;\ d_s = 0.125\ m$ three – paddle stirrer;

sufficient for the purposes of design engineering and analysis, may be assumed to be constant and equal to 0.25 (Figure 13). The independent determination of $\frac{d\varphi}{dh}$ has enabled the unambiguous determination of the coefficient of proportionality α_Γ in expression (2.22). A comparison of the experimental results for the gas holdup within the near-surface zone measured at a distance of 2 to 7 mm from the surface with those obtained by calculations (Figure 14) at α_f=0.45 indicates that the equation obtained on the basis of the model of gas entrainment from the surface is in good agreement with the measurement results. The value of coefficient α_f is reasonable since it is in good agreement with the results obtained in studying a wide range of hydrodynamic problems [4, 42, 48, 49] based on the hyposysis of the mechanics of mixing advanced by Prandtle.

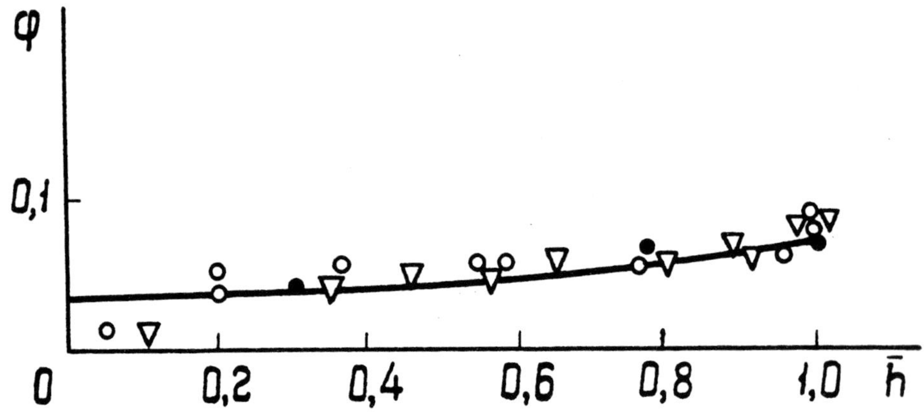

Figure 15. Gas holdup distribution in height of vessel: D=0.5 m; H=0.5 m; d_s=0.15 m; n=6 s^{-1}; ξ_s=8.4; W'=5.7·10^{-3} m/s; solid line-calculation.

Figure 15 illustrates gas holdup distribution in the vessel height obtained by experimental measurements and also shows the calculated relationship φ(h). The graphs showing the mean gas holdup as a function of ε_0 are presented in Figure 16 and 17. The comparison of the experimental and calculated values (taking into account the surface aeration) of the mean gas holdup in vessels from 15 to 100 liters in capacity is shown in Figure 18.

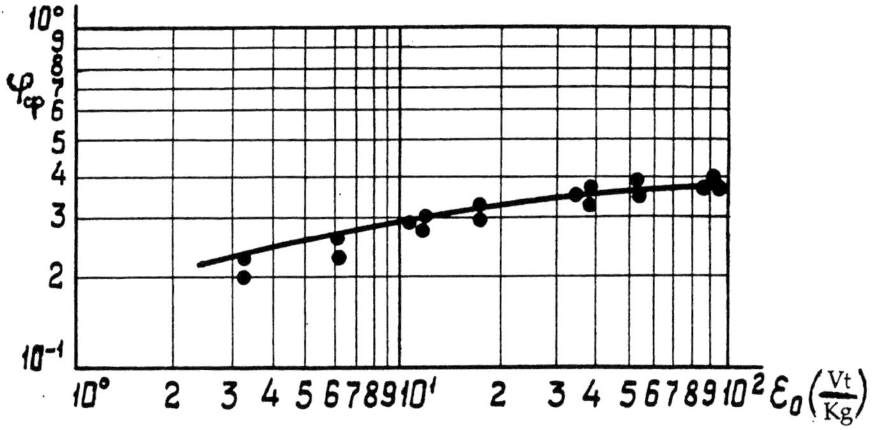

Figure 16. Mean gas holdup as a function of ronds: D=0.25 m; d_s=1.14 m; H=0.25 m; open turbine-type stirrer; G_g=8.3 m^3/h; W'=0.046 m/s; •-results obtained in [50]; - - - calculation data from Equation (2.20).

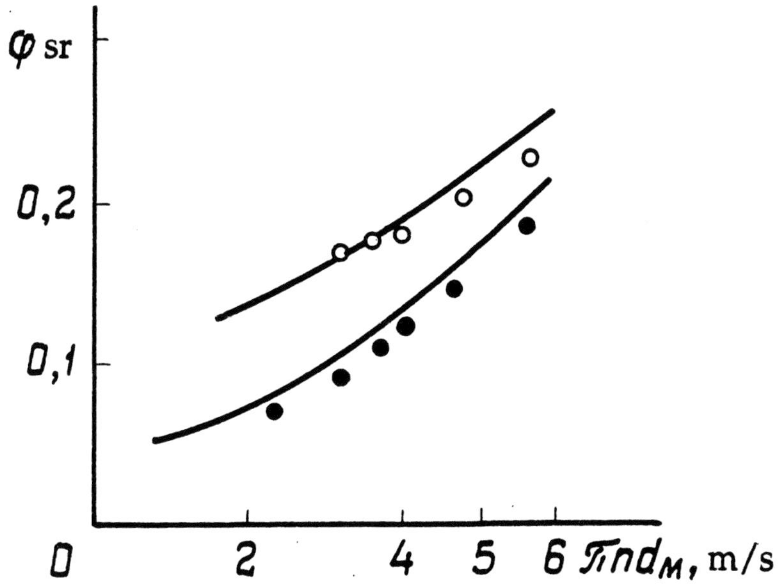

Figure 17. Mean gas holdup as a function of stirrer padolle tip speed and reduced gas velocity vessel diameter - 0.19 m; stirrer diameter - 0.076 m; open turbine-type stirrer; medium-water-air:•W'=0.014 m/s; G_g=1.44 m^3/h [51]; oW'=0.0305 m/s; G_g=3.11 m^3/h [51]. Solid lines - calculations from Equation (2.20).

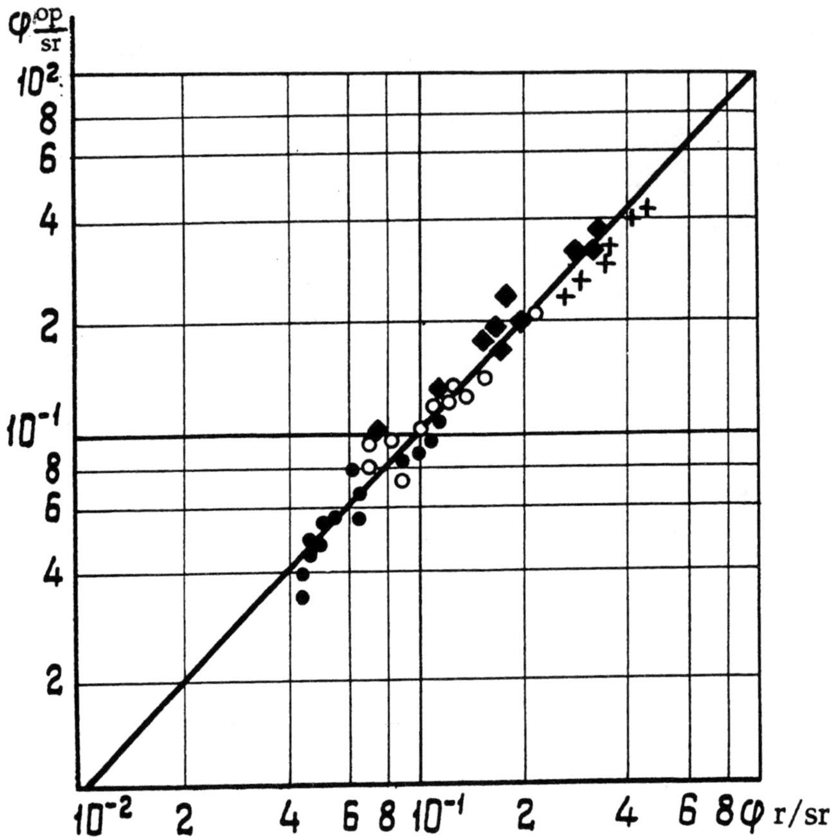

Figure 18. Comparison of analytical and experimental values of mean gas holdup: +-D=0.25 m; d_S=0.14 m; from [50]; Δ-D=0.25 m; d_S=01 m; from [50]; •-D=0.5 m; d_S=0.15 m; authors' data; o-D=0.2 m; d_S=0.1 m; authors' data.

3. THE EFFECT OF THE SPARGER DESIGN ON ENERGY CHARACTERISTICS OF AN AGITATED VESSEL

The results presented in Sectsion1 and 2 indicate that the parameter exerting the greatest effect on the size of bubbles in the absence of coalescence and on the rate of mass transfer is the rate of energy dissipation and, consequently, the power input spent for mixing. In this respect, the use of traditional gas distribution devices in the form of ring-shaped spargers mounted under the stirrer is not effective, since, at high gas flow rates, the power spent for mixing is found to decrease sharply compared to the total power input for liquid agitation (Figure 19). This effect is accounted for by the fact that at high gas flow rates the zone located under the disc of the open turbine-type stirrer becomes filled with gas and the lower half of the blades operates practically in the air medium. Thus, only the upper portion of the stirrer blades turns out to be active, which has a two-fold reduction on the amount of power spent for mixing. This power drop results in a decrease in intensity of mass transfer and efficiency of operation of the vessel as a whole.

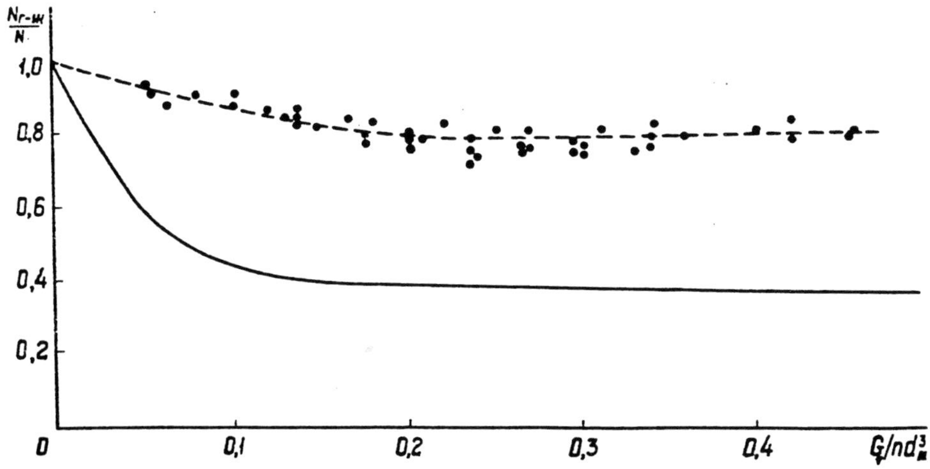

Figure 19. Function $\frac{N_{g-l}}{N_1} = f\left(\frac{G_g}{n \bullet d_s^3}\right)$ - - - traditional sparger; •••• - novel-design sparger (experimental data).

One of the possible ways to intensify the mass transfer processes in agitated gas-liquid systems is to increase the amount of the power spent for mixing, while retaining the principal design characteristics of the vessel and stirrer and the installed power of the stirrer drive. This problem can be solved by optimizing the design of the spargers so as to minimize gas localization within the stirrer zone.

This condition can be satisfied by using the ring-shaped sparger, the diameter of which is greater than that of the stirrer ((1.2 to 1.3)d_s) and which is located at the same height above the vessel bottom as the stirrer (Figure 20).

The analysis of the bubble breakup process in agitated vessels (Section 1) shows that in cases where use is made of the working media containing SAS or the electrolytes suppressing coalescence it is advisable that the gas outlet orifices be located on the inner radius of the sparger in the immediate vicinity of the stirrer blades, (the zone of maximum values of the rates of energy dissipation), since in these cases the size of the bubbles within the working volume of the vessel depends on the peculiarities of the bubbles breakup within this zone.

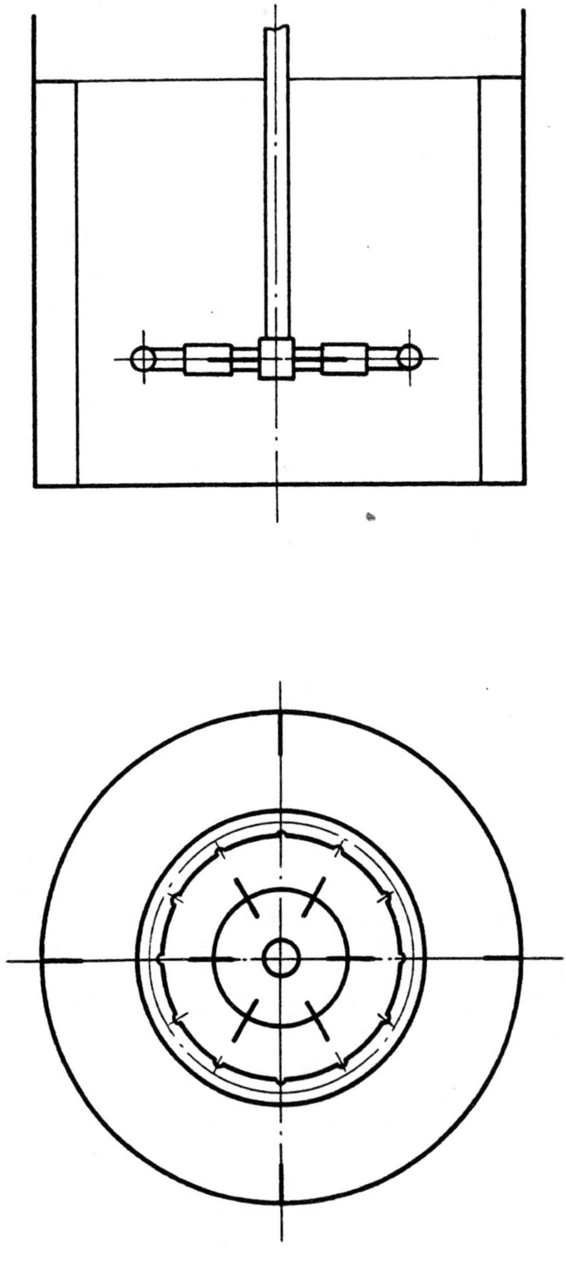

Figure 20. Schematic diagram of agitated vessel.

With the orifices located in this manner, the zone of the stable operation of the vessel, (the gas is uniformly distributed throughout the volume of the vessel), is limited by operating conditions in which the dynamic head of the working medium ΔP_1 developed by the stirrer exceeds the dynamic head of the gas ΔP_g on its exit from the orifices. In the cases where $\Delta P_g > \Delta P_g$ the so-called zoning begins when there is practically no gas in the under-stirrer zone and the gas is only concentrated within the upper portion of the vessel.

Assuming that the entire flow of the liquid caused by the meridianal circulation passes through the stirrer zone, the height of which is equal to $2h_b$, the head of the liquid within this zone may be determined as

$$\Delta P_l = \frac{\rho v_l^2}{2} = \frac{\rho\left(\frac{q}{\pi d : hs}\right)^2}{2} \tag{3.1}$$

where: 2_q is the total flow rate of the medium passing through the stirrer, m3/s;

d_s is the stirrer diameter, m;

h_b is th height of the stirrer blade, m.

The dynamic head of the gas delivered through the sparger can be determined from the formula

$$\Delta P_g = \frac{\rho_g v_g^2}{2} = \frac{\rho_g\left(\frac{G_g}{Z_{02}S_{02}}\right)^2}{2} \tag{3.2}$$

where Z_{or} is the number of sparger orifices;

S_{or} is the area of one orifice, m^2;

G_g is the rate of gas flow through the sparger, m^3/s.

Thus, the condition of the stable oepration of the vessel assumes the form

$$\frac{\rho_l\left(\frac{q}{\pi d_s h_B}\right)^2}{2} \geq \frac{\rho_g\left(\frac{G_g}{Z_{or}S_{or}}\right)^2}{2} \tag{3.3}$$

Determining the circulation flow rate q as detailed in [52] from (3.3) we can obtain the limit of the rational use of a novel-design sparger (without zoning)

$$\frac{G_g}{nd_s^3} \leq \frac{K_q Z_{or} S_{or}}{\pi d_s h_s}\sqrt{\rho/\rho_g} \tag{3.4}$$

where k_q is the flow coefficient [52].

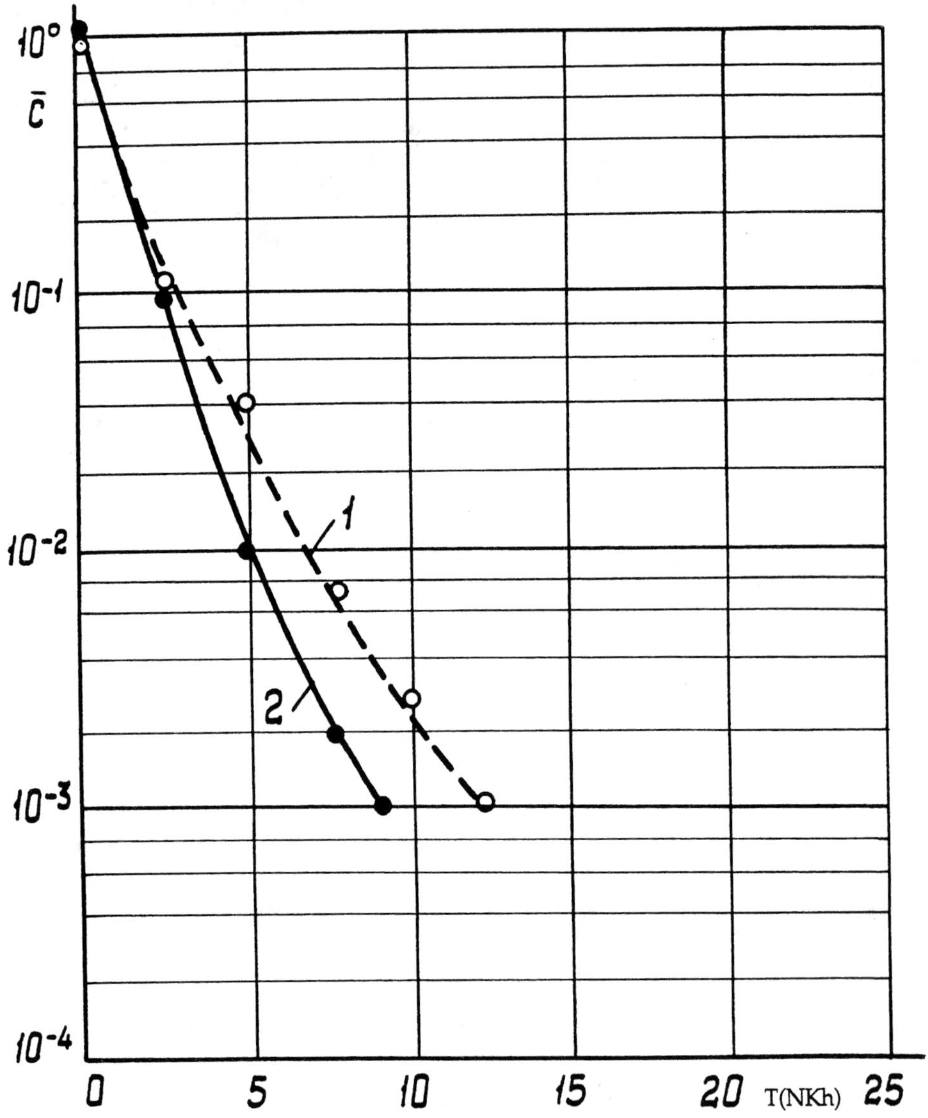

Figure 21. Kinetic curves of CO_2 desorption in vessels with novel and traditional sparger: D=0.5 m; ds=0.15 m; ξ_s=8.4; n=300 rpm; G_g=6.5 m^3/h; •, o-experimental data; 1- traditional - design stirrer; 2 - novel - design stirrer.

In the presence of coalescence the bubble size in the bulk liquid is practically unaffected by the power input introduced by the stirrer (section 1). In such a case, the orifices of the sparger are recommended to be arranged around its outer radius, which eliminates the possibility of zoning and considerably extends the zone of the stable operation of the vessel.

This sparger enables the useful power to be almost doubled (Figure 19), which increases the rate of the mass transfer (Figure 21) and, consequently, the capacity of the vessel.

4. MASS TRANSFER IN AGITATED GAS-LIQUID SYSTEMS

In determining the rate of mass transfer in the gas-liquid systems, attention is focused mainly on the limiting stage of this process, mass transfer from the liquid phase.

A few quantitative estimates [53] and results of experiments [54-56] show that in the use of the gases with low solubility the mass transfer coefficient in the gas phase is a value of the order of 0.4 m/s, whereas the mass transfer in the liquid phase at intensive agitation does not exceed 0.001 m/s.

Most often the rate of mass transfer in the case of gas dispersion in a liquid is described by the equation:

$$\frac{dq}{d\tau} = \beta_v (C_{eq} - C) \tag{4.1}$$

where: q is the quantity of the absorbed gaseous component, kg/m^3;

τ is the time, s;

C_{eq}, C are the equilibrium concentration of the component being absorbed and its mean concentration in the solution, kg/m^3;

$\beta_V = \beta \bullet F_{sp}$ is the volumetric mass transfer coefficient, s^{-1};

F_{sp} is the specific interfacial area;

β is the linear mass transfer coefficient, m/s.

For determining the value of the volumetric mass transfer coefficient β_V a large number of empirical relationships are proposed in the literature [53, 56-58] which deal with the effect of properties of working media and mixing conditions on the rate of mass transfer.

For instance, in the publication [53] for the case of agitation by a turbine-type stirrer the following relationships have been obtained:

$$\beta_v \sim \left(\frac{N_g}{V}\right)^{\alpha_1} (W')^{\alpha_2} \tag{4.2}$$

The values of the exponent α_1 measured by many authors [53] range from 0.43 to 0.95 and the value of the exponents α_2, from 0.4 to 0.84. Therefore, considerable differences are accounted for by a number of causes, the major one among them is that the effect of the value of the power input per unit volume (Ng/V) on the rate of mass transfer decreases with are increase in the intensity of mixing. For instance, according to [56] at W=0.016 m/s

$$\beta_v = 0,06\left(\frac{N_g}{v}\right)^{0,84} \quad \frac{N_g}{v} < 8 \div 10\frac{kW}{m^3} \tag{4.3}$$

$$\beta_v = 0,153\left(\frac{N_g}{v}\right)^{0,39} \quad \frac{N_g}{v} < 8 \div 10\frac{kW}{m^3} \tag{4.4}$$

Relationships (4.3) and (4.4) have been obtained for the reaction of oxidation of sodium sulphite in the diluted aqueous solution by the atmospheric oxygen in the presence of the catalyst Cu SO_4 or $CoSO_4$, which results in greater values of mass transfer coefficients as compared with those measured during physical absorption [58].

$$\beta_v = 3805\left(\frac{N_g}{v}\right)^{0,44} \varphi^{0,67} D_b^{0,5} \mu^{-0,15} \tag{4.5}$$

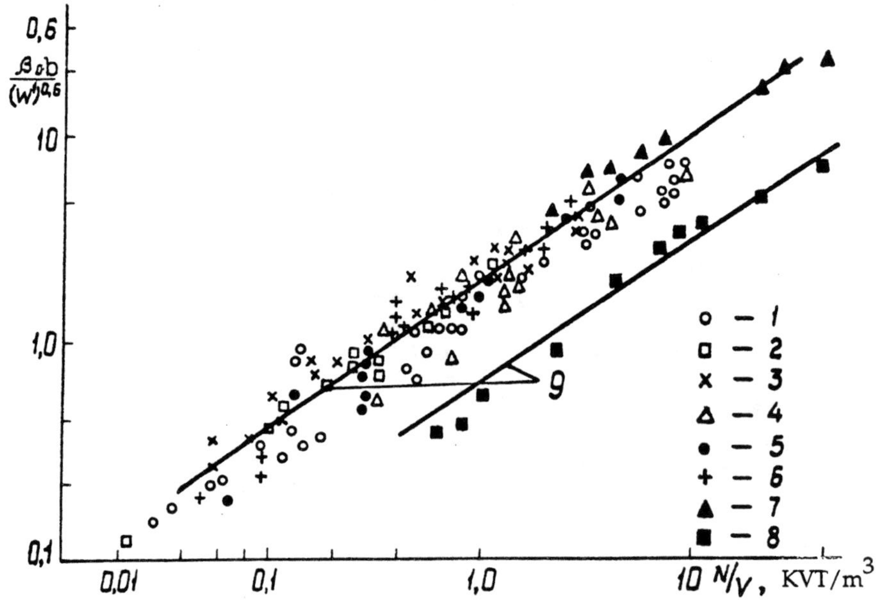

Figure 22. Volumetric mass transfer coefficient as a function of power input per unit volume (N-power consumed without gas introduction).

Points	Stirrer type	Vessel volume, m^3	Measurement conditions
1	2	3	4
1	Turbine-type	0.02-0.25	Oxidation of
2	Turbine-type	0.2	sodium sulphate
3	Closed-tubine-type	0.02-0.25	
4	Plate	0.05	
5	Paddle	0.02-0.05	
6	Impleller	0.02-0.05	
7	Three-paddle	0.01-0.005	
8	Three-paddle	0.01-0.005	Physical absorption in air-sodium sulphase system
9	Theoretical analysis		

Figure 22 illustrates the results of measurements of the volumetric mass transfer coefficient in the oxidation of sodium sulphite with the atmospheric oxygen (catalyst- $CuSO_4$) [59]. The diameter of the stirrers ranged from 75 to 100 mm, the number of the stirrers on the shaft - from 1 to 4 and the reduced gas velocity was up to 0.05 m/s. In this Figure the data [60] obtained in the conditions of oxidation of sodium sulphite and in the case of physical absorption of oxygen are presented.

The use of the notion of volumetric mass transfer coefficient considerably simplified the analysis of the experimental data on mass transfer, since there is no need for determining the specific interfacial surface area in this case as well as the gas holdup and bubble size. However, it should benoted that just as in any other empirical relationahip, these relationships are characterized by a relatively narrow scope of application limited by variations of the basic design and operating parameters for which they have been obtained.

In order to derive the generalized design formulae it is necessary to conduct a more detailed analysis of the mechanism of the phenomena being studied. In cases of mass transfer with the presence of bubbles, several mechanisms may take place concurrently. Presently, there are two principal approaches to the analysis of the effect of the turbulence on mass transfer.

In the first case, the determination of the mass transfer coefficients is based on the use of the equation for the steady-state diffusion in the boundary layer taking the laws governing tubulence decay in vicinity of the bubble surface into account [14, 61]. The second approach is based on the use of the model of non-steady-state diffusion in the vicinity of the boundary between the phases [62, 63].

The universally-accepted concepts and methods for analysis of the mass transfer from bubbles have not yet been completely developed since, at present, it is impossible to assess the characteristic time limits determining the duration of existence of the boundary layer on the interfacial surface and the duration of the interval between two consecutive renewals within the framework of the second approach with sufficient accuracy. The mass transfer at the boundary between the gas and liquid phases may be additionally complicated by the effect of admixtures, particularly SAS, and also by the spontaneous interfacial turbulence, (the so-called Marangoni effect), occurring as a result of variations in the interfacial tension.

From the data published in Reference [64] the duration of boundary layer relaxation proves somewhat shorter than the duration of the surface renewal estimated for the non-steady-state diffusion model with proper allowance made for the experimental values of the mass transfer coefficient.

Since the duration of the formation and breakup of the bubbles proves to be rather small, we shall consider the process of mass transfer from the bubbles rising in the bulk volume of the vessel.

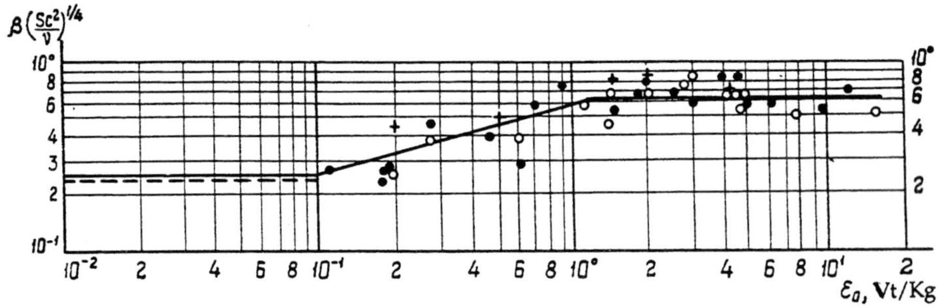

Figure 23. Function $\beta\left(\frac{Sc^2}{\nu}\right)^{1/4}$ =f(e0): ——— - calculations; - - - - experimental data for individual bubbles [65]; o - experimental data for desorption (oxygen-water) [50]; • - experimental data for desorption (carbon dioxide-water) [50]; + - experimental data for oxidation of sodium sulphite [50].

For this purpose, we shall analyze the data on the linear mass transfer coefficients β (Figure 23) obtained from the experimental values of the volumetric mass transfer coefficients in cases of desorption in the oxygen-water and carbon dioxide-water systsms [50]. The necessary data on the gas holdup and bubble size were obtained on the basis of the results described in Sections 1 and 2.

As seen from Figure 23 the three most characteristic zones can be distinguished for the variation of the mass transfer coefficient with the intensity of mixing. At average rates of energy dissipation in the vessel not exceeding 0.1-0.2 W/kg (zone I) the mass transfer coefficient is not affected by the energy introduced into the agitated vessel and is determined by the laws governing mass transfer from indiviual bubbles rising in an undisturbed fluid flow.

As the rates of energy dissipation increase in the vessel up to about 1 W/kg, the mass transfer coefficients are noted to grow in magnitude (zone II).

At the rates of energy dissipation exceeding 1 W/kg the automodel zone is observed where the mass transfer coefficient remains practically unchanged (zone III).

With a view to obtaining design relationships for determining mass transfer coefficients we shall consider the mechanism of mass transfer in each zone with proper allowance made for the continuous phase flow near the bubble surface and distribution of bubbles within the working volume of the vessel.

Zone III shall be considered in greater detail since this zone is the most characteristic in terms of practical applications. Taking the approximation of the steady-state boundary layer model into account along with the laws in accordance with which turbulence dies down near the deformable surface [14], the expression for the flow of the substance from the bubble surface can be written as:

$$j = (D_m + \frac{v_0^1}{\delta_0}) \frac{dc}{dy} \tag{4.6}$$

where δ_0 is the thickness of the layer where the major portion of diffusion resistance is concentrated, m;

D_m is the molecular diffusion coefficient, mg/s;

y is the distance from the surface, m;

v'_0 is the pulsation velocity of the scale δ_0, m/s.

Hence, the diffusion resistance is equal to:

$$R_D = \int_0^{\delta_0} \frac{dc}{j} = \int_0^{\infty} \frac{dc}{j} = \frac{\pi}{2}\left(\frac{\delta_0}{v_0^1}\right)^{1/2} \frac{1}{D_m^{1/2}} \tag{4.7}$$

and the mass transfer coefficient -

$$\beta = 1/R_D = \frac{2}{h}\left(\frac{v_0^1 Dm}{\delta_0}\right)^{1/2} \tag{4.8}$$

Determining the pulsation velocity of the scale δ_0 according to the Kolmogorov's-Obukhov's law of "two thirds" [32, 33] as

$$v_0^1 = \left(\varepsilon_0^l \delta_0\right)^{1/3} \tag{4.9}$$

and determining the value of δ_0 from the dimensionless relationship

$$\frac{v_0^1 \delta_0}{v} = A \tag{4.10}$$

we can obtain the expression for the mass transfer coefficient

$$\beta = \frac{2}{\pi \sqrt[4]{A}} \frac{\left(\varepsilon_0^l v\right)^{0,25}}{Sc^{0,5}} \tag{4.11}$$

where ε_0^l is the local value of the rate of energy dissipation, W/kg;

$\mathrm{Sc} = \dfrac{v}{D_M}$ is the Schmidt number.

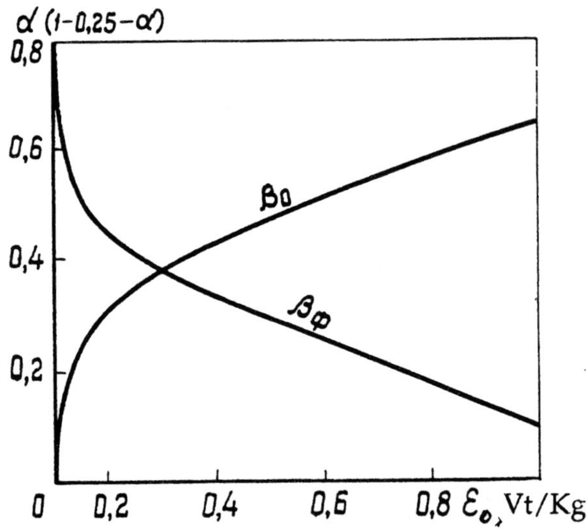

Figure 24. Variation of bubble surface share for various mass transfer mechanisms.

The distinguishing feature of zone III, shown in Figure 23, is that, with the parameter $Gg/n \cdot d_s^3$ being within the commonly used limits of magnitudes, the value of the volumetric gas fraction in the vessel as a rule exceeds 5 to 7% which corresponds to the situation when the distance between the bubbles proves to be of the same order of magnitude as the bubble size. In this case, as has been noted (Section 1), the bubble is located within the zones of the makes formed by the neighboring bubbles most of the time, where the value determined from the relationship (1.8) can be used, as the local value of the rate of energy dissipation in (4.11). For water-air systems without additives preventing coalescence where the bubble size is equal to 4.5 mm, the value of ε_0^l in accordance with (1.8) may be assumed to be constant and equal to 1.6 W/kg.

Thus, the only unknown value in (4.11) is the constant A characterizing the dimensionless thickness of the layer with the major portion of diffusion resistance. Since information is available in the literature about this hydrodynamic characteristic as applied to cases of deformable surfaces, the value of A may be found by comparing the experimental and estimated values of the mass transfer coefficients, just as it has been done, for instance, in [61].

In processing the experimental data presented in Figure 23 the value of A was assumed to be constant and equal to 2. The quantitative value of A is in good agreement with the data well-known from the literature about the thickness of the layer with the major portion of the diffusion resistance for the case of flows about hard surfaces [4] (A_T=11.5).

Actually, since the thickness of this layer depends on the intensity of the turbulence flow near the surface and on the deformable surface, turbulence dies awa at slower rates than at rigid ones, the value of A should be less than that of A_T.

With an allowance made for the value of A, the expression (4.11) shall assume the following final form:

$$\beta = \frac{0,535(\varepsilon_0^l \nu)^{0,25}}{Sc^{0,5}} \tag{4.12}$$

The obtained information makes it possible to proceed to consideration of mass transfer within zones I and II in Figure 23.

Zone I is characterized by a rather low-intensity of mixing and low values of the gas holdup, thus making these bubbles practically unaffected by one another. It is most likly that under these conditions the laws governing the mass exchange are identical to those governing the mass transfer from individual rising bubbles in an undisturbed stream. In considering this case, a distinction should be made between the mass transfer in the back portion of the bubble surface located within the wak zone (Figure 3) and the mass transfer from the front portion of the bubble surface. Taking into account the fact that for the 4.5-mm diam. bubbles the Reynolds number Re=$\frac{v_r \cdot d_b}{\nu}$=$10^3$ and the fraction of the bubble surface located in the wake α_b amounts to 25% of the total bubble surface [30], the overall mass transfer coefficient may be determined from the relationship

$$\beta = \beta_8\alpha_8 + \beta_f(1-\alpha_8) \tag{4.13}$$

The value of the mass transfer coefficient in the back portion of the surface bb may be determined from the relationship (4.12) at the value of ε_0^l=1.6 W/kg.

For determininng β_f use can be made of the expression which

$$\beta_f = \frac{0,65 \cdot D_M}{d_8} \sqrt{\mathrm{Re} \cdot Sc} \tag{4.14}$$

is well known in the literature [64].

As seen in Figure 23, the value of mass transfer B estimated from the equation (4.13), with the equations (4.12) and (4.14) properly accounted for, is in considerably good agreement with the experimental data obtained in zone I.

The most complex zone, in terms of interpretation of the experimental results and deriving the design formulae, is zone II where the mean rates of energy dissipation in the vessel δ_0 vary within wide limits from 0.1 to 1.0 W/kg.

At the values of ε_0 exceeding 0.1 W/kg the turbulence intensity in the vessel markedly increases and the rate of mass transfer from the front portion of the bubble surface becomes affected not only by the average flow around the bubble, but also by the turbulent pulsations penetrating into the layer with the major portion of diffusion resistance.

In the bulk of zone II, just as within zone I, the gas holdup in the vessel is so small that the reciprocal influence of the bubbles and their wakes may be neglected.

Thus, within zone II the coefficient of mass transfer from the bubble shall be determined:

a) in the back portion of the bubble surface accounting for 25% of the overall bubble surface, by the turbulent pulsations whose velocity depends on the rate of energy dissipation in the bubble wake;
b) over some portion of the frontal surface of the bubble α, by the averaged motion around the bubble;
c) over the remaining portion of the bubble surface (1-0.25-α), by the pulsation motion whose intensity is determined by the mean value of the rate of energy dissipation.

As the value of ε_0 approaches 1 W/kg, a certain portion of the front surface shall find itself within the wakes formed by the neighboring bubbles where the mechanism of mass transfer will be the same as that in the back portion of the surface. It should be noted, however, that for the values of ε_0 1 W/kg the proposed design scheme does not allow for this contribution to the mass transfer because of the complexity and amgiguity of the solution.

The portions of the front surface characterized by some mechanism have been determined (Figure 24) from the comparison of the experimental values of the mass transfer coefficients with the results of the estimation with the use of the relationship

$$\beta = 0.25\beta_8 + \alpha\beta_f + (1 - 0,25 - \alpha)\beta \tag{4.15}$$

where β_0 is the mass transfer coefficient determined from the equation (4.12) with the mean value of the rate of energy dissipation in the vessel taken as the local one.

The results presented in Figure 23 indicate that the linear mass transfer coefficient within the practically-employed range of variations of the energy dissipation rates (ε_0>1 W/kg) remains unchanged, and the variations in the volumetric mass transfer coefficient (Figure 22) is caused solely by the changes of the specific interfacial area.

$$F_{sp} = \frac{\sigma \varphi_m}{d_8} \qquad (4.16)$$

It is interesting to note that in water-air systems with coalescence, the value of F_{sp} depends solely on the volumetric gas holdup, since the bubble size in this case remains unchanged and equals 4.5 mm. In systems containing SAS or electrolytes suppressing coalescence the variation of the specific interfacial area is considerably affected not only by the volumetric gas holdup but also by the bubble size in the working volume of the vessel, depending on the agitation intensity. On the basis of the relationships for calculating the bubble sizes and gas holdup in the agitated vessels, (Section I and II), and the data on the linear mass transfer coefficient (Figure 23) the values of the volumetric mass transfer coefficients have been obtained which are in good agreement with the experimental results.

CONCLUSION

1. It has been shown that the information on local values of the rates of energy dissipation and the data on the mechanisms underlying the behavior of the turbulent flow near the boundary between phases are of primary importance for determining the hydrodynamic characteristics and intensity of mass transfer in the agitated gas-liquid systems.

2. Expressions have been obtained for a quantitative assessment of the bubble size in mixing gas-liquid systems. It has been shown that in the systems with coalescence the diameter of the bubble in the bulk volume of the agitated vessel does not vary and is only affected by the properties of the continuous phase. For instance, in the water-air systems the bubble diameter is constant and equal to 4.5 mm. In the presence of the system of coalescence-reducing additives the size of the bubbles in the working volume proves to be the same as that of the bubbles within the stirrer zone whose diameter depends on the intensity of agitation.

3. A technique has been proposed for calculating the distribution of the gas holdup and its mean value over the volume of the vessel with forced delivery of gas through the sparger, taking the phenomenon of the natural surface aeration into account.

4. Relationships have been obtained for calculating the quantity of the gas entrained from the surface as a function of agitation conditions, intensity of the forced aeration through the sparger and properties of working media.

5. The operating parameters of the ring-shaped sparger have been analyzed and its optimum design has been proposed to increase its efficiency and reliability.

6. A technique has been proposed for calculating the mass transfer coefficients in an agitated gas-liquid system based on the real hydrodynamics of a rising bubble and also accounting for the specific features of the turbulent flow near the phase boundary and the conclusions concerning the local isotropic turbulence.

REFERENCES

1. V.V. Kafarov, *Fundamentals of mass transfer*. M.: Vysshaya shkola, 1979, p. 439.
2. V.V. Kafarov, A.Yu. Vinarov, L.S. Gordeev, *Modelling biochemical reactors,* M.: Lesnaya promyshlennost, 1979, pp. 430.
3. L.N. Braginsky, M.A. Evilevich, et al. *Modelling waste purifying aerators* - L.: Khimiya, 1980, p. 144.
4. L.N. Braginsky, V.I. Begachev, V.M. Barabash *Mixing in lqiuid phases*. L. Khimiya, 1984, pp. 336.
5. T. Shidhar, O.E. Potter, *Gas holdup and bubble diameters in pressurized gas stirred vessels,* Ind. Eng. Chem. Fundam. 1980, V. 19, N. 1, p. 21-36.
6. R.D. Soyfer, V.V. Kafarov, *Energy consumption for mixing in gas-liquid systems*. Khimicheskoe i neftyanoe mashinostroenie. 1967, No. 3, pp. 16-18.
7. T. Vermeulen. *Chem. Eng. Progr.,* 1955, V. 51, N. 1, p. 85-95.
8. D.N. Miller "Scale-up of agitated vessel gas-liquid mass transfer," *A.I. ch. E.J.,* 1974, V. 20, N. 3, p. 448-453.
9. R.S. Smirnov, Falkov, Bulinkina, E.S., Planovsky, A.N. "On determining energy consumption in mixing gas-liquid systems." *Abstracts of report on the 2nd All-Union Conference* "Theory and practice of mixing in liquid phases." Zelenogorsk. Oct. 12-14, 1982. M. NIITEKhIM, 1982, pp. 82-83.
10. Z. Shterbachek, P. Tausk, *Mixing in chemical industry,* L. GOSKHIMIZDAT, 1963.
11. V.N. Sokolov, I.V. Domansky, *Gas-liquid reactors*. l.: Mashinostroenie, 1976, p. 216.
12. I.M. Zhernovaya, V.V. Kafarov, V.A. Klipinitser, *"Gas holdup in agitated vessel." Thes. of report at 2nd All-Union Conf.* "Theory and practice of mixing in liquid phases." Cherkassy. May 29 - June 1, 1973. M. NIITEKhIM, 1973, pp. 145-150.
13. R.S. Smirnov, "Hydrodynamics and mass transfer in large-scale agitated vessels with spragers." *Thes. Dissert. Cdt. Sc.* (Eng.), M., 1985, p. 20.
14. V.G. Levich, *Physico-chemical hydrodynamics*. M.: Fizmatgiz, 1959, p. 799.

15. P. Breadshow, *Introduction into turbulence and its measurement.* M.: Mir, 1974, p. 277.
16. V.M. Barabash, L.N. Braginsky, "On effect of near-surface turbulence on gas entrainment in mechanically agitated bubbled volume." *Thes. of report VII-th All-Union Conf. Two-phase flow in power installations and apparatuses.* Oct. 23-25, 1985, Leningrad. L: 1985, Vol. 3, pp. 187-189.
17. V.M. Barabash, L.N. Braginsky, G.V. Gorbacheva "On analysis of gas holdup in agitated vesselles." *Teor. osnovy khim. tekhnol,* 1987, Vol. 21, N. 5, pp. 654-660.
18. Z.Z. Van Dierendonck, J.M.H. Fortuit, D. Vanderboss, "The specific contact area in gas-liquid reactor," *4th European Symp on the Chem. Reaction Engng Brussel.* 1968, p. 205-211.
19. K.R. Westerterp, Z.Z. Van Dierendonck, "DeDraq J.A.", *Chem. Eng. Sci.* 1963, V. 18, N. 1, p. 157-162.
20. H.C. Foust, D.E. Mack, J.H. Rushton, "Gas-liquid contacting by mixers" *Ind. Eng. Chem.* 1944, V. 36, N. 5, p. 517-525.
21. V.M. Barabash, Yu.V. Korotov "On operating conditions of apparatuses for mixing gas-liquid systems." *Index of VINITI* "Dep. rukopisi," 1979, No. 9 (35), pp. 82-83.
22. P.H. Calderbank, M.B. Moo-Young, "The continuous phase heat and mass transfer properties of dispersions" *Trans. Inst. Chem. Engrs.* 1958, V. 36, N. 5, p. 443-448.
23. A.N. Kolmogorov, "On drops breakup in turbulent flow." *DAN USSR,* 1949, Vol. 66, N. 5, pp. 825-828.
24. G.K. Patterson, H. Wu, "Distribution of turbulence energy dissipation rates in mixers," *5th European conference on Mixing.* Wurzburg, West Germany, June 10-12, 1985, BHRA, p. 355-364.
25. H.D. Laufhutter, "A.B. Mersmann, Dissipation of power in stirred vessels" *5th European conference on mixing.* Wurzburg, West Germany, June 10-12, 1985, BHRA, 1985, p. 331-339.
26. F. Liepe, H.O. Mockel, H. Winkler, "Untersuchungen uber Homogenisieren und uber Turbulen in Ruhrmashinen" *Chem. Techn.,* 1971, B. 23, N. 412, S. 231-237.
27. H.O. Mockel "Die verteilung der ortlichen Energudissipation in einen Ruhrwerk" *Chem. Techn.* 1980, B. 32, N. 3, S. 127-129.
28. S.S. Kurateladze, M.A. Styrikovich, *Hydrodynamics of gas-liquid systems.* M.: Energiya, 1976, p. 296.
29. *Album of liquid and gas flows* (compiled and commented by M. Van-Dike. Trsl. from Engl. by L.V. Sokolovskaya under editorship of G.I. Barenblatt and V.P. Shidlovsky) M. Mir, 1986, p. 180, ill. photos.
30. Sen. P. *Separation flows.* Trsl from Engl., M.: Mir. 1982, p. 279.
31. Ya.Ye. Geguzin, *Bubbles,* M. Nauka, 1985, p. 1974.
32. A.N. Kolmogorov, "Energy dissipation under local isotropic turbulence conditions," *Reports of the USSR,* Acad. of Sciences, 1941, Vol. 32, No. 1, pp. 19-23.

33. A.M. Obukhov, "On energy distribution in turbulent flow spectrum." *Proceedings of USSR Acad.* Sc. Serial: geography and geophysics, 1941, Nos 4-5, pp. 452-458.
34. I.O. Protodyakonov, V.A. Ulyanov, *Hydrodynamics and mass transfer in dispersed liquid-liquid systems*. M. Nauka, 1986.
35. A.A. ZHukauskas, *Convective transport to heat exchanger*. M.: Nauka, 1982, p. 472.
36. N. Donald, Miller. *Scale-up of Agitated Vessels Gas-liquid mass transfer,* "AIchE J." 1974, V. 20, N. 3, p. 445-453.
37. J.H. Rushton, J.B. Gallagher, J.J. Oldshoe, *Chem. Eng. Progr.*, 1956, V. 52, N. 2, p. 319-326.
38. J.B. Gallagher, W. Resnich, *Gas residence time in agitated gas-liquid contactor. Experimental test of mass transfer model,* "Ind. Eng. Chem. Fundame." 1966, V. 5, N. 1, p. 15-21.
39. F. Joshida, J. Miure, "Gas absorption in agitated gas-liquid contactor" *Ind. Eng. Chem. Fund. Dev.*, 1963, V. 2, N. 6, p. 263-266.
40. S.V. Boltsezhak, G.P. Solomakha, "Mean gas holdup in bubbled layer of two- and three-phase systems in stirred vessles." *Thes. of report at IV-th All-Union Conf.* "Theory and practice of mixing in liquid media." Zelengorsk, Oct. 12-14, 1982, *M. NIITEKhIM*, 1982, pp. 86-87.
41. N.N. Falkov, "Studies of hydrodynamics and mass transfer in sparger apparatuses with high liquid level." *Thes. of dissert... Cdt. Sc. Eng.).* M.: 1973, p. 22.
42. L.N. Braginsky, V.I. Begachev, R.Z. Kofman, "Distribution of solid particles in mixing," *Teor. of osnovy khim. tekhnol.*, 1968, Vol. 2, N. 1, p. 128-133.
43. B.G. Tov, L.N. Braginsky, L.N. Novichkov, L.N. Dobroserdov, "Determining eddy-diffusion coefficient in sparger-mixing apparatuses," *Thes. of report at IInd All-Union Conf. Theory and practice of mixing in liquid media.* Zelenogorsk, Oct. 12-14, 1982. M.: NIITEKhIM, 1982, pp. 83-86.
45. M. Tanaka, T. Izumi, "Gas entrainment in stirred-tank reactors," *Chem. Eng. Res. Des.* 1987, V. 65, N. 2, p. 195-198.
46. R. Friedrich, E.J. Plate, "Die wirkung der Oberflachennahen Turbulenz auf den Sauerstoffeintrag" *Ind. J. Heat Mass Transfer,* 1983, V. 26, N. 4, p. 521-530.
47. T. Korn, T. Korn, *Reference book in mathematics,* M.: Nauka, 1974, p. 831.
48. P. Breadshow, T. Sebisi, G.G. Fernholtz, et al. *Turbulence,* M. Mashinostroeniye, 1980, p. 342.
49. Bai-Shi-I. *Turbulent gas and liquid flow.* Il., 1962, p. 342.
50. B.L. Efimov, "Researches into hydrodynamics and mass transfer in liquid phase in mechanically agitated gas-liquid systems," *Thes. of diss...Cdt. Sc.* (Eng.) M.: 1973, p. 22
51. T. Reich, *Physical Aspects of bubble dispersions in liquids.* University of Debft, Holland. 1968.
52. L.N. Braginsky, V.M. Barabash, V.B. Yaroshenko, "Meridional liquid circulation in agitated baffled vessels." *Theor. osnovy khim. tekhnol.* 1988, Vol. 22, No. 6, pp. 787-793.

53. F. Strenk, *Mixing and agitated vessels.* L.: Khimiya, 1975, p. 348.
54. P.H. Calderbank, *Trans. Inst. Chem. Eng..*, 1950, V. 37, N. 2, p. 131-137.
55. C.M. Cooper, G.A. Fernstrom, S.A. Miller, *Ind. Eng. Chem.* 1944, V. 36, N. 6, p. 504-514.
56. B.A. Litmans, I.S. Kukucherenko, "Yu.V. Tumanov Studies of mass transfer in liquid phase in cases of large amounts of power per unit volume spent for mixing." *Thes. of report at II-nd All-Union Conf. Theory and practice of mixing in liquid media.* Cherkassy, May 29 - June 1, 1973. M.: NIITEKhIM, 1973, pp. 137-140.
57. N.A. Yakovlev, G.P. Solomakha, "On allowing for gas flow energy in studying mass transfer in stirred vessels." *Thes. report at II-nd All-Union Conf. Theory and practice of mixing in liquid media.* Cherkassy, May 29 - June 1, 1973. M.: NIITEKhIM, 1973, pp. 137-140.
58. B.L. Efimov, G.P. Solomakha, "On same laws of mass transfer in liquid phase in agitated vessels with spargers." *Thes. of report at II-nd All-Union Conf. Theory and practice of mixing in liquid media.* Cherkassy, May 20 - June 1, 1973. M.: NIITEKhIM< 1973, pp. 131-137.
59. I.S. Pavlushnko, L.N. Braginsky, V.N. Brylov, *Zhurnal prikoadnoy khimii,* 1972, Vol. 6, No. 5, pp. 821-826.
60. B.A. Litmans, I.S. Kukucherenko, I.D. Boyko, Yu.V. Tumanov, *Zhurnal prikladnoy khimii,* 1972, Vol. 6, No. 5, pp. 821-826.
61. V.A. Klipinitser, Yu.M. Dubinsky, "On estimation of coefficient of mass transfer from gas bubble in agitated vessel." *Thes. of report at V-th All-Union Conf.* "Theory and practice of mixing in liquid media." Zelenogorsk., L. 1986, pp. 105-107.
62. R. Higbi, *Trans. AIchE.* 1935, V. 35, p. 365-368.
63. P.V. Dankverts, *Gas-liquid reactions.* M.: Khimiya, 1973, p. 295.
64. S. Sow *Hydrodynamics of multiphase systems,* M.: Mir, 1971, p. 295.
65. R.I. Nigmatullin, *Dynamics of multiphase media.* Part I., M.: Nauka, Gl. red. fiz. -mat. lit. 1987, p. 464.

Subject Index